CHEMICAL ENGINEERING HANDBOOK

CHEMICAL ENGINEERING HANDBOOK

Matthew Parker
Audrey Holland
Aiden Jenkins

Kruger Brentt
Publishers
2026

Kruger Brentt Publishers UK. LTD.
Company Number 9728962

Regd. Office: 68 St Margarets Road, Edgware, Middlesex HA8 9UU

ISBN: 978-1-78715-635-7

For information on all our publications visit our website at http://krugerbrentt.com/

PREFACE

Chemical Engineering Handbook is designed as a comprehensive reference that provides both foundational knowledge and practical insights for students, researchers, and professionals in the field of chemical engineering. The discipline of chemical engineering bridges science and industry, encompassing the design, operation, and optimization of processes that transform raw materials into valuable products, from chemicals and fuels to food, pharmaceuticals, and advanced materials.

This handbook covers the core principles of chemical engineering, including thermodynamics, fluid mechanics, heat and mass transfer, reaction engineering, process control, and separation processes. Each section is designed to provide a clear understanding of theoretical concepts while demonstrating their application in real-world industrial scenarios.

Special emphasis is placed on modern developments in chemical engineering, such as process intensification, green and sustainable engineering, process simulation, and advanced materials. The book also addresses safety, environmental, and ethical considerations, recognizing the responsibility of chemical engineers to design processes that are not only efficient but also safe, environmentally sound, and socially responsible.

Practical examples, case studies, and data tables are included throughout to aid understanding and provide quick references for professional use. The handbook also serves as a guide for problem-solving, design calculations, and decision-making in both academic and industrial contexts.

We hope that Chemical Engineering Handbook will serve as a reliable resource for learning, teaching, and professional practice, supporting the development of knowledge, skills, and innovation in the field of chemical engineering.We hope that Carbon Nanotubes: Properties and Applications will serve as an essential resource for students, researchers, and industry professionals, providing both a solid theoretical foundation and practical guidance for exploring and applying the remarkable potential of carbon nanotubes.

Matthew Parker | Audrey Holland | Aiden Jenkins

CONTENTS

Chapter-1

Chebyshev Polynomials and SAFT EOS for Predicting Asphaltene Precipitation

1. Introduction: An Overview

Equations of state are fundamental instruments for modeling physical and chemical processes involving fluids. The predominant PVT computations for oil and gas mixtures utilize a cubic equation of state (EoS). This category of equations originates almost a century ago with the renowned Van der Waals equation (Van der Waals, 1873). The cubic equations of state predominantly employed in the petroleum business are closely akin to the Van der Waals equation; nonetheless, it required nearly a century for the industry to recognize this equation as a significant engineering instrument. The Redlich and Kwong equation of state (Redlich & Kwong, 1949) was derived from the Van der Waals equation, incorporating a distinct attractive term while retaining the original repulsive term. Since 1949, when Redlich and Kwong (RK) developed their two-parameter cubic equation of state (EoS), other researchers have proposed various adjustments to enhance the efficacy of the RK EoS. Two prominent cubic equations are the Soave-Redlich-Kwong (SRK) equation (Soave, 1972) and the Peng-Robinson (PR) equation (Peng & Robinson, 1976), which have distinct formulations of the attractive term and are widely utilized in the oil industry for thermodynamic modeling of hydrocarbon fluids.

Numerous cubic equations of state exist, along with several noncubic equations. Noncubic equations, like the Benedict-Webb-Rubin equation (Benedict et al., 1942) and its version by Starling (Starling, 1973), contain numerous constants and accurately characterize the volumetric behavior of pure substances. However, due to the difficulties of mixing rules, hydrocarbon mixtures and crude oils may not be appropriate (Katz & Firoozabadi, 1978). Cubic equations containing many constants may not enhance the predictive accuracy of volumetric behavior for complicated reservoir fluids. Indeed, the majority of cubic equations have comparable accuracy in predicting the phase behavior of intricate hydrocarbon systems, with simpler models frequently outperforming them (Firoozabadi, 1999).

The vapor-liquid equilibrium properties of hydrocarbons and other non-polar fluids can be effectively modeled using a symmetric approach to represent both the vapor and liquid phase fugacity, employing a Van der Waals type equation model, with the Soave-Redlich-Kwong and Peng-Robinson equations being the most prevalent. Activity coefficient models are better appropriate for representing the liquid phase when polar fluids are present at modest pressures. In scenarios where elevated pressure ranges are pertinent, a symmetric equation of state technique utilizing intricate mixing rules, inclusive of an excess Gibbs energy term derived from an activity coefficient model, can yield favorable outcomes. Regrettably, these methodologies exhibit constraints for complicated fluids and may significantly falter in proximity to the key region, unless a particular intervention is incorporated (Llovell et al., 2004, 2008).

Since the early 1980s, there has been heightened interest in formulating an equation of state for pure fluids and mixtures of large polyatomic molecules that does not depend on a lattice representation of molecular arrangements. A comprehensive statistical-mechanical theory for extensive polyatomic molecules in continuous space is challenging due to its asymmetric architecture, numerous internal degrees of freedom, and robust coupling between intra- and intermolecular interactions. Nonetheless, a rather straightforward model depicts chain-like structures as freely linked tangent hard spheres (Chapman et al., 1984; Song et al., 1994; Wertheim, 1984). A hard-sphere-chain (HSC) equation of state can serve as the reference system instead of the hard-sphere reference employed in the majority of current equations of state for simple fluids. Although they are simplistic, hard-sphere-chain models incorporate essential characteristics of actual fluids with chain-like molecules, such as omitted volume effects and chain connectedness. To characterize the properties of fluids composed of massive polyatomic molecules, it is essential to incorporate attractive forces by introducing a perturbation to a hard sphere chain equation of state. Assuming that the impact of attractive forces on fluid structure is minimal, a Van der Waals-type or other mean-field term (e.g., square-well fluids) is typically employed to incorporate attractive forces into the reference hard-sphere-chain equation of state (Prausnitz & Tavares, 2004).

Molecular-based equations of state, grounded in statistical mechanics, remain relevant in chemical engineering calculations due to their applicability across a broad range of thermodynamic circumstances and substances, while being far less computationally intensive than molecular simulations. The Statistical Associating Fluid Theory (SAFT) equation of state has gained significant popularity due to its proficiency in forecasting the thermodynamic features of several complex fluids, including chain hydrocarbons, aromatic hydrocarbons, chlorinated hydrocarbons, esters, alkanols, and carboxylic acids (Huang & Radosz, 1990). SAFT was conceived as an application of Wertheim's association theory (Wertheim, 1984, 1986) utilizing first-order thermodynamic perturbation theory (TPT) to develop a physically grounded equation of state (Chapman et al., 1990; Huang & Radosz, 1991). The aspiration to refine SAFT into a precise equation for engineering applications has led to the creation of various iterations aimed at addressing the shortcomings of the original formulation (Economou, 2002; Muller & Gubbins, 1995).

SAFT resembles group contribution theories in that the fluid of interest is originally seen as a combination of unlinked groups or segments. SAFT incorporates a chain connectivity term to consider the bonding of diverse groups in polymer formation, as well as an explicit term for intermolecular hydrogen bonding. A theory grounded in statistical mechanics presents numerous advantages. The primary advantage is that every approximation employed in the construction of SAFT has been validated against molecular simulation outcomes. The application range of each phrase in the EoS has been established. The second benefit is that the EoS can be methodically enhanced. Weak approximations in SAFT can be detected, leading to enhancements in the EoS through improved approximations or the extension of the theory. Similar to most thermodynamic models, SAFT methodologies necessitate the assessment of many parameters that connect the model to the experimental system. A third advantage of SAFT-type equations over alternative methodologies is their foundation in statistical mechanics, which endows parameters with distinct physical significance; when meticulously calibrated, they possess predictive capabilities to investigate regions of the phase diagram distant from the data and operating conditions employed in parameter regression, demonstrating superior performance compared to other models for interacting compounds, such as activity coefficient models (Prausnitz et al., 1999). In SAFT, a chain molecule is defined by the diameter or volume of a segment, the quantity of segments in the chain, and the segment-segment dispersion energy. For a molecule exhibiting association or hydrogen bonding, two additional physical characteristics are required: the association energy, which pertains to the change in enthalpy during association, and the bond volume, which is associated with the change in entropy upon association. The SAFT equation has demonstrated significant technical applications in fluids exhibiting chain and hydrogen bonding (Chapman et al., 2004).

Asphaltenes are defined as the fraction of crude oil that is insoluble in light standard alkanes (such as n-heptane or n-pentane) yet soluble in aromatic solvents (such as benzene or toluene). The solubility class classification of asphaltenes indicates a wide range of molecular configurations that differ significantly across crude oil sources. Asphaltenes generally exhibit fused ring aromaticity, diminutive aliphatic side chains, and polar functional groups comprising heteroatoms that can donate or take protons both inter- and intra-molecularly. Despite the complexity of asphaltene fractions as mixes of molecular species, they exhibit a clear chemical resemblance overall, regardless of their geographic origin (Ting, 2003). The stability of asphaltenes is influenced by various elements, including pressure, temperature, and fluid composition, which encompasses the introduction of light gases, solvents, and other oils, as well as contamination from mixed operations or charges. Asphaltene aggregate production is found during pressure depletion at constant temperature, within a range above and below the bubble point. As production pressure decreases from reservoir levels, asphaltene precipitation may occur because to alterations in the solubility of asphaltene in crude oil. Asphaltene precipitation reaches its peak at or near the bubble point pressure. Below the bubble point, light gases are released from the solution, hence enhancing asphaltene solubility once again (Ting, 2003). Temperature variations also influence asphaltene precipitation. For hydrocarbons situated in shallow formations, the wellhead flowing temperatures are generally moderate, ranging from 110 to 140 °F. Nonetheless, the temperature of the seabed in deep waters is frigid, frequently approaching or falling below 40°F, even in equatorial regions. The cooling of flow streams during transit may result in asphaltene precipitation (Huang & Radosz, 1991). Elevated temperatures at steady pressure typically stabilize the asphaltene in crude oil. Depending on the oil's composition, there are instances where precipitation initially diminishes and subsequently escalates with rising temperature (Verdier et al., 2006). Moreover, considerable temperature impacts can be found contingent upon the temperature level (Buenrostro-Gonzales & Lira-Galeana, 2004). Composition alterations transpire during gas injection procedures utilized in Enhanced Oil Recovery (EOR). Gas injection include methods such as miscible flooding utilizing CO_2, N_2, or natural gas, as well as artificial gas lifting. The presence of dissolved gas reduces asphaltene solubility, rendering the asphaltene increasingly unstable (Verdier et al., 2006).

The propensity of petroleum asphaltenes to aggregate in solution and adhere to surfaces can lead to substantial challenges during the extraction, recovery, pipeline conveyance, and refining of crude oil. Consequently, it is essential to forecast the conditions under which precipitation transpires and the quantity of precipitate formed. We have adopted the SAFT EoS for modeling, as it explicitly incorporates association interactions and chain connectedness to account for the bonding of different groups. Consequently, the equation can elucidate the asphaltene precipitation behavior. Through algebraic manipulations of this equation, we obtain a simpler expression for the compressibility factor or pressure as a function of density. In response to pressure, an approximation method utilizing Chebyshev polynomials is employed to derive the explicit form of the SAFT equation of state, facilitating the calculation of density and, consequently, fugacity necessary for phase equilibrium computations. A binary system consisting of ethanol and toluene is utilized to illustrate the efficacy of the SAFT EoS. The density of the aforementioned system is computed using Chebyshev polynomial approximation at various temperatures ranging from 283.15 K to 353.15 K and pressures up to 45 MPa. Assessing fugacity is an essential component in phase equilibrium computations. Consequently, fugacity is calculated utilizing the SAFT equation of state. The model is employed to forecast the phase behavior of oil-asphaltene systems.

2. Articulation of the issue

2.1 SAFT equation of state

The Statistical Associating Fluid Theory (SAFT) (Chapman et al., 1990) is founded on the first-order perturbation theory proposed by Wertheim (Wertheim, 1987). This theory posits that the residual Helmholtz energy is represented as a summation of terms that encompass short-range repulsions, long-range dispersion forces, chemically bonded aggregation (such as the formation of stable chains), and association or solvation (such as hydrogen bonding) among distinct molecules or chains. A three-step process for the development of stable aggregates (e.g., chains) and their subsequent association is illustrated in Figure 1 for a pure component. Initially, a fluid consists of equal-sized, single hard spheres.

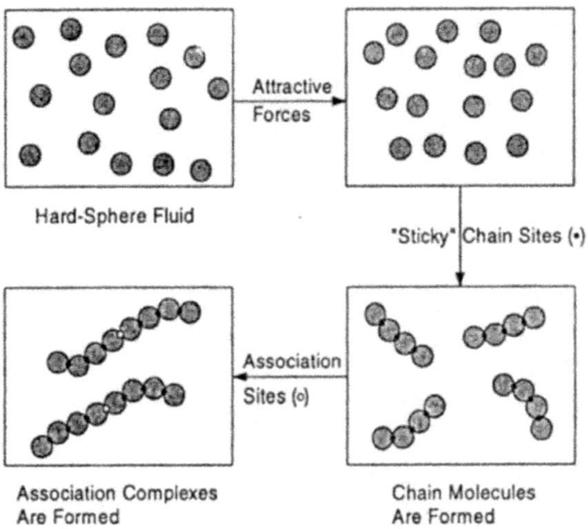

Fig. 1. Three steps to form chain molecules and association complexes from hard spheres in the SAFT model (Prausnitz et al., 1999).

Intermolecular attractive forces are added which are described by an appropriate potential function, such as the square-well potential. Next, each sphere is given one, two or more "sticky spots", such that the spheres can stick together (covalent bonding) to form dimmers, trimers and higher stable aggregates as chains. Finally, specific interaction sites are introduced at some position in the chain to form association complex through some attractive interaction (e.g. hydrogen bonding). Each step provides a contribution to the Helmholtz energy.

Using SAFT EoS, the residual molar Helmholtz energy A^R contributes from formation of hard spheres, chains, dispersion (attraction), and association which would be in the form of:

$$A^R = A_{hs} + A_{ch} + A_{disp} + A_{assoc} \tag{1}$$

Here the sum of the first two terms is the hard-sphere-chain reference system accounting for molecular repulsion and chain connectivity (chemical bonding); the sum of the last two terms is the perturbation accounting for molecular attraction and for association due to specific interactions like hydrogen bonding. Application of the relation between molar Helmholtz energy, A, and the equation of state, gives the SAFT EoS for pure fluids (Prausnitz et al., 1999). We can write for compressibility factor of a real fluid:

$$Z = \frac{P}{\rho RT} = Z^{id} + Z_{hs} + Z_{ch} + Z_{disp} + Z_{Assoc} \tag{2}$$

with $Z^{id} = 1.0$ and for mixtures,

$$Z_{hs} = \frac{6}{\pi N_A \rho}\left[\frac{\xi_0 \xi_3}{1-\xi_3} + \frac{3\xi_1\xi_2}{(1-\xi_3)^2} + \frac{(3-\xi_3)\xi_2^3}{(1-\xi_3)^2}\right] \tag{3}$$

With

$$\xi_k = \frac{\pi N_A \rho}{6}\sum_{i=1}^{N_c} z_i r_i (d_i)^k \; k = 0,1,2,3 \tag{4}$$

$$d_i = \sigma_i[1 - 0.12\exp\left(-3\varepsilon_i/kT\right)] \tag{5}$$

$$Z_{hs} = \frac{hs_1\rho + hs_2\rho^2 + hs_3\rho^3}{1 + hs_4\rho + hs_5\rho^2 + hs_6\rho^3} \tag{6}$$

here ρ is the total molar density, z_i, is the mole fraction of component i, r_i is the number of segments per molecule i, and d_i is the temperature dependent segment diameter. The parameters T, N_A, k, ε_i and σ_i are temperature, Avogadro's and Boltzmann's constants, segment energy and diameter, respectively. By simple algebraic manipulation on Eq. (3), we arrive at the following simplified form of the hard sphere term:

where,

$$hs_1 = (\bar{r}S_3 + 3S_1S_2)P_{na} \tag{7}$$

$$hs_2 = (-2\bar{r}S_3^2 - 3S_1S_2S_3 + 3S_2^3)P_{na}^2 \tag{8}$$

$$hs_3 = (\bar{r}S_3^3 - S_2^3S_3)P_{na}^3 \tag{9}$$

$$hs_4 = -3S_3P_{na} \tag{10}$$

$$hs_5 = 3S_3^2P_{na}^2 \tag{11}$$

$$hs_6 = -S_3^3P_{na}^3 \tag{12}$$

The parameters used in Eqs. (7) - (12) are defined as the following:

$$\bar{r} = \sum_{i=1}^{N_c} z_i r_i \tag{13}$$

$$S_1 = \sum_{i=1}^{N_c} z_i r_i d_i \tag{14}$$

$$S_2 = \sum_{i=1}^{N_c} z_i r_i d_i^2 \tag{15}$$

$$S_3 = \sum_{i=1}^{N_c} z_i r_i d_i^3 \tag{16}$$

$$P_{na} = \frac{\pi N_A}{6} \tag{17}$$

The contribution accounting for the formation of chain molecules of the various components in the mixture is

$$Z_{ch} = \sum_{i=1}^{N_c} z_i(1 - r_i)L(d_i) \tag{18}$$

$$L(d_i) = \frac{2\xi_3 + 3d_i\xi_2 - 4\xi_3^2 + 2d_i^2\xi_2^2 + 2\xi_3^3 + d_i^2\xi_2^2\xi_3 - 3d_i\xi_2\xi_3^2}{(1 - \xi_3)(2 - 4\xi_3 + 3d_i\xi_2 + 2\xi_3^2 + d_i^2\xi_2^2 - 3d_i\xi_2\xi_3)} \tag{19}$$

where $\xi_k(k = 2,3)$ is given by Eq. (4). It is remarkable that no mixing rules are necessary in Eq. (3) and (18). After some arithmetic operations on Eq. (18), the following simplified density dependent equation for the chain term of SAFT EoS is presented as:

$$Z_{ch} = \sum_{i=1}^{N_c} z_i(1 - r_i)\frac{ch_1(d_i)\rho + ch_2(d_i)\rho^2 + ch_3(d_i)\rho^3}{2 + ch_4(d_i)\rho + ch_5(d_i)\rho^2 + ch_6(d_i)\rho^3 + ch_7(d_i)\rho^4} \tag{20}$$

where,

$$ch_1(d_i) = [2S_3 + 3d_iS_2]P_{na} \tag{21}$$
$$ch_2(d_i) = [-4S_3^2 + 2d_i^2S_2^2]P_{na}^2 \tag{22}$$
$$ch_3(d_i) = [d_i^2S_2^2S_3 + 2S_3^3 - 3d_iS_2S_3^2]P_{na}^3 \tag{23}$$
$$ch_4(d_i) = [-6S_3 + 3d_iS_2]P_{na} \tag{24}$$
$$ch_5(d_i) = [6S_3^2 + d_i^2S_2^2 - 9d_iS_2S_3]P_{na}^2 \tag{25}$$
$$ch_6(d_i) = [9d_iS_2S_3^2 - 2S_3^3 - d_i^2S_2^2S_3]P_{na}^3 \tag{26}$$
$$ch_7(d_i) = [-3d_iS_2S_3^3]P_{na}^3 \tag{27}$$

SAFT uses the following expression for the dispersion contribution to the compressibility factor (Pedersen & Christensen, 2007):

$$Z_{\text{disp}} = -2\pi\rho\frac{\partial(\xi_3 I_1)}{\partial(\xi_3)}\overline{r^2\varepsilon\sigma^3} - \pi\rho\bar{r}\left[C_1\frac{\partial(\xi_3 I_2)}{\partial(\xi_3)} + C_2\xi_3 I_2\right]\overline{r^2\varepsilon^2\sigma^3} \tag{28}$$

where

$$C_1 = 1 + \bar{r}\frac{8\xi_3 - 2\xi_3^2}{(1-\xi_3)^4} + (1-\bar{r})\frac{20\xi_3 - 27\xi_3^2 + 12\xi_3^3 - 2\xi_3^4}{[(1-\xi_3)(2-\xi_3)]^2} \tag{29}$$

$$C_2 = -C_1^2\left[\bar{r}\frac{-4\xi_3^2 + 20\xi_3 + 8}{(1-\xi_3)^5} + (1-\bar{r})\frac{2\xi_3^3 + 12\xi_3^2 - 48\xi_3 + 40}{[(1-\xi_3)(2-\xi_3)]^3}\right] \tag{30}$$

$$\overline{r^2\varepsilon\sigma^3} = \sum_{i=1}^{N_c}\sum_{j=1}^{N_c} z_i z_j r_i r_j\left(\frac{\varepsilon_{ij}}{kT}\right)\sigma_{ij}^3 \tag{31}$$

$$\overline{r^2\varepsilon^2\sigma^3} = \sum_{i=1}^{N_c}\sum_{j=1}^{N_c} z_i z_j r_i r_j\left(\frac{\varepsilon_{ij}}{kT}\right)^2\sigma_{ij}^3 \tag{32}$$

$$I_1 = \sum_{j=0}^{6} a_j(\bar{r})\xi_3^j \tag{33}$$

$$I_2 = \sum_{j=0}^{6} b_j(\bar{r})\xi_3^j \tag{34}$$

In equations (31) and (32):

$$\varepsilon_{ij} = \sqrt{\varepsilon_i \varepsilon_j}(1 - k_{ij}) \tag{35}$$

$$\sigma_{ij} = \frac{1}{2}(\sigma_i + \sigma_j) \tag{36}$$

where k_{ij} is a binary interaction parameter similar to that in the mixing rule for the a parameter of a cubic EoS (Pedersen & Christensen, 2007). In equationa (33) and (34):

$$a_j(\bar{r}) = a_{0j} + \frac{\bar{r}-1}{\bar{r}} a_{1j} + \frac{\bar{r}-1}{\bar{r}} \cdot \frac{\bar{r}-2}{\bar{r}} a_{2j}, j = 0,1,\ldots,6 \tag{37}$$

$$b_j(\bar{r}) = b_{0j} + \frac{\bar{r}-1}{\bar{r}} b_{1j} + \frac{\bar{r}-1}{\bar{r}} \cdot \frac{\bar{r}-2}{\bar{r}} b_{2j}, j = 0,1,\ldots,6 \tag{38}$$

The universal constants for $a_{0j}, a_{1j}, a_{2j}, b_{0j}, b_{1j}$ and b_{2j} are given in Table 1.

Table 1. The universal constants for $a_{0j}, a_{1j}, a_{2j}, b_{0j}, b_{1j}$ and b_{2j} parameters used in SAFT EoS (Pedersen & Christensen, 2007).

J	a_{0j}	a_{1j}	a_{2j}	b_{0j}	b_{1j}	b_{2j}
0	0.9105631	- 0.3084017	-0.0906148	0.7240947	0.5755498	0.0976883
1	0.6361281	0.1860531	0.4527843	2.2382792	0.6995095	-0.2557575
2	2.6861348	-2.5030047	0.5962701	-4.0025849	3.8925674	-9.1558561
3	-26.547362	21.419793	-1.7241829	-21.003577	-17.215472	20.642076
4	97.759209	-65.255885	-4.1302112	26.855641	192.67226	-38.804430
5	-159.59154	83.318680	13.776632	206.55134	-161.82646	93.626774
6	91.297774	-33.746923	-8.6728470	-355.60235	-165.20769	-29.666905

Again, simplification of Eq. (28), would yield the following density dependent form of the dispersion term in SAFT EoS:

$$
\begin{aligned}
Z_{disp} &= \left[dI_1.\rho + dI_2.\rho^2 + dI_3.\rho^3 + dI_4.\rho^4 + dI_5.\rho^5 + dI_6.\rho^6 + dI_7.\rho^7\right] - \left(6P_{na}\overline{r}.Adisp_2\right) \\
&\cdot\left[\left(\frac{\rho + disp_1.\rho^2 + disp_2.\rho^3 + disp_3.\rho^4 + disp_4.\rho^5 + disp_5.\rho^6 + disp_6.\rho^7}{4 + disp_7.\rho + disp_8.\rho^2 + disp_9.\rho^3 + disp_{10}.\rho^4 + disp_{11}.\rho^5 + disp_{12}.\rho^6}\right)\right. \\
&\times\left(b_0 + dI_8.\rho + dI_9.\rho^2 + dI_{10}.\rho^3 + dI_{11}.\rho^4 + dI_{12}.\rho^5 + dI_{13}.\rho^6\right) - \left(P_{na}S_3\rho^2\right) \\
&\times\left(\frac{1 + disp_1.\rho + disp_2.\rho^2 + disp_3.\rho^3 + disp_4.\rho^4 + disp_5.\rho^5 + disp_6.\rho^6}{4 + disp_7.\rho + disp_8.\rho^2 + disp_9.\rho^3 + disp_{10}.\rho^4 + disp_{11}.\rho^5 + disp_{12}.\rho^6}\right)^2 \\
&\times\left(\frac{disp_{13} + disp_{14}.\rho + disp_{15}.\rho^2 + disp_{16}.\rho^3 + disp_{17}.\rho^4 + disp_{18}.\rho^5}{8 + disp_{19}.\rho + disp_{20}.\rho^2 + disp_{21}.\rho^3 + disp_{22}.\rho^4 + disp_{23}.\rho^5 + disp_{24}.\rho^6 + disp_{25}.\rho^7 + disp_{26}.\rho^8}\right) \\
&\left.\times\left(b_0 + Idisp_1.\rho + Idisp_2.\rho^2 + Idisp_3.\rho^3 + Idisp_4.\rho^4 + Idisp_5.\rho^5 + Idisp_6.\rho^6\right)\right]
\end{aligned}
\tag{39}
$$

where the parameters used in the equation are given below,

$$disp_1 = 4(3\bar{r} - 4)S_3P_{na} \tag{40}$$

$$disp_2 = (27\bar{r} - 6)S_3^2P_{na}^2 \tag{41}$$

$$disp_3 = (82 - 70\bar{r})S_3^3P_{na}^3 \tag{42}$$

$$disp_4 = (51\bar{r} - 52)S_3^4P_{na}^4 \tag{43}$$

$$disp_5 = 16(1 - \bar{r})S_3^5P_{na}^5 \tag{44}$$

$$disp_6 = 2(\bar{r} - 1)S_3^6P_{na}^6 \tag{45}$$

$$disp_7 = -20_3S_3P_{na} \tag{46}$$

$$disp_8 = 41S_3^2P_{na}^2 \tag{47}$$

$$disp_9 = -44S_3^3P_{na}^3 \tag{48}$$

$$disp_{10} = 26S_3^4P_{na}^4 \tag{49}$$

$$disp_{11} = -8S_3^5P_{na}^5 \tag{50}$$

$$disp_{12} = S_3^6P_{na}^6 \tag{51}$$

$$disp_{13} = 40 + 24\bar{r} \tag{52}$$

$$disp_{14} = (192\bar{r} - 128)S_3P_{na} \tag{53}$$

$$disp_{15} = (148 - 372\bar{r})S_3^2P_{na}^2 \tag{54}$$

$$disp_{16} = (230\bar{r} - 70)S_3^3P_{na}^3 \tag{55}$$

$$disp_{17} = (8 - 52\bar{r})S_3^4P_{na}^4 \tag{56}$$

$$\mathrm{disp}_{18} = (2 + 2\bar{r})S_3^5 P_{na}^5 \quad (57)$$

$$\mathrm{disp}_{19} = -52 S_3 P_{na} \quad (58)$$

$$\mathrm{disp}_{20} = 146 S_3^2 P_{na}^2 \quad (59)$$

$$\mathrm{disp}_{21} = -231 S_3^3 P_{na}^3 \quad (60)$$

$$\mathrm{disp}_{22} = 225 S_3^4 P_{na}^4 \quad (61)$$

$$\mathrm{disp}_{23} = -138 S_3^5 P_{na}^5 \quad (62)$$

$$\mathrm{disp}_{24} = 52 S_3^6 P_{na}^6 \quad (63)$$

$$\mathrm{disp}_{25} = -11 S_3^7 P_{na}^7 \quad (64)$$

$$\mathrm{disp}_{26} = S_3^8 P_{na}^8 \quad (65)$$

$$I\mathrm{disp}_1 = b_1 S_3 P_{na} \quad (66)$$

$$I\mathrm{disp}_2 = b_2 S_3^2 P_{na}^2 \quad (67)$$

$$I\mathrm{disp}_3 = b_3 S_3^3 P_{na}^3 \quad (68)$$

$$I\mathrm{disp}_4 = b_4 S_3^4 P_{na}^4 \quad (69)$$

$$I\mathrm{disp}_5 = b_5 S_3^5 P_{na}^5 \quad (70)$$

$$I\mathrm{disp}_6 = b_6 S_3^6 P_{na}^6 \quad (71)$$

$$\mathrm{dIdisp}_1 = 2a_1 S_3 P_{na} \quad (72)$$

$$\mathrm{dIdisp}_2 = 3a_2 S_3^2 P_{na}^2 \quad (73)$$

$$\mathrm{dIdisp}_3 = 4a_3 S_3^3 P_{na}^3 \quad (74)$$

$$\mathrm{dIdisp}_4 = 5a_4 S_3^4 P_{na}^4 \quad (75)$$

$$\mathrm{dIdisp}_5 = 6a_5 S_3^5 P_{na}^5 \quad (76)$$

$$\mathrm{dIdisp}_6 = 7a_6 S_3^6 P_{na}^6 \quad (77)$$

$$\mathrm{dI}_1 = -12P_{na} \cdot Adisp_1 \cdot a_0 \quad (78)$$

$$\mathrm{dI}_2 = -12P_{na} \cdot Adisp_1 \cdot dIdisp_1 \quad (79)$$

$$dI_3 = -12P_{na} \cdot Adisp_1 \cdot dIdisp_2 \quad (80)$$

$$dI_4 = -12P_{na} \cdot Adisp_1 \cdot dIdisp_3 \quad (81)$$

$$dI_5 = -12P_{na} \cdot Adisp_1 \cdot dIdisp_4 \quad (82)$$

$$dI_6 = -12P_{na} \cdot Adisp_1 \cdot dIdisp_5 \quad (83)$$

$$dI_7 = -12P_{na} \cdot Adisp_1 \cdot dIdisp_6 \quad (84)$$

$$dI_8 = 2b_1S_3P_{na} \quad (85)$$

$$dI_9 = 3b_2S_3^2P_{na}^2 \quad (86)$$

$$dI_{10} = 4b_3S_3^3P_{na}^3 \quad (87)$$

$$dI_{11} = 5b_4S_3^4P_{na}^4 \quad (88)$$

$$dI_{12} = 6b_5S_3^5P_{na}^5 \quad (89)$$

$$dI_{13} = 7b_6S_3^6P_{na}^6 \quad (90)$$

$$Adisp_1 = \overline{r^2\varepsilon\sigma^3} = \sum_{i=1}^{N_C}\sum_{j=1}^{N_C} z_iz_jr_ir_j(\varepsilon_{ij}/kT)\sigma_{ij}^3 \quad (91)$$

$$Adisp_2 = \overline{r^2\varepsilon^2\sigma^3} = \sum_{i=1}^{N_C}\sum_{j=1}^{N_C} z_iz_jr_ir_j(\varepsilon_{ij}/kT)^2\sigma_{ij}^3 \quad (92)$$

Similarly, Z_{assoc} can be derived rigorously from statistical mechanics (Chapman et al., 1990). The relation is a mole fraction average of the corresponding pure-component equations:

$$Z_{\text{assoc}} = \rho\sum_{i=1}^{N_C} z_i\left[\sum_{S_i}\left(\frac{1}{X^{S_i}}-\frac{1}{2}\right)\frac{\partial X^{S_i}}{\partial\rho}\right] \quad (93)$$

where X^{S_i}, the mole fraction of component i in the mixture not bonded with other components at site S, is given by:

$$X^{S_i} = \left(1 + N_A\sum_{j=1}^{N_C}\sum_{Y_j} z_i\rho X^{Y_j}W_{ij}\right)^{-1} \quad (94)$$

with

$$W_{ij} = \left[\frac{1}{1-\xi_3} + \frac{3d_id_j}{d_i+d_j}\frac{\xi_2}{(1-\xi_3)^2} + 2\left(\frac{d_id_j}{d_i+d_j}\right)^2\frac{\xi_2^2}{(1-\xi_3)^3}\right]\left(\sigma_{ij}k^{S_iY_j}\right)\left[\exp\left(\frac{\varepsilon^{S_iY_j}}{kT}\right)-1\right]\#(95)$$

In Eq. (94), summation $\sum_{Y_j}$ is over all specific interaction sites on molecule j and summation $\sum_j$ is over all N_c components. The association/salvation $\varepsilon^{S_iY_j}$ and the dimensionless parameter $k^{S_iY_j}$ characterize, respectively, the association ($i = j$) and solvation ($i \neq j$) energy and volume for the specific interaction between sites S and Y. These parameters are adjustable. Equation (93) requires no mixing rules. As it can be seen in Eq. (94), X^{S_i} 's satisfy a non-linear system of equations which can be solved using any iterative technique such as Gauss-Seidel, Successive-Over-Relaxation (SOR) or Jacobi iterative method. The derivative of the function X^{S_i} with respect to ρ yields the following equation:

$$\left(\frac{\partial X^{S_i}}{\partial \rho}\right) = -(X^{S_i})^2 N_A \left[\begin{array}{c} \sum_{j=1}^{N_c}\sum_{Y_j} z_j X^{Y_j} W_{ij} + \sum_{j=1}^{N_c}\sum_{Y_j} z_j \rho X^{Y_j}\left(\frac{\partial W_{ij}}{\partial \rho}\right) \\ + \sum_{j=1}^{N_c}\sum_{Y_j} z_j \rho W_{ij} X^{Y_j}\left(\frac{\partial X^{Y_j}}{\partial \rho}\right) \end{array}\right] \tag{96}$$

where,

$$\left(\frac{\partial W_{ij}}{\partial \rho}\right) = \left(\sigma_{ij} k^{S_i Y_j}\right)\left[\exp\left(\frac{\varepsilon^{S_i Y_j}}{kT}\right) - 1\right] \times \left[\begin{array}{c} (1-\xi_3)^{-2}\frac{\partial \xi_3}{\partial \rho} + \frac{3 d_i d_j}{d_i + d_j}\left[(1-\xi_3)^{-2}\left(\frac{\partial \xi_2}{\partial \rho}\right) + 2(1-\xi_3)^{-3}\xi_2\left(\frac{\partial \xi_3}{\partial \rho}\right)\right] \\ +2\left(\frac{d_i d_j}{d_i + d_j}\right)^2\left[2\xi_2\left(\frac{\partial \xi_2}{\partial \rho}\right)(1-\xi_3)^{-3} + 3(1-\xi_3)^{-4}\left(\frac{\partial \xi_3}{\partial \rho}\right)\xi_2^2\right] \end{array}\right] \tag{97}$$

As it can be seen from Eq. (96), $(\partial X^{S_i}/\partial \rho)'$ s are solutions of a linear system of equations which can be estimated using a known technique such as Gaussian Elimination, GaussJordan or Least Square method (Burden et al., 1981).

2.2 Fugacity derivation utilizing the SAFT equation of state

The fugacity of component i in terms of independent variables V and T is given by the following equation for a given phase (Danesh, 1998; Prausnitz et al., 1999; Tabatabaei Nejad, & Khodapanah, 2009):

$$R\ln \varphi_k^\alpha = = R\ln \frac{f_k^\alpha}{z_k^\alpha P} = \int_{V^\alpha}^{\infty}\left[\left(\frac{\partial P}{\partial n_k}\right)_{T,V,n_{j\neq k}} - \frac{RT}{V}\right] dV - R\ln Z^\alpha \tag{98}$$

$$k = 1,2,\dots,N_c \ \alpha = L,V \qquad (98)$$

where $f_k, \varphi_k, n_k, V, Z$ and P are fugacity, fugacity coefficient and the number of moles of component k, volume, compressibility factor, and pressure, respectively. The superscript α denotes liquid (L) and vapor phases (V).

The compressibility factor is related to the volume by the following equations:

$$Z^{\alpha} = \frac{P}{\rho^{\alpha} RT} \tag{99}$$

$$\rho^{\alpha} = \frac{n^{\alpha}}{V^{\alpha}} \tag{100}$$

where n refers to the total number of moles of the known phase. To use equation (98), we require a suitable EoS that holds for the entire range of possible mole fractions z at the system temperature and for the density range between 0 and $\left(\frac{n}{V}\right)$. Application of the SAFT EoS in Eq. (98) yields the following equation for calculating the fugacity of the components:

$$RT\ln\left(Z^{\alpha}\varphi_k^{\alpha}\right) = \int_V^{\infty} \frac{RT}{V}\left[\frac{\partial(nZ_{hs})}{\partial n_k}\right]_{T,V,n_{j\neq k}} dV + \int_V^{\infty} \frac{RT}{V}\left[\frac{\partial(nZ_{ch})}{\partial n_k}\right]_{T,V,n_{j\neq k}} dV$$

$$+\int_V^{\infty} \frac{RT}{V}\left[\frac{\partial(nZ_{disp})}{\partial n_k}\right]_{T,V,n_{j\neq k}} dV + \int_V^{\infty} \frac{RT}{V}\left[\frac{\partial(nZ_{assoc})}{\partial n_k}\right]_{T,V,n_{j\neq k}} dV \tag{101}$$

$$k = 1,2,\ldots,N_c$$

The following equations are derived for the first term in Eq. (101) accounting for the hard sphere contribution of SAFT:

$$IR\_HS = \int_V^{\infty} \frac{RT}{V}\left[\frac{\partial(nZ_{hs})}{\partial n_k}\right]_{T,V,n_{j\neq k}} dV = \int_0^{\rho} \frac{RT}{\rho}\left[\frac{\partial(nZ_{hs})}{\partial n_k}\right]_{T,V,n_{j\neq k}} d\rho \tag{102}$$

$$k = 1,2,\ldots,N_c$$

$$\left[\frac{\partial(nZ_{hs})}{\partial n_k}\right]_{T,V,n_{j\neq k}} = \left[(hs_{1k} + 2hs_1).\rho + (hs_{2k} + 3hs_2).\rho^2 + (hs_{3k} + 4hs_3).\rho^3\right]$$

$$\times\left[1 + hs_4.\rho + hs_5.\rho^2 + hs_6.\rho^3\right]^{-1} - \left[1 + hs_4.\rho + hs_5.\rho^2 + hs_6.\rho^3\right]^{-2} \tag{103}$$

$$\times\left[(hs_{4k} + hs_4).\rho + (hs_{5k} + 2hs_5).\rho^2 + (hs_{6k} + 3hs_6).\rho^3\right]\times\left[hs_1.\rho + hs_2.\rho^2 + hs_3.\rho^3\right]$$

The parameters used in Eq. (103) are given by equations (7) - (17) and the following equations:

$$hs_{1k} = (\bar{r}_k S_3 + \bar{r} S_{3k} + 3S_{1k}S_2 + 3S_1 S_{2k})P_{na} \tag{104}$$

$$hs_{2k} = (-2\bar{r}_k S_3^2 - 4\bar{r} S_{3k} S_3 - 3S_{1k}S_2S_3 - 3S_{2k}S_1S_3 - 3S_1S_2S_{3k} + 9S_{2k}S_2^2)P_{na}^2 \tag{105}$$

$$hs_{3k} = (\bar{r}_k S_3^3 + 3\bar{r} S_3^2 S_{3k} - 3S_2^2 S_{2k} S_3 - S_2^3 S_{3k})P_{na}^3 \tag{106}$$

$$hs_{4k} = -3S_{3k}P_{na} \tag{107}$$

$$hs_{5k} = 6S_3 S_{3k} P_{na}^2 \tag{108}$$

$$hs_{6k} = -3S_3^2 S_{3k} P_{na}^3 \tag{109}$$

where,

$$\bar{r}_k = (-\bar{r} + r_k) \tag{110}$$

$$S_{1k} = -S_1 + r_k d_k \tag{111}$$

$$S_{2k} = -S_2 + r_k d_k^2 \tag{112}$$

$$S_{3k} = -S_3 + r_k d_k^3 \tag{113}$$

$$k = 1,2, \dots, N_c \tag{113}$$

The non-ideality of the mixture due to formation of chain molecules of the various components which was described using the second term in Eq. (101) is derived as the following form:

$$IR\_Chain = \int_V^{\infty} \frac{RT}{V}\left[\frac{\partial(nZ_{ch})}{\partial n_k}\right]_{T,V,n_{j\neq k}} dV = \int_0^{\rho} \frac{RT}{\rho}\left[\frac{\partial(nZ_{ch})}{\partial n_k}\right]_{T,V,n_{j\neq k}} d\rho \tag{114}$$

$$k = 1, 2, \dots, N_c$$

$$\left[\frac{\partial(nZ_{ch})}{\partial n_k}\right]_{T,V,n_{j\neq k}}$$

$$\times\sum_{i=1}^{N_c}\left[\begin{array}{c} \delta_{ik}.(1-r_i)\times\left[ch_{1k}(d_i).\rho + ch_{2k}(d_i).\rho^2 + ch_{3k}(d_i).\rho^3\right] \\ \times\left[2 + ch_{4k}(d_i).\rho + ch_{5k}(d_i).\rho^2 + ch_{6k}(d_i).\rho^3 + ch_{7k}(d_i).\rho^4\right]^{-1} \\ +z_i(1-r_i)\times\left[2 + ch_{4k}(d_i).\rho + ch_{5k}(d_i).\rho^2 + ch_{6k}(d_i).\rho^3 + ch_{7k}(d_i).\rho^4\right]^{-1} \\ \left[\begin{array}{c} (ch_{1k}(d_i) + ch_1(d_i)).\rho \\ +(ch_{2k}(d_i) + 2ch_2(d_i)).\rho^2 + (ch_{3k}(d_i) + 3ch_3(d_i)).\rho^3 \end{array}\right] \\ -z_i(1-r_i)\times\left[ch_1(d_i).\rho + ch_2(d_i).\rho^2 + ch_3(d_i).\rho^3\right] \\ \times\left[2 + ch_4(d_i).\rho + ch_5(d_i).\rho^2 + ch_6(d_i).\rho^3 + ch_7(d_i).\rho^4\right]^{-2} \\ \left[\begin{array}{c} (ch_{4k}(d_i) + ch_4(d_i)).\rho + (ch_{5k}(d_i) + 2ch_5(d_i)).\rho^2 \\ +(ch_{6k}(d_i) + 3ch_6(d_i)).\rho^3 + (ch_{7k}(d_i) + 4ch_7(d_i)).\rho^4 \end{array}\right] \end{array}\right] \tag{115}$$

$$k = 1,2,\dots,N_c$$

where the parameters are described using equations (21) - (27) and the equations given below:

$$ch_{1k}(d_i) = (2S_{3k} + 3d_iS_{2k})P_{na} \quad (116)$$
$$ch_{2k}(d_i) = (-8S_3S_{3k} + 4d_i^2S_2S_{2k})P_{na}^2 \quad (117)$$
$$ch_{3k}(d_i) = (2d_i^2S_2S_{2k}S_3 + d_i^2S_2^2S_{3k} + 6S_3^2S_{3k} - 3d_iS_{2k}S_3^2 - 6d_iS_2S_3S_{3k})P_{na}^3 \quad (118)$$
$$ch_{4k}(d_i) = (-6S_{3k} + 3d_iS_{2k})P_{na} \quad (119)$$
$$ch_{5k}(d_i) = (12S_3S_{3k} + 2d_i^2S_2S_{2k} - 9d_iS_{2k}S_3 - 9d_iS_2S_{3k})P_{na}^2 \quad (120)$$
$$ch_{6k}(d_i) = (9d_iS_{2k}S_3^2 + 18d_iS_2S_3S_{3k} - 6S_3^2S_{3k} - 2d_i^2S_2S_{2k}S_3 - d_i^2S_2^2S_{3k})P_{na}^3 \quad (121)$$
$$ch_{7k}(d_i) = (-3d_iS_{2k}S_3^3 - 9d_iS_2S_3^2S_{3k})P_{na}^4 \quad (122)$$
$$k = 1,2,\dots,N_c \quad (122)$$

The dispersion contribution to the non-ideal behavior of the mixture (the third term in the right hand side of Eq. (101) is derived as the following forms:

$$IR\_Disp = \int_V^\infty \frac{RT}{V}\left[\frac{\partial(nZ_{disp})}{\partial n_k}\right]_{T,V,n_{j\neq k}} dV = \int_0^\rho \frac{RT}{\rho}\left[\frac{\partial(nZ_{disp})}{\partial n_k}\right]_{T,V,n_{j\neq k}} d\rho \quad (123)$$
$$k = 1,2,\dots,N_c \quad (123)$$

$$\left[\frac{\partial(nZ_{disp})}{\partial n_k}\right]_{T,V,n_{j\neq k}} = Z_{disp} + (Z_{disp})_k \quad (124)$$

$$(Z_{disp})_k = A_{1k} + A_{2k}A_3 + A_2A_{3k}, k = 1,2,\dots,N_c \quad (124)$$

where,

$$A_1 = dI_1.\rho + dI_2.\rho^2 + dI_3.\rho^3 + dI_4.\rho^4 + dI_5.\rho^5 + dI_6.\rho^6 + dI_7.\rho^7 \quad (125)$$

$$\begin{gathered} A_{1k} = (dI_{1k} + dI_1).\rho + (dI_{2k} + 2dI_2).\rho^2 + (dI_{3k} + 3dI_3).\rho^3 \\ +(dI_{4k} + 4dI_4).\rho^4 + (dI_{5k} + 5dI_5).\rho^5 + (dI_{6k} + 6dI_6).\rho^6 + (dI_{7k} + 7dI_7).\rho^7 \\ k = 1,2,\dots,N_c \end{gathered} \quad (126)$$

$$A_2 = -6P_{na}\overline{r}Adisp_2 \quad (127)$$

$$\begin{gathered} A_{2k} = -6P_{na}\left[\overline{r}_k.Adisp_2 + \overline{r}.Adisp_{2k}\right] \\ k = 1,2,\dots,N_c \end{gathered} \quad (128)$$

$$A_3 = \left(\frac{B_1}{B_2}\right) \cdot B_3 - \left(\frac{B_1}{\rho B_2}\right)^2 \cdot \left(\frac{B_4}{B_5}\right) \cdot B_6 \cdot B_7 \tag{129}$$

$$B_1 = 4\rho + disp_1 \cdot \rho^2 + disp_2 \cdot \rho^3 + disp_3 \cdot \rho^4 + disp_4 \cdot \rho^5 + disp_5 \cdot \rho^6 + disp_6 \cdot \rho^7 \tag{130}$$

$$\begin{gathered} B_{1k} = 4\rho + \left(d_{1k} + 2disp_1\right) \cdot \rho^2 + \left(d_{2k} + 3disp_2\right) \cdot \rho^3 + \left(d_{3k} + 4disp_3\right) \cdot \rho^4 \\ + \left(d_{4k} + 5disp_4\right) \cdot \rho^5 + \left(d_{5k} + 6disp_5\right) \cdot \rho^6 + \left(d_{6k} + 7disp_6\right) \cdot \rho^7 \\ k = 1, 2, \ldots, N_c \end{gathered} \tag{131}$$

$$B_2 = 4 + disp_7 \cdot \rho + disp_8 \cdot \rho^2 + disp_9 \cdot \rho^3 + disp_{10} \cdot \rho^4 + disp_{11} \cdot \rho^5 + disp_{12} \cdot \rho^6 \tag{132}$$

$$\begin{gathered} B_{2k} = \left(d_{7k} + disp_7\right) \cdot \rho + \left(d_{8k} + 2disp_8\right) \cdot \rho^2 + \left(d_{9k} + 3disp_9\right) \cdot \rho^3 \\ + \left(d_{10k} + 4disp_{10}\right) \cdot \rho^4 + \left(d_{11k} + 5disp_{11}\right) \cdot \rho^5 + \left(d_{12k} + 6disp_{12}\right) \cdot \rho^6 \end{gathered} \tag{133}$$

$$B_3 = b_0 + dI_8 \cdot \rho + dI_9 \cdot \rho^2 + dI_{10} \cdot \rho^3 + dI_{11} \cdot \rho^4 + dI_{12} \cdot \rho^5 + dI_{13} \cdot \rho^6 \tag{134}$$

$$\begin{gathered} B_{3k} = b_{0k} + \left(dI_{8k} + dI_8\right) \cdot \rho + \left(dI_{9k} + 2dI_9\right) \cdot \rho^2 + \left(dI_{10k} + 3dI_{10}\right) \cdot \rho^3 \\ + \left(dI_{11k} + 4dI_{11}\right) \cdot \rho^4 + \left(dI_{12k} + 5dI_{12}\right) \cdot \rho^5 + \left(dI_{13k} + 6dI_{13}\right) \cdot \rho^6 \end{gathered} \tag{135}$$

$$B_4 = disp_{13} + disp_{14} \cdot \rho + disp_{15} \cdot \rho^2 + disp_{16} \cdot \rho^3 + disp_{17} \cdot \rho^4 + disp_{18} \cdot \rho^5 \tag{136}$$

$$\begin{gathered} B_{4k} = d_{13k} + \left(d_{14k} + disp_{14}\right) \cdot \rho + \left(d_{15k} + 2disp_{15}\right) \cdot \rho^2 \\ + \left(d_{16k} + 3disp_{16}\right) \cdot \rho^3 + \left(d_{17k} + 4disp_{17}\right) \cdot \rho^4 + \left(d_{18k} + 5disp_{18}\right) \cdot \rho^5 \end{gathered} \tag{137}$$

$$\begin{gathered} B_5 = 8 + disp_{19} \cdot \rho + disp_{20} \cdot \rho^2 + disp_{21} \cdot \rho^3 + disp_{22} \cdot \rho^4 + disp_{23} \cdot \rho^5 \\ + disp_{24} \cdot \rho^6 + disp_{25} \cdot \rho^7 + disp_{26} \cdot \rho^8 \end{gathered} \tag{138}$$

$$\begin{gathered} B_{5k} = \left(d_{19k} + disp_{19}\right) \cdot \rho + \left(d_{20k} + 2disp_{20}\right) \cdot \rho^2 \\ + \left(d_{21k} + 3disp_{21}\right) \cdot \rho^3 + \left(d_{22k} + 4disp_{22}\right) \cdot \rho^4 + \left(d_{23k} + 5disp_{23}\right) \cdot \rho^5 \\ + \left(d_{24k} + 6disp_{24}\right) \cdot \rho^6 + \left(d_{25k} + 7disp_{25}\right) \cdot \rho^7 + \left(d_{26k} + 8disp_{26}\right) \cdot \rho^8 \end{gathered} \tag{139}$$

$$B_6 = b_0 \cdot \rho^2 + Idisp_1 \cdot \rho^3 + Idisp_2 \cdot \rho^4 + Idisp_3 \cdot \rho^5 + Idisp_4 \cdot \rho^6 + Idisp_5 \cdot \rho^7 + Idisp_6 \cdot \rho^8 \tag{140}$$

$$\begin{gathered} B_{6k} = \left(b_{0k} + 2b_0\right) \cdot \rho^2 + \left(Idisp_{1k} + 3Idisp_1\right) \cdot \rho^3 \\ + \left(Idisp_{2k} + 4Idisp_2\right) \cdot \rho^4 + \left(Idisp_{3k} + 5Idisp_3\right) \cdot \rho^5 \\ + \left(Idisp_{4k} + 6Idisp_4\right) \cdot \rho^6 + \left(Idisp_{5k} + 7Idisp_5\right) \cdot \rho^7 + \left(Idisp_{6k} + 8Idisp_6\right) \cdot \rho^8 \end{gathered} \tag{141}$$

$$B_7 = P_{na} S_3 \tag{142}$$

$$B_{7k} = P_{na} S_{3k} \quad k = 1,2,\ldots,N_c \tag{143}$$

The parameters used in Eqs. (125) - (143) are given through Eqs. (40) - (92) and the following equations:

$$d_{1k} = 12[\bar{r}_k S_3 + \bar{r} S_{3k}] P_{na} \tag{144}$$
$$d_{2k} = [27\bar{r}_k S_3^2 + 2(27\bar{r} - 26) S_3 S_{3k}] P_{na}^2 \tag{145}$$
$$d_{3k} = [70\bar{r}_k S_3^3 + 3S_3^2 S_{3k}(70\bar{r} + 42)] P_{na}^3 \tag{146}$$
$$d_{4k} = [51\bar{r}_k S_3^4 + 3S_3^3 S_{3k}(51\bar{r} - 27)] P_{na}^4 \tag{147}$$
$$d_{5k} = [-16\bar{r}_k S_3^5 + 5S_3^4 S_{3k}(8 - 16\bar{r})] P_{na}^5 \tag{148}$$
$$d_{6k} = [2\bar{r}_k S_3^6 + 6S_3^5 S_{3k}(2\bar{r} - 1)] P_{na}^6 \tag{149}$$
$$d_{7k} = -20 S_{3k} P_{na} \tag{150}$$
$$d_{8k} = 82 S_3 S_{3k} P_{na}^2 \tag{151}$$
$$d_{9k} = -132 S_3^2 S_{3k} P_{na}^3 \tag{152}$$
$$d_{10k} = 104 S_3^3 S_{3k} P_{na}^4 \tag{153}$$
$$d_{11k} = -40 S_3^4 S_{3k} P_{na}^5 \tag{154}$$
$$d_{12k} = 6 S_3^5 S_{3k} P_{na}^6 \tag{155}$$
$$d_{13k} = 24\bar{r}_k \tag{156}$$
$$d_{14k} = [192\bar{r}_k S_3 + (192\bar{r} - 128) S_{3k}] P_{na} \tag{157}$$
$$d_{15k} = [-372\bar{r}_k S_3^2 + 2S_3 S_{3k}(148 - 372\bar{r})] P_{na}^2 \tag{158}$$
$$d_{16k} = [230\bar{r}_k S_3^3 + 3S_3^2 S_{3k}(230\bar{r} - 70)] P_{na}^3 \tag{159}$$
$$d_{17k} = [-52\bar{r}_k S_3^4 + 4S_3^3 S_{3k}(8 - 52\bar{r})] P_{na}^4 \tag{160}$$
$$d_{18k} = [2\bar{r}_k S_3^5 + 10 S_3^4 S_{3k}(1 + \bar{r})] P_{na}^5 \tag{161}$$
$$d_{19k} = -52 S_{3k} P_{na} \tag{162}$$
$$d_{20k} = 292 S_3 S_{3k} P_{na}^2 \tag{163}$$
$$d_{21k} = -693 S_3^2 S_{3k} P_{na}^3 \tag{164}$$
$$d_{22k} = 900 S_3^3 S_{3k} P_{na}^4 \tag{165}$$
$$d_{23k} = -690 S_3^4 S_{3k} P_{na}^5 \tag{166}$$
$$d_{24k} = 312 S_3^5 S_{3k} P_{na}^6 \tag{167}$$
$$d_{25k} = -77 S_3^6 S_{3k} P_{na}^7 \tag{168}$$

$$d_{26k} = 8S_3^7 S_{3k} P_{na}^8 \tag{169}$$

$$k = 1,2, \dots, N_c \tag{169}$$

$$\mathrm{Adisp}_{1k} = 2 \times \left[-\sum_{i=1}^{N_c} \sum_{j=1}^{N_c} z_i z_j r_i r_j \left(\varepsilon_{ij}/kT \right) \sigma_{ij}^3 + \sum_{i=1}^{N_c} z_i r_i r_k (\varepsilon_{ik}/kT) \sigma_{ik}^3 \right] \tag{170}$$

$$\mathrm{Adisp}_{2k} = 2 \times \left[-\sum_{i=1}^{N_c} \sum_{j=1}^{N_c} z_i z_j r_i r_j \left(\varepsilon_{ij}/kT \right)^2 \sigma_{ij}^3 + \sum_{i=1}^{N_c} z_i r_i r_k (\varepsilon_{ik}/kT)^2 \sigma_{ik}^3 \right] \tag{171}$$

$$a_{jk}(\bar{r}) = \begin{bmatrix} \left[\frac{\bar{r}_k}{\bar{r}} - \frac{\bar{r}_k \cdot (\bar{r}-1)}{\bar{r}^2} \right] a_{1j} \\ + \left[\frac{\bar{r}_k \cdot (\bar{r}-2)}{\bar{r}^2} + \frac{\bar{r}_k \cdot (\bar{r}-1)}{\bar{r}^2} - 2 \frac{\bar{r}_k \cdot (\bar{r}-1) \cdot (\bar{r}-2)}{\bar{r}^3} \right] a_{2j} \end{bmatrix} \tag{172}$$

$$b_{jk}(\bar{r}) = \begin{bmatrix} \left[\frac{\bar{r}_k}{\bar{r}} - \frac{\bar{r}_k \cdot (\bar{r}-1)}{\bar{r}^2} \right] b_{1j} \\ + \left[\frac{\bar{r}_k \cdot (\bar{r}-2)}{\bar{r}^2} + \frac{\bar{r}_k \cdot (\bar{r}-1)}{\bar{r}^2} - 2 \frac{\bar{r}_k \cdot (\bar{r}-1) \cdot (\bar{r}-2)}{\bar{r}^3} \right] b_{2j} \end{bmatrix} \tag{173}$$

$$j = 0,1, \dots ,6 \; k = 1,2, \dots, N_c \tag{173}$$

$$\mathrm{Idisp}_{\ 1k} = [b_{1k} S_3 + b_1 S_{3k}] P_{na} \tag{174}$$

$$\mathrm{Idisp}_{\ 2k} = [b_{2k} S_3^2 + 2b_2 S_3 S_{3k}] P_{na}^2 \tag{175}$$

$$\mathrm{Idisp}_{\ 3k} = [b_{3k} S_3^3 + 3b_3 S_3^2 S_{3k}] P_{na}^3 \tag{176}$$

$$Idisp_{4k} = [b_{4k} S_3^4 + 4b_4 S_3^3 S_{3k}] P_{na}^4 \tag{177}$$

$$\mathrm{Idisp}_{\ 5k} = [b_{5k} S_3^5 + 5b_5 S_3^4 S_{3k}] P_{na}^5 \tag{178}$$

$$\mathrm{Idisp}_{\ 6k} = [b_{6k} S_3^6 + 6b_6 S_3^5 S_{3k}] P_{na}^6 \tag{179}$$

$$\mathrm{dIdisp}_{\ 7k} = 2[a_{1k} S_3 + a_1 S_{3k}] P_{na} \tag{180}$$

$$\mathrm{dIdisp}_{\ 2k} = 3[a_{2k} S_3^2 + 2a_2 S_3 S_{3k}] P_{na}^2 \tag{181}$$

$$\mathrm{dIdisp}_{\ 3k} = 4[a_{3k} S_3^3 + 3a_3 S_3^2 S_{3k}] P_{na}^3 \tag{182}$$

$$\mathrm{dIdisp}_{\ 4k} = 5[a_{4k} S_3^4 + 4a_4 S_3^3 S_{3k}] P_{na}^4 \tag{183}$$

$$\mathrm{dIdisp}_{\ 5k} = 6[a_{5k} S_3^5 + 5a_5 S_3^4 S_{3k}] P_{na}^5 \tag{184}$$

$$dIdisp_{6k} = 7[a_{6k} S_3^6 + 6a_6 S_3^5 S_{3k}] P_{na}^6 \tag{185}$$

$$dI_{1k} = -12[Adisp_{1k} \cdot a_0 + Adisp_1 \cdot a_{0k}]P_{na} \tag{186}$$

$$dI_{2k} = -12\left[Adisp_{1k} \cdot dIdisp_1 + Adisp_1 \cdot dI \text{ disp }_{1k}\right]P_{na} \tag{187}$$

$$dI_{3k} = -12\left[Adisp_{1k} \cdot dIdisp_2 + Adisp_1 \cdot dI \text{ disp }_{2k}\right]P_{na} \tag{188}$$

$$dI_{4k} = -12\left[Adisp_{1k} \cdot dIdisp_3 + Adisp_1 \cdot dI \text{ disp }_{3k}\right]P_{na} \tag{189}$$

$$dI_{5k} = -12[Adisp_{1k} \cdot dIdisp_4 + Adisp_1 \cdot dIdisp_{4k}]P_{na} \tag{190}$$

$$dI_{6k} = -12[Adisp_{1k} \cdot dIdisp_5 + Adisp_1 \cdot dIdisp_{5k}]P_{na} \tag{191}$$

$$dI_{7k} = -12[Adisp_{1k} \cdot dIdisp_6 + Adisp_1 \cdot dIdisp_{6k}]P_{na} \tag{192}$$

$$dI_{8k} = 2[b_{1k}S_3 + b_1 S_{3k}]P_{na} \tag{193}$$

$$dI_{9k} = 3[b_{2k}S_3^2 + 2b_2 S_3 S_{3k}]P_{na}^2 \tag{194}$$

$$dI_{10k} = 4[b_{3k}S_3^3 + 3b_3 S_3^2 S_{3k}]P_{na}^3 \tag{195}$$

$$dI_{11k} = 5[b_{4k}S_3^4 + 4b_4 S_3^3 S_{3k}]P_{na}^4 \tag{196}$$

$$dI_{12k} = 6[b_{5k}S_3^5 + 5b_5 S_3^4 S_{3k}]P_{na}^5 \tag{197}$$

$$dI_{13k} = 7[b_{6k}S_3^6 + 6b_6 S_3^5 S_{3k}]P_{na}^6 \tag{198}$$

The association contribution to the non-ideal behavior of mixtures containing associating compounds (the last term given by the right hand side of Eq. (101) is derived as the following forms:

$$IR\_Assoc = \int_V^\infty \frac{RT}{V}\left[\frac{\partial(nZ_{\text{assoc}})}{\partial n_k}\right]_{T,V,n_{j\neq k}} dV = \int_0^\rho \frac{RT}{\rho}\left[\frac{\partial(nZ_{\text{assoc}})}{\partial n_k}\right]_{T,V,n_{j\neq k}} d\rho \tag{199}$$

$$k = 1,2,\ldots,N_c \tag{199}$$

$$\left[\frac{\partial(nZ_{\text{assoc}})}{\partial n_k}\right]_{T,V,n_{j\neq k}} = Z_{\text{assoc}} + \rho(A_k) + \rho^2 \sum_{i=1}^{N_c} z_i \left(\frac{\partial A_i}{\partial n_k}\right)_{T,V,n_{j\neq k}} \tag{200}$$

$$k = 1,2,\ldots,N_c \tag{200}$$

where,

$$\left(\frac{\partial A_i}{\partial n_k}\right)_{T,V,n_{j\neq k}} = \sum_{S_i} -\frac{1}{(X^{S_i})^2}\left(\frac{\partial X^{S_i}}{\partial n_k}\right)_{T,V,n_{j\neq k}} \cdot \left(\frac{\partial X^{S_i}}{\partial \rho}\right) + \left(\frac{1}{X^{S_i}} - \frac{1}{2}\right)\left[\frac{\partial}{\partial n_k}\left(\frac{\partial X^{S_i}}{\partial \rho}\right)\right]_{T,V,n_{j\neq k}} \tag{201}$$

$$k = 1,2,\ldots,N_c$$

$$A_i = \sum_{S_i} \left(\frac{1}{X^{S_i}} - \frac{1}{2}\right)\left(\frac{\partial X^{S_i}}{\partial \rho}\right) \tag{202}$$

where ($\partial X^{S_i}/\partial \rho$) is given by Eq. (96), other derivatives of X^{S_i} used in Eq. (202) are given below:

$$\left(\frac{\partial X^{S_i}}{\partial n_k}\right)_{T,V,n_j\neq k} = -(X^{S_i})^2 N_A \left[\sum_{j=1}^{N_c}\sum_{Y_j} \frac{1}{V}\delta_{jk} X^{Y_j} W_{ij} + \sum_{j=1}^{N_c}\sum_{Y_j} z_j \rho X^{Y_j} \left(\frac{\partial W_{ij}}{\partial n_k}\right)_{T,V,n_j\neq k} + \sum_{j=1}^{N_c}\sum_{Y_j} z_j \rho W_{ij} \left(\frac{\partial X^{Y_j}}{\partial n_k}\right)_{T,V,n_j\neq k} \right] \tag{203}$$

$$k = 1, 2, \ldots, N_c$$

$$\left[\frac{\partial}{\partial n_k}\left(\frac{\partial X^{S_i}}{\partial \rho}\right)\right]_{T,V,n_j\neq k} = 2\left(X^{S_i}\right)^{-1}\left(\frac{\partial X^{S_i}}{\partial n_k}\right)_{T,V,n_{j\neq k}} \cdot \left(\frac{\partial X^{S_i}}{\partial \rho}\right)$$
$$-\left(X^{S_i}\right)^2 N_A \left[\begin{aligned} &\sum_{j=1}^{N_c}\sum_{Y_j} \frac{\delta_{jk} - z_j}{n} X^{Y_j} W_{ij} + \sum_{j=1}^{N_c}\sum_{Y_j} z_j W_{ij} \left(\frac{\partial X^{Y_j}}{\partial n_k}\right)_{T,V,n_{j\neq k}} + \sum_{j=1}^{N_c}\sum_{Y_j} z_j X^{Y_j} \left(\frac{\partial W_{ij}}{\partial n_k}\right)_{T,V,n_{j\neq k}} \\ &+ \sum_{j=1}^{N_c}\sum_{Y_j} \frac{\delta_{jk}}{V} X^{Y_j} \left(\frac{\partial W_{ij}}{\partial \rho}\right) + \sum_{j=1}^{N_c}\sum_{Y_j} z_j \rho \left(\frac{\partial X^{Y_j}}{\partial n_k}\right)_{T,V,n_{j\neq k}} \cdot \left(\frac{\partial W_{ij}}{\partial \rho}\right) \\ &+ \sum_{j=1}^{N_c}\sum_{Y_j} z_j \rho X^{Y_j} \left(\frac{\partial}{\partial n_k}\left(\frac{\partial W_{ij}}{\partial \rho}\right)\right)_{T,V,n_{j\neq k}} + \sum_{j=1}^{N_c}\sum_{Y_j} \frac{\delta_{jk}}{V} W_{ij} \left(\frac{\partial X^{Y_j}}{\partial \rho}\right) \\ &+ \sum_{j=1}^{N_c}\sum_{Y_j} z_j \rho \left(\frac{\partial W_{ij}}{\partial n_k}\right)_{T,V,n_{j\neq k}} \cdot \left(\frac{\partial X^{Y_j}}{\partial \rho}\right) + \sum_{j=1}^{N_c}\sum_{Y_j} z_j \rho W_{ij} \left(\frac{\partial}{\partial n_k}\left(\frac{\partial X^{Y_j}}{\partial \rho}\right)\right)_{T,V,n_{j\neq k}} \end{aligned} \right] \tag{204}$$

where $\left(\partial W_{ij}/\partial\rho\right)$ is given by Eq. (97), other derivatives of W_{ij} are given as the following

$$\left(\frac{\partial W_{ij}}{\partial n_k}\right)_{T,V,n_{j\neq k}} = \left(\sigma_{ij}k^{S_iY_j}\right)\left[exp\left(\frac{\varepsilon^{S_iY_j}}{kT}\right)-1\right]$$
$$\times\left[\begin{array}{c}(1-\xi_3)^{-2}\frac{\partial\xi_3}{\partial n_k}+3\frac{d_id_j}{d_i+d_j}\left[(1-\xi_3)^{-2}\left(\frac{\partial\xi_2}{\partial n_k}\right)+2(1-\xi_3)^{-3}\xi_2\left(\frac{\partial\xi_3}{\partial n_k}\right)\right]\\ +2\left(\frac{d_id_j}{d_i+d_j}\right)^2\left[2(1-\xi_3)^{-3}\xi_2\left(\frac{\partial\xi_2}{\partial n_k}\right)+3(1-\xi_3)^{-4}\left(\frac{\partial\xi_3}{\partial n_k}\right)\xi_2^2\right]\end{array}\right] \quad (205)$$
$$k=1,2,\ldots,N_c$$

$$\left(\frac{\partial}{\partial n_k}\left(\frac{\partial W_{ij}}{\partial\rho}\right)\right)_{T,V,n_{j\neq k}}$$
$$=\left(\sigma_{ij}k^{S_iY_j}\right)\left[exp\left(\frac{\varepsilon^{S_iY_j}}{kT}\right)-1\right]$$
$$\times\left[\begin{array}{c}2\left(1-\xi_3\right)^{-3}\left(\frac{\partial\xi_3}{\partial n_k}\right)\left(\frac{\partial\xi_3}{\partial\rho}\right)+\left(1-\xi_3\right)^{-2}\frac{\partial}{\partial n_k}\left(\frac{\partial\xi_3}{\partial\rho}\right)\\ +\frac{3d_id_j}{d_i+d_j}\left[\begin{array}{c}2\left(1-\xi_3\right)^{-3}\left(\frac{\partial\xi_3}{\partial n_k}\right)\left(\frac{\partial\xi_2}{\partial\rho}\right)+\left(1-\xi_3\right)^{-2}\frac{\partial}{\partial n_k}\left(\frac{\partial\xi_2}{\partial\rho}\right)\\ +6\left(1-\xi_3\right)^{-4}\left(\frac{\partial\xi_3}{\partial n_k}\right)\xi_2\left(\frac{\partial\xi_3}{\partial\rho}\right)+2\left(1-\xi_3\right)^{-3}\left(\frac{\partial\xi_2}{\partial n_k}\right)\left(\frac{\partial\xi_3}{\partial\rho}\right)\\ +2\left(1-\xi_3\right)^{-3}\xi_2\frac{\partial}{\partial n_k}\left(\frac{\partial\xi_3}{\partial\rho}\right)\end{array}\right]\\ +2\left(\frac{d_id_j}{d_i+d_j}\right)^2\left[\begin{array}{c}2\left(1-\xi_3\right)^{-3}\left(\frac{\partial\xi_2}{\partial n_k}\right)\left(\frac{\partial\xi_2}{\partial\rho}\right)+2\xi_2\left(1-\xi_3\right)^{-3}\frac{\partial}{\partial n_k}\left(\frac{\partial\xi_2}{\partial\rho}\right)\\ +6\left(1-\xi_3\right)^{-4}\left(\frac{\partial\xi_3}{\partial n_k}\right)\xi_2\left(\frac{\partial\xi_2}{\partial\rho}\right)+12\left(1-\xi_3\right)^{-5}\left(\frac{\partial\xi_3}{\partial n_k}\right)\left(\frac{\partial\xi_3}{\partial\rho}\right)\xi_2^2\\ +3\left(1-\xi_3\right)^{-4}\frac{\partial}{\partial n_k}\left(\frac{\partial\xi_3}{\partial\rho}\right)\xi_2^2+6\left(1-\xi_3\right)^{-4}\left(\frac{\partial\xi_2}{\partial n_k}\right)\xi_2\left(\frac{\partial\xi_3}{\partial\rho}\right)\end{array}\right]\end{array}\right] \quad (206)$$
$$k=1,2,\ldots,N_c$$

here,

$$\frac{\partial \xi_2}{\partial \rho} = P_{na} S_2 \tag{207}$$

$$\frac{\partial \xi_3}{\partial \rho} = P_{na} S_3 \tag{208}$$

$$\left(\frac{\partial \xi_2}{\partial n_k}\right)_{T,V,n_{j\neq k}} = \frac{1}{V} P_{na} r_k d_k^2 \tag{209}$$

$$\left(\frac{\partial \xi_3}{\partial n_k}\right)_{T,V,n_{j\neq k}} = \frac{1}{V} P_{na} r_k d_k^3 \tag{210}$$

$$\left(\frac{\partial}{\partial n_k}\left(\frac{\partial \xi_2}{\partial \rho}\right)\right)_{T,V,n_{j\neq k}} = P_{na} \cdot \sum_{i=1}^{N_c} \left(\frac{\delta_{ik} - z_i}{n}\right) r_i d_i^2 = \frac{1}{V} \cdot P_{na} \cdot \frac{1}{\rho} [r_k d_k^2 - S_2] \tag{211}$$

$$\left(\frac{\partial}{\partial n_k}\left(\frac{\partial \xi_3}{\partial \rho}\right)\right)_{T,V,n_{j\neq k}} = P_{na} \cdot \sum_{i=1}^{N_c} \left(\frac{\delta_{ik} - z_i}{n}\right) r_i d_i^3 = \frac{1}{V} \cdot P_{na} \cdot \frac{1}{\rho} [r_k d_k^3 - S_3] \tag{212}$$

$$kk = 1,2, \ldots, N_{c_c} \qquad (212)$$

In the above equations, δ_{jk} refers to dirac delta function which is defined as following:

$$\delta_{jk}\delta_{ik} = \begin{cases} 0 & j \neq k \\ 1 & j = k \end{cases} \tag{213}$$

3. Utilization of Chebyshev polynomials for density computation

The integration of the terms in equations (102), (114), (123), and (119) for computing the fugacity coefficients is executed numerically utilizing the Gaussian quadrature method. The five-point quadrature approach yields results with satisfactory precision. Equation (101) illustrates that the fugacity coefficient depends on temperature, pressure, composition, and the characteristics of the components. To determine the fugacity coefficient of each component, we must first compute the mixture's density at a certain pressure, temperature, and composition using Equation (2). The equation indicates that the density as a function of pressure is not clearly defined. Consequently, the determination of density at a certain pressure must be executed through an iterative process, commencing with initial approximations due to the existence of several solutions. A solution derived via an iterative method is contingent upon the selection of the first estimate. Consequently, iterative methods cannot encompass all valid roots unless the quantity of roots and the estimated values of the solutions (i.e., starting estimates) are already known. Therefore, a robust, rapid, and precise alternative technique capable of predicting all possible solutions is given. The suggested approach relies on numerical interpolation utilizing Chebyshev polynomials inside a finite interval (Burden et al., 1981).

It should be pointed out that Chebyshev series provide high accuracy and can be transformed to power series which are suitable for root finding procedure. More general accounts of root finding through Chebyshev approximations are given in (Boyd, 2006). The aforementioned method enables us to calculate all possible solutions and select among them those which are physically interpretable.

It should be considered that using Chebyshev polynomials to approximate a given function will become more efficient when it has non-zero values at both end points of the interval. It can be shown that the pressure vs. density function in SAFT EoS linearly goes to zero for negligible values of the density. In order to avoid this problem, P/ρ vs. density using Chebyshev polynomials was interpolated.

Another advantage of using Chebyshev polynomials for approximating a function is that for a specific number of basic functions, it always leads to a well-conditioned matrix during the calculation of the unknown coefficients of the basis functions, which is more accurate than the other interpolation techniques.

Figure 2 shows the interpolation error using Chebyshev polynomials of degree 15 for approximating pressure vs. density of a binary mixture of ethanol and toluene containing 37.5 mole% ethanol. Figure 3 shows the error in interpolation for another system (oil sample) for which the composition is given in Table 2 (Jamaluddin et al., 2000).

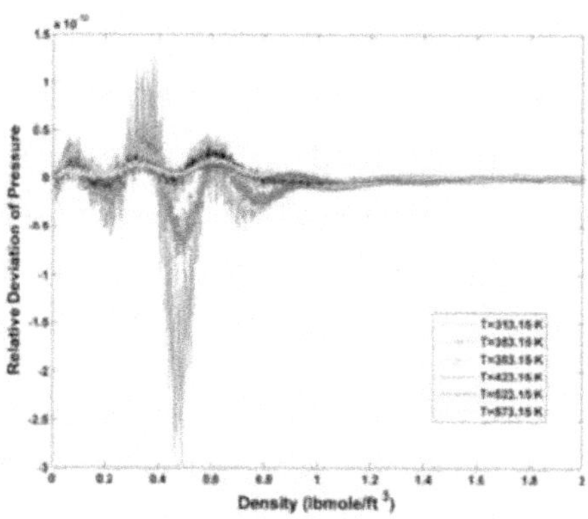

Fig. 2. Interpolation error using Chebyshev polynomials for approximating preesure vs. density of a binary mixture of ethanol and toluene containing 37.5 mole% of ethanol at different temperatures.

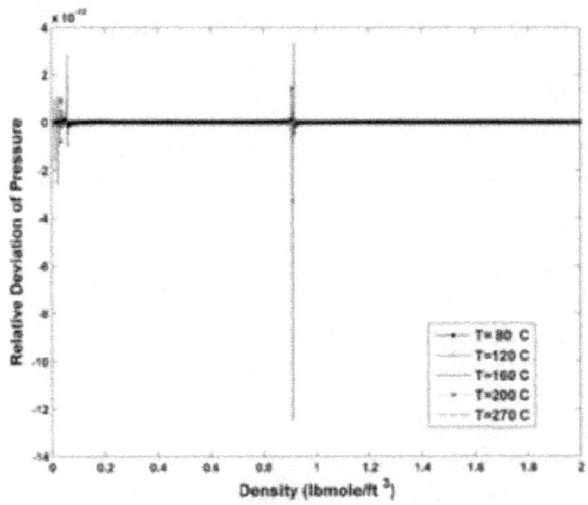

Fig. 3. Interpolation error using Chebyshev polynomials for approximating preesure vs. density at different temperatures for an oil sample of the composition given in Table 2.

After approximating the $P(\rho)$ function using Chebyshev polynomials, it is necessary to find solutions for density values at the given pressure(s) and select those which are physically interpretable. In doing so, the complex and negative solutions and those which make $\partial P/\partial \rho$ negative, are discarded because they have no physical meaning.

Table 2. Composition (mole%) and properties of the oil sample used to investigate the effect of temperature and pressure on asphaltene precipitation (Jamaluddin et al., 2000).

Component and Properties	Oil
N_2	0.49
CO_2	11.37
H_2S	3.22
C_1	27.36
C_2	9.41
C_3	6.70
iC_4	0.81
nC_4	3.17

iC_5	1.22
nC_5	1.98
C_6	2.49
C_{7+}	31.79
C_{7+} molecular weight	248.3
C_{7+} density (g/cm^3)	0.877

Figure 4 shows a typical plot of pressure versus density for SAFT EoS in the positive region of density. As it can be seen in Figure 4, the derivative of pressure with respect to density ($\partial P/\partial\rho$) has two zeros in this region for different values of the shown temperatures. For pressures between the maximum and minimum of the $P(\rho)$ function (e.g. the pressure region between two parallel lines passing through the maximum and minimum of the middle curve), the system has three zeros one of which is not acceptable. The smaller root corresponds to the vapor phase density and the larger root corresponds to the liquid phase density. At pressures below the minimum of $P(\rho)$, the function has only a single root which is identified as the vapor phase density. At pressures above the maximum of $P(\rho)$, only a single zero is detected for the function which is identified as the liquid phase density. By increasing the temperature (Figure 5), the roots of $\partial P/\partial\rho$ approaches to each other. At some temperature they coincide above which $\partial P/\partial\rho$ has not any zero. At these temperatures the system has only a single root for any value of the pressure which is identified as the vapor phase density. Therefore, the procedure for finding roots of the SAFT EoS at the given pressure can be summarized as the following:

1. The pressure versus density of SAFT EoS is approximated using Chebyshev polynomials.

2. The derivative of pressure with respect to density is calculated to find zeros of $\partial P/\partial\rho$. The complex and negative zeros are eliminated.

3. The roots of the fitting polynomial are estimated at the given pressure using a proper root finding algorithm for polynomials. The negative and complex roots and those which make $\partial P/\partial\rho$ negative are eliminated.

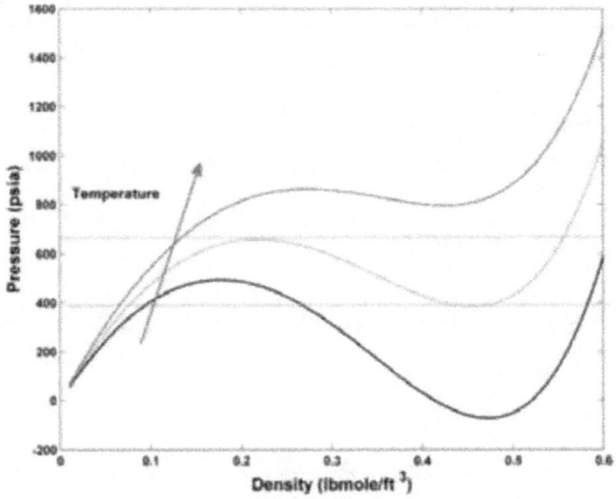

Fig. 4. A typical plot of pressure versus density of SAFT EoS at different temperatures.

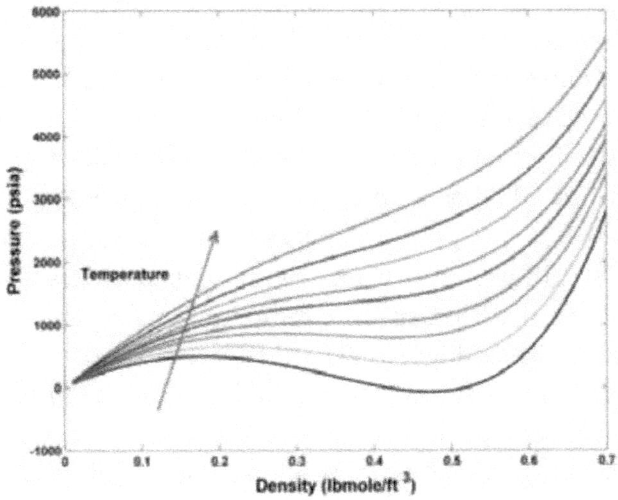

Fig. 5. A typical plot of pressure versus density of SAFT EoS at different temperatures.

4. If two physically meaning roots are obtained at the given pressure, the smaller root corresponds to vapor phase density and the larger one corresponds to the liquid phase density.

5. If the system has only a single root at the given pressure and $\partial P/\partial \rho$ has two zeros, if the obtained root is larger than the larger root of $\partial P/\partial \rho$, it is identified as the

liquid phase density, otherwise, if the estimated root is smaller than the smaller root of $\partial P/\partial \rho$, it corresponds to the vapor phase density.

6. If the system has only a single root at the given pressure and $\partial P/\partial \rho$ has not any zero, the calculated root is identified as the vapor phase density.

4. Findings and discussion

4.1 Calculation of density for binary mixtures of ethanol and toluene

The SAFT EoS was first applied to calculate densities of the asymmetrical binary systems composed of ethanol and toluene. Experimental liquid densities for ethanol (1) and toluene (2) and seven of their binary mixtures in the temperature range 283.15-353.15 K at each 10 K and for pressures up to 45 MPa in steps of 5 MPa are given in (Zeberg-Mikkelsen et al., 2005). No density measurements were performed at 353.15 K and 0.1 MPa for ethanol as well as for mixtures containing more than 25 mole% ethanol, since ethanol and all mixtures with a composition higher than 25 mole% ethanol is either located in the two phase region or the gaseous phase (Zeberg-Mikkelsen et al., 2005). A comparison of the experimental density values of the aforementioned binary mixtures and pure compounds with the values calculated using SAFT EoS has been performed in this work. Figure 6 shows plots of the compressibility factor (Z-factor) of ethanol for different pressures of 0.1,25 and 45 MPa using the SAFT EoS. As can be seen in this figure the contribution from the hard chain term ($Z_{hc} = Z_{hs} + Z_{ch}$), the dispersion term (Z_{disp}), and the association term (Z_{assoc}) are shown at different pressures versus density. Every point on a constant pressure curve is associated with a specific temperature. As the temperature rises, the density of the liquid diminishes. Figures 7-10 illustrate a comparison of experimental and estimated densities utilizing the SAFT equation, plotted against pressure at various temperatures. The average absolute values of the relative deviations (AAD) between experimental and estimated densities for various compositions of the binary mixes of ethanol and toluene at differing pressures and temperatures is 0.143%. Figure 11 illustrates the relative deviations for various combinations of ethanol and toluene in a three-dimensional graphic.

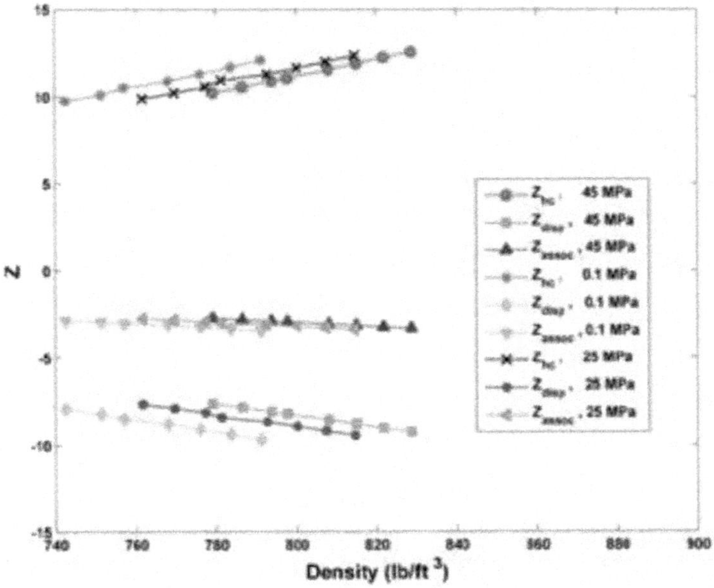

Fig. 6. Contributions to Z -factor of ethanol at different pressures and temperatures according to SAFT EoS.

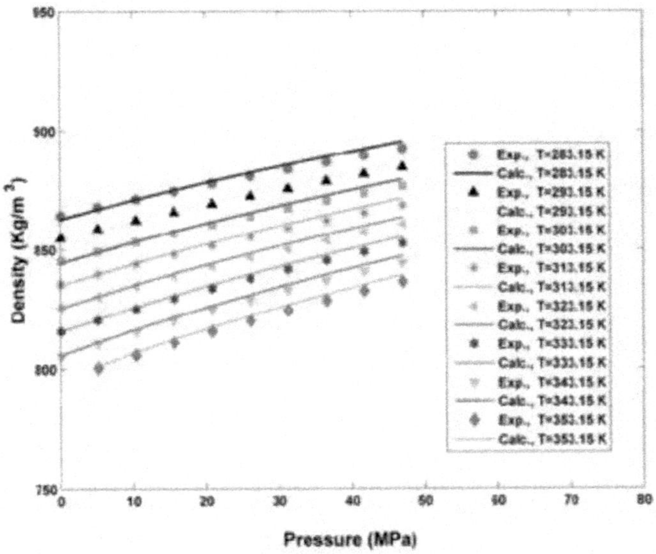

Fig. 7. Experimental and calculated densities versus pressure at different temperatures using SAFT EoS for binary system composed of ethanol and toluene at 25.0 mole% **of ethanol.**

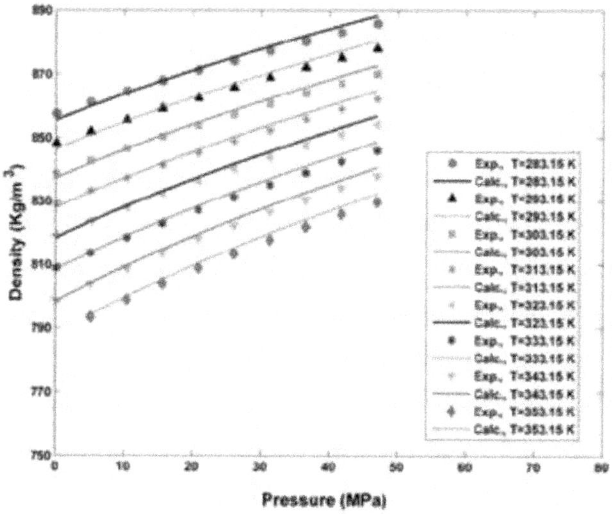

Fig. 8. Experimental and calculated densities versus pressure at different temperatures using SAFT EoS for binary system composed of ethanol and toluene at 37.5 mole% **of ethanol.**

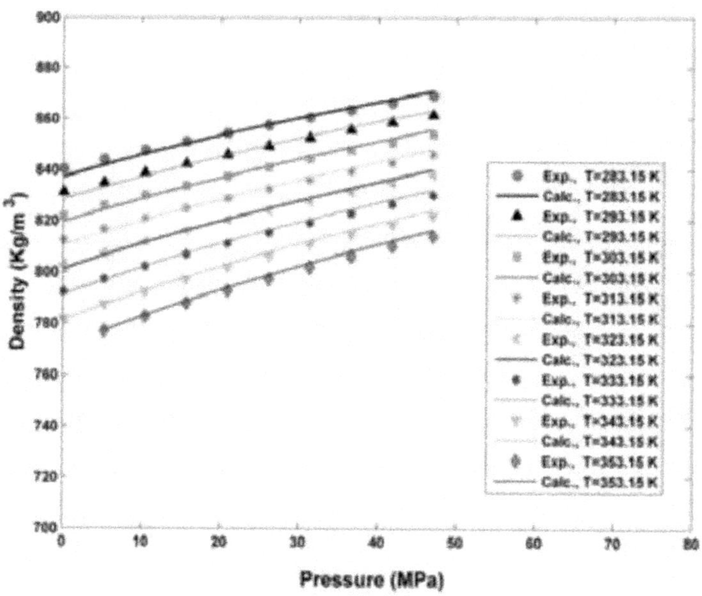

Fig. 9. Experimental and calculated densities versus pressure at different temperatures using SAFT EoS for binary system composed of ethanol and toluene at 62.5 mole% **of ethanol.**

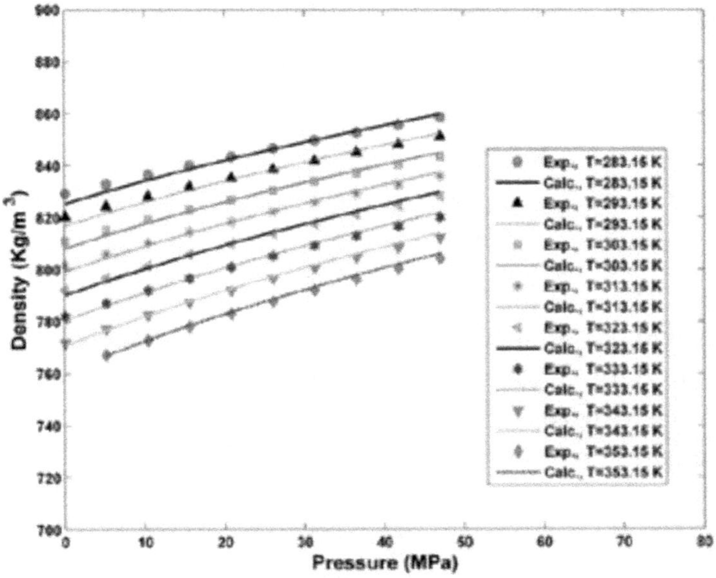

Fig. 10. Experimental and calculated densities versus pressure at different temperatures using SAFT EoS for binary system composed of ethanol and toluene at 75.0 mole% **of ethanol.**

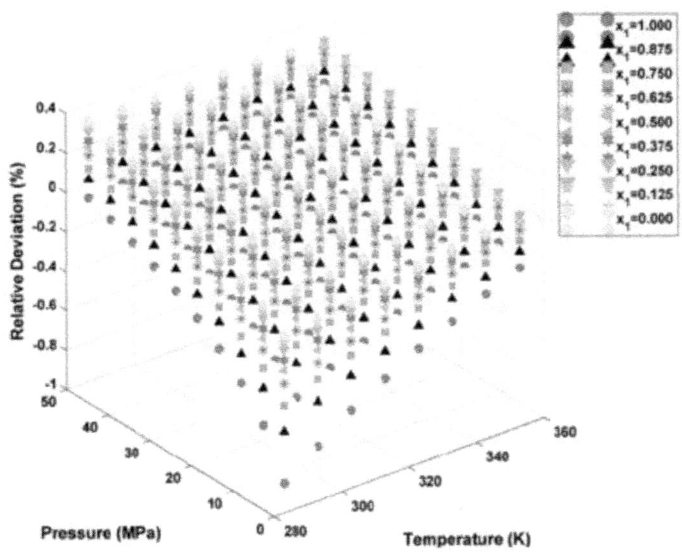

Fig. 11. Percent relative deviation of the calculated densities using SAFT EoS for different compositions of ethanol.

4.2 Calculations of phase equilibrium for asphaltene-oil systems

Given a pressure, P, a temperature, T, and a mixture with global composition $z_i(i = 1, \dots, n_c)$, flash calculations determine the phase molar fraction and composition by solving equations (214) and (215),

$$f_i^V = f_i^L \; i = 1, \dots, N_c \tag{214}$$

$$f_{N_c}^L = f_{N_c}^S \tag{215}$$

f_i^V, f_i^L are the fugacity of component i in the vapor and liquid phase, respectively. $f_{N_c}^L$ and $f_{N_c}^S$ represent the fugacity of the asphaltene component in the liquid and solid phases respectively. The fugacity of the components in the vapor and liquid phases are calculated using the SAFT model as mentioned previously. In order to calculate the fugacity of the asphaltene component in the solid phase an experimental value of the asphaltene precipitation amount at a given temperature and pressure is required. Then, the thermodynamic equality of the fugacities at equilibrium conditions is used to calculate the reference value of the fugacity of the asphaltene component. The Quasi Newton Successive Substitution (QNSS) method, is used to solve the equilibrium equations. In conjunction with the above equations, the following material balance equations can be derived:

$$\sum_{i=1}^{N_c} \frac{(K_i^{VL} - 1)z_i}{1 + V(K_i^{VL} - 1) + S(K_i^{SL} - 1)} = 0 \; i = 1, \dots, N_c \tag{216}$$

$$\sum_{i=1}^{N_c} \frac{(K_i^{SL} - 1)z_i}{1 + V(K_i^{VL} - 1) + S(K_i^{SL} - 1)} = 0 \; i = 1, \dots, N_c \tag{217}$$

where, V and S are the mole fraction of the gas and solid phases, respectively. K_i^{VL} and K_i^{SL} are the equilibrium ratios of the vapor-liquid and liquid-solid phase equilibrium, respectively. In order to find the values of V and S using the above system of equations, we used the NewtonRaphson method. The complexity of multiphase flash calculations is due to the fact that the number of phases in equilibrium is not known a priori. The stability test for performing multiphase flash calculations has been performed using the stage-wise method developed by Michelsen (Michelsen, 1982a, 1982b). The parameters of the SAFT EoS including r, σ and ε have been given by Gross and Sadowski (2001) for N_2, CO_2 and hydrocarbons from C_1 to C_{20}. The values of these parameters for heavier and lumped component, also the volume and energy parameters used in association term have been estimated using a tuning approach. The thermodynamic model was applied to predict the precipitation behavior of petroleum fluids. The SAFT model has been used to describe the oil and gas phases. Table 3 presents the composition of an oil sample and a solvent from Burke et al. (1990). The oil sample was mixed with various amounts of solvent. Table 4 shows the precipitation and saturation pressure data for different concentrations of solvent in the oil sample for which the compositions are given in Table 3. The weight percent corresponds to the percentages with respect to the original mass of the oil. The last column in Table 4 shows the total amount of precipitates. To estimate the fugacity of pure asphaltene phase at a reference state, data at 0 mole% of solvent and $P^* = 3014.7$ psia and $T = 218°F$ was used. The amount of precipitate, 0.14 weight %, was removed from the feed. The fugacity of the asphaltene component in the remaining mixture was then calculated using SAFT equation of state and equated to f_s^* (reference state fugacity of the asphaltene). The amounts of asphaltene precipitation and saturation pressures are also calculated using WinProp (CMG software) in which the fluid phases are described with a cubic equation of state and the fugacities of components in the solid phase are predicted using the solid model desribed in (Nghiem & Heidemann, 1982). In this work the Peng-Robinson equation of state has been used to describe the fluid phases non-ideality. Figures 12-13 represent a comparison between experimental and calculated asphaltene precipitation amounts and saturation pressures at different concentrations of solvent in the oil sample using SAFT EoS and WinProp software. As it can be seen in Figure 12 and Table 4, for solvent concentration above 78 mole%, the measured values show a substantial decrease in the precipitate. For this case, WinProp does not show a drop in the amount of precipitate. Howevere, it shows that the amount of precipitate levels off at high concentration of the solvent. Burke et al. (1990), and Chaback (1991) attributed the decrease in the precipitate at high solvent concentration to the switching of the mixture from bubble point fluid to a dew point fluid. WinProp does not show a decrease in the precipitate, while SAFT EoS based on the developed method for the calculation of densities and fugacities shows a very good agreement with measured data. The saturation pressures calculated using the proposed model at solvent concentrations above 78 mole % correspond to the upper dew point pressures while those calculated using WinProp are the buuble point pressues (Figure 13). The mean relative variations of saturation pressure and asphaltene precipitation quantity utilizing the SAFT EoS are 4.6% and 3.8%, respectively. The AAD values derived by WinProp exceed 17% for both parameters.

Table 3. Composition (mole%) and properties of oil ad solvent from Burke et al. (1990).

Component and Properties	Oil	Solvent
N_2	0.51	3.17
CO_2	1.42	17.76
C_1	6.04	30.33
C_2	7.00	26.92
C_3	6.86	13.09
iC_4	0.83	1.26
nC_4	3.35	4.66
iC_5	0.70	0.77
nC_5	3.46	1.26
C_6	3.16	0.78
C_{7+}	66.68	
C_{7+} molecular weight 281		
C_{7+} specific gravity	0.9020	
Oil molecular weight	202.4	
API gravity of stock tank oil	24.0	
Reservoir temperature, °F	218	
Saturation pressure, psia	600	

Table 4. Experimental and calculated values of the amount of asphaltene precipitation and saturation pressure for different mixtures of the oil sample and solvent given in Table 3.

Solvent (mol%)	Mixture Saturation Pressure (psia)		Test Pressure (psia)	Precipitates from oil (wt%)		Total Precipitaes (wt%)
	Exp.	Calc.		Exp.	Calc.	
0	600	596.14	3014.7	0.14	0.1578	8.97
20	1050	1053.3	3014.7	0.27	0.2667	7.83
50	2310	2142	3014.7	1.46	1.4700	6.96
70	3750	3467	4214.7	1.65	1.6503	4.34
78	4510	4565	5014.7	3.21	3.4025	7.84
85	5000	5395	5014.7	1.29	1.2714	8.02
90	4250	4578	5014.7	1.10	1.0428	7.17

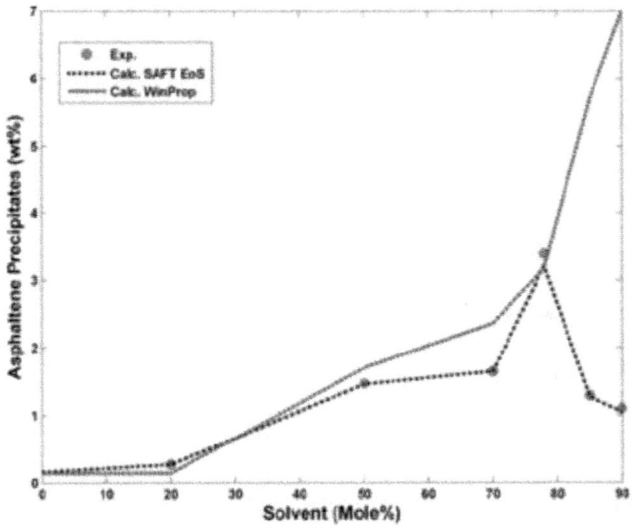

Fig. 12. Comparison between experimental and calculated values of the amount of asphaltene precipitation for different mixtures of the oil sample and solvent given in Table 3 using SAFT EoS and WinProp.

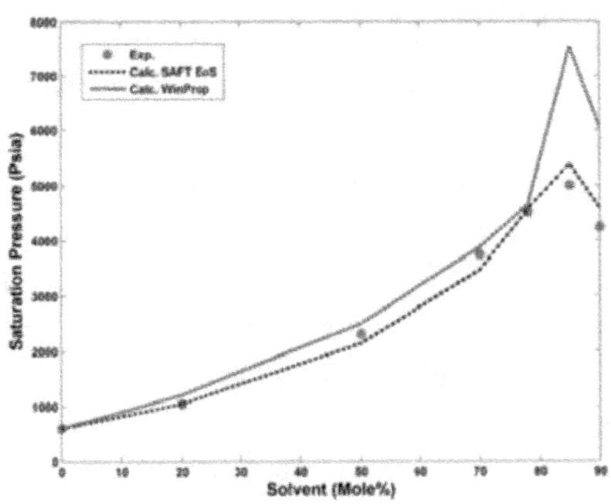

Fig. 13. Comparison between experimental and calculated values of the saturation pressure for different mixtures of the oil sample and solvent given Table 3 using SAFT EoS and WinProp.

The effect of temperature and pressure on the solid model prediction results has also been investigated. To do so, different hydrocarbon mixtures have been used. Figure 14 shows a comparison between the calculated and experimental values of the bubble point pressure and upper asphaltene onset pressure (AOP) for an oil sample (Jamaluddin et al., 2002). As it can be seen in this figure, excellent agreement is observed between experimental and predicted values of the upper AOP using SAFT EoS. The average values of the relative

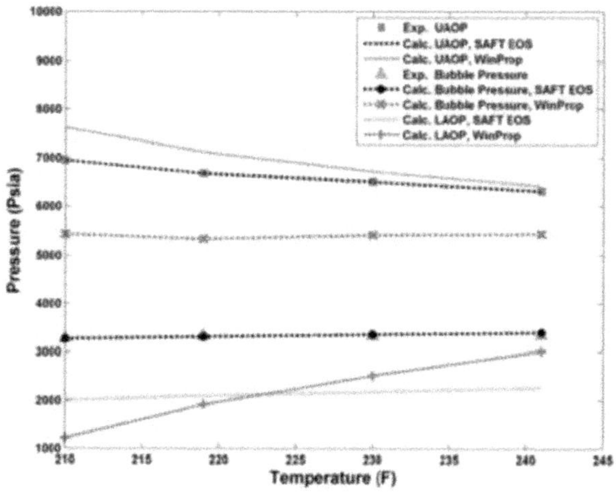

Fig. 14. Comparison between experimental and calculated values of the saturation pressure, upper and lower AOP's for the oil sample given in (Jamaluddin et al., 2002) using SAFT EoS and WinProp.

deviation in calculating the upper AOP using SAFT EoS and WinProp are 0.002% and 5.26%, respectively. Also, shown in Figure 14 are the calculated values of the lower AOP at different temperatures. The amounts of asphaltene precipitation vs. pressure at different temperatures is presented in Figure 15 for SAFT EoS. As it shown (Fig. 15) the maximum values of the asphaltene precipitation occures at the bubble point pressure of the mixture above which the amount of precipitaes decreases by increasing pressure up to upper AOP. In the pressure range below the bubble point pressure, decreasing pressure leads to a decrease in the amount of asphaltene precipitation and becomes infinitesimal at lower AOP. The effect of the injection gas on the asphaltene precipitation conditions has also been investigated for an oil sample given in (Rydahl et al., 1997) (Figures 16-17). Figure 17 shows that by increasing the amount of the injection gas added to the initial oil, the pressure interval of the asphaltene precipitation increases and shifts to the right side of the figure. Also, the amount of asphaltene precipitation increases by increasing the amount of the injection gas. A similar scenario happens when using WinProp.

The amount of asphalltene precipitation at different dilution ratios of normal heptane (nC_7) and stock tank conditions, are also calculated using both models for two Iranian oil samples, Sarvak oil A and Fahliyan oil (Bagheri et al., 2009). Figures 18-19 shows a comparison between experimental and calculated amount of asphaltene precipitation vs. dilution ration of nC_7 using SAFT EoS and WinProp. Again an excellent agreement has been observed using SAFT EoS with the experimental data. As it can be seen in these figues using WinProp, the amount of precipitaes increases rapidly at low dilution ratios after which the slope of the curve does not change considerably. The results show that SAFT EoS with average relative deviations of 2.32% and 1.73% for Sarvak oil A and Fahliyan oil, agrees well with the experimental data in comparison with the results obtained using WinProp (10.26% and 13.19%).

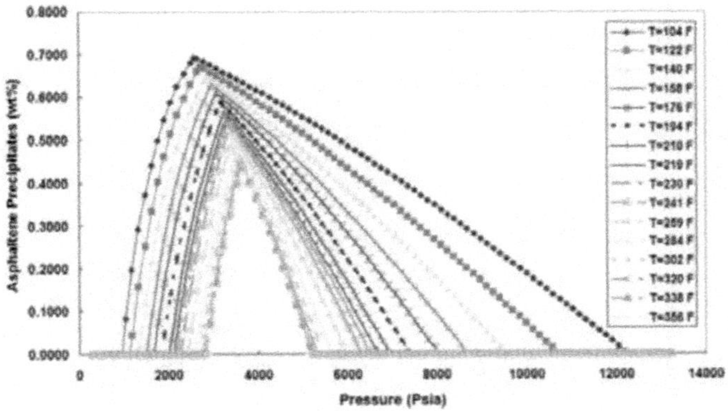

Fig. 15. The amount of asphaltene precipitation vs. pressure at different temperatures for the oil sample given in (Jamaluddin et al., 2002) using SAFT EoS.

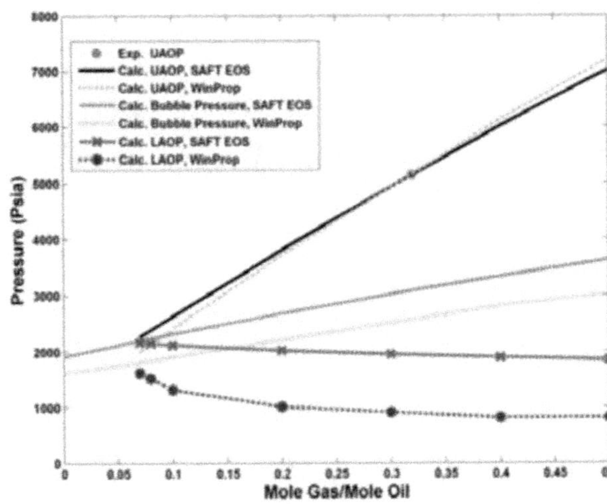

Fig. 16. Calculated values of the saturation pressure, upper and lower asphaltene AOP's vs. the amount of injection gas for the oil sample given in (Rydahl et al., 1997) using SAFT EoS and WinProp.

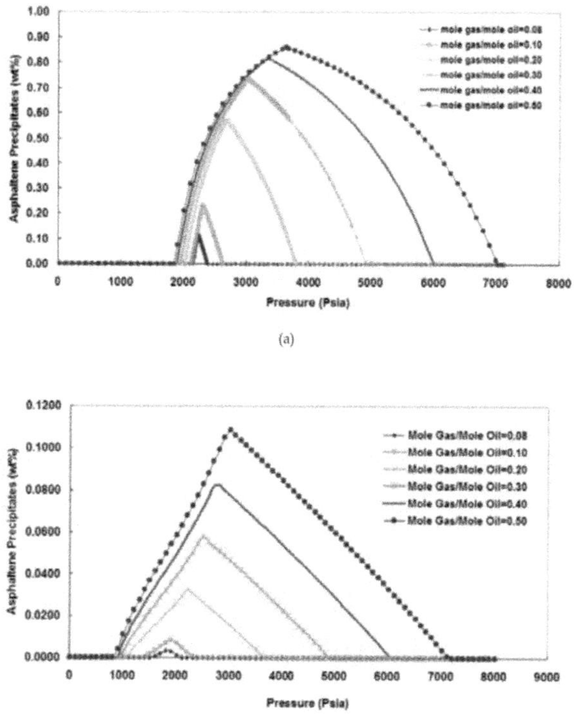

(a)

(b)

Fig. 17. The amount of asphaltene precipitation vs. pressure at different values of the injection gas for the oil and gas samples given in (Rydahl et al., 1997) using (a) SAFT EoS and (b) WinProp.

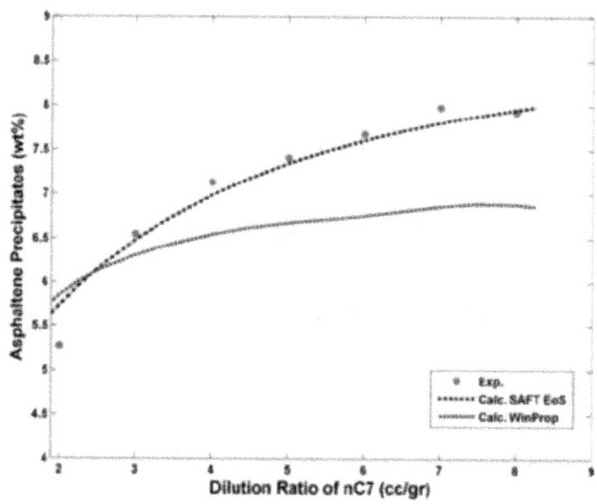

Fig. 18. Comparison between experimental and calculated values of the amount of asphaltene precipitation vs. dilution ration of nC_7 for Sarvak Oil A (Bagheri et al., 2009) using SAFT EoS and WinProp.

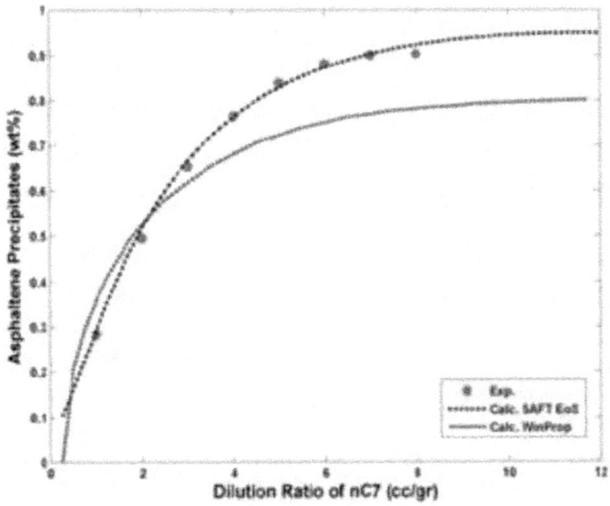

Fig. 19. Comparison between experimental and calculated values of the amount of asphaltene precipitation vs. dilution ration of nC_7 for Fahliyan Oil (Bagheri et al., 2009) using SAFT EoS and WinProp.

5. Concise Summary

This paper presents a model utilizing statistical associating fluid theory (SAFT) to predict the phase behavior of hydrocarbon systems that include asphaltene and associating components. A reliable, rapid, and precise technique utilizing Chebyshev polynomial approximation was presented to determine density using the SAFT EoS, which is crucial for calculating fugacity coefficients. The model was initially assessed utilizing binary mixtures of ethanol and toluene. An excellent correlation between experimental and estimated liquid densities across various pressures, temperatures, and compositions was achieved. The proposed model was utilized to examine the impact of solvent addition on the quantity of asphaltene precipitate. The results demonstrated a strong correlation between the experimental and estimated values of precipitate quantity for various solvent-oil mixes. The influence of temperature on the initiation pressure of asphaltene precipitation and bubble point pressure was examined. A strong correlation was noted between the experimental and predicted values of asphaltene onset pressure across various temperatures.

6. References

Bagheri, MB; Kharrat, R. & Mirzabozorg, A. (2009). A Novel Method to Develop a New Scaling Equation for Modeling of Asphaltene Precipitation, Proceedings of the 2009 SPE/EAGE Reservoir Characterization and Simulation, Abu Dhabi, UAE, October 1921, 2009.

Benedict, M.; Webb, G.B. & Rubin, L.C. (1942). An Empirical Equation for Thermodynamic Properties of Light Hydrocarbons and Their Mixtures II: Mixtures of Methane, Ethane, Propane, and n-Butane. J. Chem. Physics, Vol. 10, pp. 747.

Boyd, J.P. (2006). Computing Real Roots of a Polynomial in Chebyshev Series form Through Subdivision. Appl. Numer. Math., Vol. 56, No. 8, pp. 1077-1091.

Buenrostro-Gonzales, E. & Lira-Galeana, C. (2004). Asphaltene Precipitation in Crude Oils, Theory and Experiments. AIChE J., Vol. 50, pp. 2552-2570.

Burden, R.L.; Faures, J.D. & Reynolds, A.C. (1981). Numerical analysis (Second Edition), Prindle, Weber & Schmidt, Boston.

Burke, N.E.; Hobbs, R.E. & Kashou, S.F.R. (1990). Measurement and Modeling of Asphaltene Precipitation. J. Petroleum Tech., Vol. 42, pp.1440.

Chaback JJ. (1991). Discussion on Measurement and Modeling of Asphaltene Precipitation. Journal of Petroleum Technology. Vol. 43, pp. 1519-1520.

Chapman, W.G.; Jackson, G. & Gubbins, K.E. (1988). Phase Equilibria for Associating Fluids Chain Molecules with Multiple Bonding Sites. Molec. Phys., Vol. 65, pp. 1057.

Chapman, W.G.; Gubbins, K.E., Jackson, G. & Radosz, M. (1990). New Reference Equation of State for Associating Liquids. Ind. Eng. Chem. Res., Vol. 29, pp. 1709-1721.

Chapman, W.G.; Sauer, Sh.G.; Ting, D. & Ghosh, A. (2004). Phase Behavior Applications of SAFT Based Equations of State from Associating Fluids to Polydisperse, Polar Copolymers. Fluid Phase Equilibria, Vol. 217, pp. 137-143.

Danesh, A. (1998). PVT and Phase Behavior of Petroleum Reservoir Fluids, Elsevier Science, ISBN 0444 821961, Netherlands.

Economou, I.G. (2002).A Successful Model for the Calculation of Thermodynamics and Phase Equilibria Propeties of Complex Fluid Mixtures. Ind. Eng. Chem. Res., Vol. 41, pp. 953-962.

Firoozabadi. (1999). Thermodynamics of Hydrocarbon Reservoirs, McGraw-Hill, NewYork.

Gross, J. & Sadowski, G. (2001). Perturbed-Chain SAFT : An Equation of State Based on Perturbation Theory for Chain Molecules. Ind. Eng. Chem. Res., Vol. 40, pp. 12441260.

Huang, S.H. & Radosz, M. (1990). Equation of State for Small, Large, Polydisperse and Associating molecules, Ind. Eng. Chem. Res., Vol. 29, No. 11, pp. 2284-2294.

Huang, S.H. & Radosz, M. (1991). Equation of State for Small, Large, Polydisperse and Associating molecules, Ind. Eng. Chem : Extension to Fluid Mixtures. Ind. Eng. Chem. Res., Vol. 30, pp. 1994-2005.

Jamaluddin, A.K.M.; Joshi, N.; Joseph, M.T.; D'Cruz, D.; Ross, B.; Greek, J.; Kabir. C.S. & Fadden, J.D. (2000). Laboratory Techniques to Defines the Asphaltene Deposition Envelop, Proceedings of the Petroleum Society's Canadian International Petroleum Conference, Calgary, Canada, Jane 4-8, 2000.

Jamaluddin, K.M.; Joshi, N.; Iwere, F. & Gurpinar, F. (2002). An Investigation of Asphaltene Instability under Nitrogen Injection, Proceedings of SPE International Petroleum Conference and Exhibition, SPE 74393, Villahermosa, Mexico, February 10-12, 2002.

Katz, D.L. & Firoozabadi, A. (1978). Predicting Phase Behavior of Condensate/Crude Oil Systems Using Methane Interaction Coefficients. J. Pet. Tech., Vol. 27, pp. 1649.

Llovell, F.; Pamies, J.C. & Vega, L.F. (2004). Thermodynamic Properties of Lennard-Jones Chain Molecules: Renormalization Group Corrections to a Modified Statistical Associated Fluid Theory. J. Chem. Phys., Vol. 121, No. 21, pp. 10715-10724.

Llovell, F.; Vega, L.F.; Seiltgens, D.; Mejia, A. & Segura, H. (2008). An Accurate Direct Technique for Parametrizing Cubic Equations of State: Part III. Application of a Crossover Treatment. Fluid Phase Equilibria, Vol. 264, No. 1-2, pp. 201-210.

Michelsen, M.L. (1982a). The Isothermal Flash Problem, I. Stability. Fluid Phase Equilibria, Vol. 9, pp. 1-19.

Michelsen, M.L. (1982b). The Isothermal Flash Problem, II. Phase Split Calculation. Fluid Phase Equilibria, Vol. 9, pp. 21-40.

Muller, E.A. & Gubbins, K.E. (1995). An Equation of State for Water from a Simplified Intermolecular Potential. Ind. End. Chem. Res., Vol. 34, pp. 3662-3673.

Nghiem, L.X., Heidemann, R.A. (1982). General Accelaration Procedure for Multiphase Flash Calculation with Application to Oil-Gas-Water Systems, Proceedings of the 2nd European Symposium on Enhanced Oil Recovery, Paris, France, November 8-10, 1982.

Pedersen, K.S. & Christensen, P.L. (2007). Phase Behavior of Petroleum Reservoir Fluids, CRC Press, Talor & Francis Group LLC.

Peng, D.Y. & Robinson, D.B. (1976). A New Two Constant Equation of State. Ind. Eng. Chem. Fundam., Vol. 15, pp. 59-64.

Prausnitz, J.M.; Lichtenthaler, R.N. & de Azevedo, E.G. (1999). Molecular Thermodynamics of Fluid Phase Equilibria (Third Edition), Prentice Hall PTR, pp. 390.

Prausnitz, J.M. & Tavares, F.W. (2004). Thermodynamics of Fluid-Phase Equilibria for Standard Chemical Engineering Operations. AIChE J., Vol. 50, No.4, pp. 739-761.

Redlich, O. & Kwong, J.N.S. (1949). On the Thermodynamics of Solutions. V. An Equation of Sate. Fugacities of Gaseous Solutions. Chem. Rev., Vol. 44, pp. 233-244.

Rydahl, A.K.; Pedersen, K.S. & Hjermstad, H.P. (1997). Modeling of Live Oil Asphaltene, Precipitation Proceedings of AIChE Spring Meeting, Houston, March 9-13, 1997.

Segura, H.; Seiltgens, D.; Mejia, A.; Llovell, F. & Vega, L.F. (2008). An Accurate Direct Technique for Parameterizing Cubic Equations of State. Part II. Specializing Models for Predicting Vapor Pressures and Phase Densities. Fluid Phase Equilibria, Vol. 265, pp. 155-172.

Soave, G. (1972). Equilibrium Constants from a Modified Redlich-Kwong Equation of State. Chem. Eng. Sci., Vol. 27, pp. 1197-1203.

Song, Y.; Lambert, S.M. & Prausnitz, J.M. (1994). Equation of State for Mixtures of Hard Sphere Chains Including Copolymers. Macromolecules, Vol. 27, pp. 441.

Starling, K.E. (1973). Fluid Thermodynamic Properties for Light Petroleum Systems, Gulf Publishing Company, Houston.

Tabatabaei-Nejad, S.A.R. & Khodapanah, E. (2009). An Investigation on the Sensitivity Analysis of the Parameters of Proposed Wax Precipitation Model. Journal of Petroleum Science & Engineering, Vol. 68, pp. 89-98.

Ting, D.L., (2003). Thermodynamic Stability and Phase Behavior of Asphaltenes in Oil and of Other Highly Asymmetric Mixtures, PhD Thesis, Rice University, Houston, USA.

Van der Waals, J.D. (1873). Over de Continuiteit van der Gas en Vloeistoftoestand Leiden, Doctoral Dissertation, The Netherlands.

Verdier, S.; Carrier, H.; Andersen, S.I. & Daridon, J.L. (2006). Study of Pressure and Temperature Effects on Asphaltene Stability in Presence of CO2. Energy and Fuels, Vol. 20, pp. 1584-1590.

Wertheim, M.S. (1984). Fluids with Highly Directional Attractive Forces. I. Statistical Thermodynamics. J. Stat. Phys., Vol. 35, No.1, pp. 19-47.

Wertheim, M.S. (1986). Fluids with Highly Directional Attractive Forces. III. Multiple Attraction Sites. J. Stat. Phys., Vol. 42, No.3, pp. 459-476.

Wertheim, M.S. (1987). Thermodynamic Perturbation Theory of Polymerization. J. Chem. Phys., Vol. 87 , pp. 7323.

Zeberg-Mikkelsen, C.K.; Lugo, L.; Garcia, J. & Fernandez, J. (2005). Volumetric Properties under Pressure for the Binary System Ethanol + Toluene, Fluid Phase Equilibria, Vol. 235, pp. 139-151.

Chapter-2

Colorimetric Analysis of Individual Oil–Gas Inclusions via Inverted Microscopy

1. Introduction: An Overview

Hydrocarbon fluid inclusions are found in diverse geological settings, predominantly inside carbonate rocks of petroliferous sedimentary origin (Stasiuk & Snowdon, 1997). Oil-gas inclusions (OGIs) are classified as hydrocarbon inclusions. The dimensions of OGIs typically range from 5 to 20 micrometers. Micro-mass (ng-fg) palaeo-oils were confined within a singular OGI. The interaction between OGIs and external factors is rather minimal despite the passage of a vast geological period. The OGI may be seen as a closed system. An unblemished (not fractured) OGI can be regarded as a micro-oil-gas reservoir. The OGIs get extensive information regarding the palaeo-oils. Typically, as the depth of deposition in an oil-gas reservoir increases, the temperature of the strata rises. The organic macromolecules were broken down into micromolecules. The oil maturity progresses from low to high. Correspondingly, in the various digenetic stages, the types, colors, and compositions of the OGIs exhibit variation. As organisms mature from lower to higher forms, the types of OGIs transition primarily from the liquid phase to a combination of liquid and gas phases, ultimately reaching the gas phase. The colors transition from colorless, yellow, and brown to black (Liu, Y. R. et al., 2003, as referenced in Burruss R.C., 1991). The micro-fluorescence features of OGIs, mostly influenced by the aromatic qualities of hydrocarbons, signify the organic chemical makeup. This characteristic was frequently employed to differentiate OGIs from saltwater inclusions. The varying fluorescence colors of OGIs qualitatively ascertain the oil and gas charge history and the maturity of oil and gas. The prevalence of grains carrying OGIs (GOI) (Liu, K. Y. & Eadington, 2005, as referenced in Eadington, 1996) allows for a qualitative estimation of the petroleum properties of the associated strata. Micro-spectroscopy is crucial for OGIs and mostly encompasses VIS and UV-VIS spectra.

1.1 The VIS spectral method

The VIS spectral technique encompasses fluorescence microphotometry (FMP), fluorescence alteration of multiple macerals (FAMM), and laser scanning confocal microscopy (LSCM).

1.1.1 Fluorescence micro photometry (FMP)

FMP is a technique combined microscope and photometer (see Fig.1). Generally, the exciting source is UV light (365 nm , called internal light source in this chapter) from a mercury arc lamp in a microscope. The UV light is focused by an objective and incidents onto a sample. The sample absorbs the exciting photons then emits fluorescence. The wavelengths of the fluorescence are longer than that of the incident light.

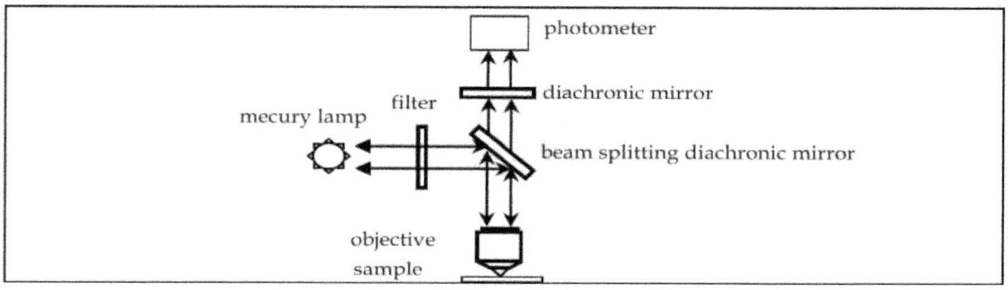

Fig. 1. A schematic diagram of a FMP. The upright microscope includes in a beam splitting diachronic mirror (with higher reflectivity to the incident light and also a higher transmittance to the fluorescence) and a diachronic mirror (absorbing the scattered light from the sample, with a higher transmission to the fluorescence).

Some fluorescence backs (upward for the upright microscope and downward for the inverted microscope) and is collected by the same microscope objective. That light enters into the aperture of the photometer and the intensity is measured. For the inverted microscope, except for the downward fluorescence, one can measure the upward (transmitted) fluorescence because there is a big space above the sample stage to connect a photometer with the body of the microscope.

According to different study contents, one can choose different filter, beam splitting diachronic mirror, diachronic mirror and detector to realize fluorescence, transmission, reflective and absorption measurement. For example, in Fig. 1, if the photometer is an energy meter, one can measure the intensity of the fluorescence. And if the diachronic mirror has a higher absorption to the fluorescence and a higher transmission to the scattered light from the sample, the reflectivity of the sample can be measured. MPV3microphotometer is an example. It was widely used to measure the vitrinite reflectivity of coal and source rock (Xiao et al., 1998, 2000).

The superiors of the above FMP are simple and low cost. But there are still some shortcomings. The first is one can only gain visible spectra (400 – 780 nm) for the exciting light is at 365 nm . But some aromatic hydrocarbons emit UV fluorescence which can not be measured by above method. The second is the pin hole (about 0.15 mm) to limit the incident light spot often larger than the under studied OGI. Not only the under studied OGI emits fluorescence, but also other OGIs, the cements around the mineral grain trapping the OGI and even the mineral grain itself. The other OGIs may not be the same generation as the under studied OGI. What' more, the intensities of the cements are usually stronger than that of the under studied OGI. That means the fluorescence of the under studied OGI often mixes with other fluorescence. This makes the OGI looks like involving into a colourful cloud. So one can see if the size of the incident light spot is larger than that of the under studied OGI, the measured fluorescence spectrum may include the fluorescence of other OGIs, the cements around the mineral grain trapping the OGI and even the mineral grain itself. This will lead to a decreasing reliability. So one should be cautious to judge the oil and gas charge history and oil-gas maturity by the fluorescence colours of the OGIs.

In 1997, Stasiuk et al. (Stasiuk & Snowdon,1997) used Zeiss MPV II to study the fluorescence of OGIs. The wavelength of the exciting light is 365 nm . A 2.5 – 5μ m light spot was obtained by a pin hole. The size of the light spot is less than that of the OGIs. The authors measured the micro-fluorescence spectra of artificially synthesized and natural hydrocarbon inclusions. The fluorescence spectra were related with the chemical constitutes, API degree and other geochemistry parameters of crude oils. The migration of oil and gas was revealed. The crude oils used to synthesize the artificial hydrocarbon inclusions composite a wide range of saturated hydrocarbons, aromatic hydrocarbons and NSO compounds. And also span a long geologic time (Ordovician, Devonian, Carboniferous, Jurassic and Cretaceous). For the artificial hydrocarbon inclusions, the spectra are the real spectra for single hydrocarbon inclusions. And for the natural inclusions, there are some organic compounds in the mineral grains difficult to be observed even under the microscope, so strictly speaking, the fluorescence spectra of natural inclusions are the addition of the inclusions and the mineral matrix. The results showed that most artificially synthesized hydrocarbon inclusions have main colours. A small amount have other colours. The authors gave a detailed analysis between the peak position $L_{\max}$, Q index (Q = Intensity @ 650 nm/ Intensity @ 500 nm) of the fluorescence spectra and saturated hydrocarbons, aromatic hydrocarbons, NSO compounds, nC_{18}/Ph、nC_{18}/Pr, API degree of the crude oils. The results showed the correlative factors between the peak position $L_{\max}$ and the saturated hydrocarbons, Q index and saturated hydrocarbons are all up to 0.84 . Q index can be used to indicate the maturities of the crude oils.

Caja et al. (Caja, 2007, 2009) measured the fluorescence spectra of hydrocarbon inclusions. The results showed that Q_{580} index ($580-700$ nm vs $400-580$ nm areas) is very sensitive to the total composites of the petroleum. The higher the Q_{580} index, the heavier the crude oil and the tighter relation with the API degree. The hydrocarbon inclusions exhibiting yellow fluorescence possess an API degree ranging from 15 to 22. The API degree for hydrocarbon inclusions exhibiting blue fluorescence ranges from 23 to 45.

1.1.2 Fluorescence Modification of Various Macerals (FAMM)

In the 1990s, CSIRO of Australia established FAMM. This technique integrated laser technology with microscopy. The laser's excellent monochromaticity and high intensity facilitate the attainment of a micrometer light spot, hence improving the spatial resolution of the apparatus. Utilizing a micro-Raman setup with an Ar+ laser as the excitation light source (488 nm), a focal spot of merely several micrometers was achieved. The emission intensities of source rocks at 625 nm were measured over time. Wilkins (Wilkins et al., 1992, 1995) investigated the correlation between vitrinite inhibition and the temporal variation in fluorescence intensity. The results indicated that FAMM may successfully rectify the inhibitory effect of vitrinite reflectivity before to the high maturity stage. Lo et al. (1997) and Veld et al. (1997) employed this technique to assess the maturity of source rock. Xiao (Xiao et al., 2002), aided by FAMM, employed laser-induced fluorescence microscopy (LIFM) to investigate the maturities of carbonate rocks exhibiting elevated maturities.

The FAMM and LIFM procedures are not particularly suitable for measuring the fluorescence spectra of individual OGIs. Upon activating the laser, one can measure the fluorescence spectra; however, direct visual observation of the material is not possible due to the laser safety threshold for human eyes being below 5 μW. (Sliney, D. H., 1995). In most instances, the absorption bands reside inside the ultraviolet spectrum. However, the wavelengths of the Ar^+ ion laser, 488 nm and 515 nm, do not align effectively with the absorption bands of the materials. Utilizing an Ar^+ ion laser to trigger the fluorescence of a single OGI is not advisable. The concentrated laser exhibits a greater radiation flux density at the sample surface, resulting in potential damage to the sample. The temporal variations in fluorescence intensity at 625 nm may pertain to the photolysis (Sanches S., 2011) and photopolymerization of vitrinite in coal and source rock.

1.1.3 Laser scanning confocal microscope (LSCM)

LSCM was established in the 1980s. Figure 2 presents a schematic representation of a Laser Scanning Confocal Microscope (LSCM). The illuminating pinhole and the detecting pinhole are conjugate to the objective's focal plane. A point on the focal plane of the objective is simultaneously focused on the illuminating and detecting pinholes. The other points outside the focus plane of the objective do not produce pictures at the detection pinhole. This signifies confocal. A point light source and a point picture are created by an illuminating pinhole positioned before the laser and a detecting pinhole located before the detector. The light that traverses the detecting pinhole will be captured by a PMT or a cCCD and promptly photographed on a computer. The photos include superior spatial resolution and great quality. By regulating the movement of the objective, LSCM may provide continuous optical sectioning through tomography akin to CT. A 3D profile of a micro sample can be obtained following computer 3D imaging.

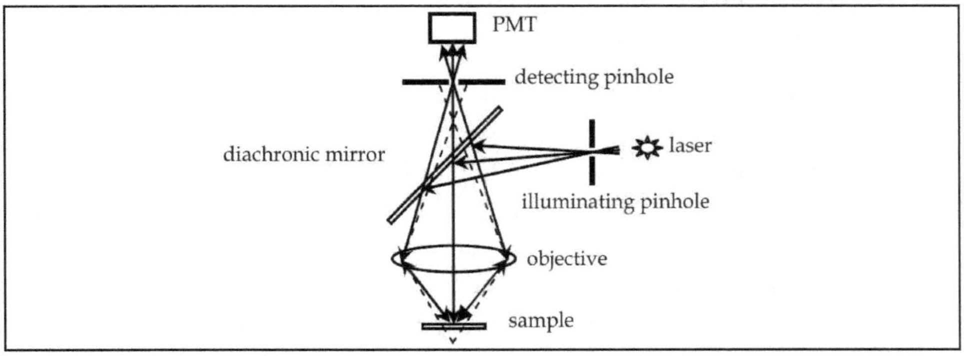

Fig. 2. The schematic diagram of a LSCM.

LSCM has been widely used in cytobiology, cytophysiology, neurophysiology and other modern medicine and biology relative to cells (Damaskinos et al., 1995; Kevin, 2003). It has been a forceful tool in the field of bio-science. Recently, LSCM was used to obtain precise volume ratios between liquid and gas phases in OGIs. Combining homogenous temperature, GC-MS analysis and PVTsim software, the pressure of palaeo-liquid can be obtained (Aplin et al., 1999; Thiéry et al. 2002, Liu, D. H. et al., 2003). Further more, the depths of the strata existing palaeo-liquids can also be obtained by the pressure. This is meaningful to study the migration of oil, gas and reservoir formation.

LSCM possesses unique advantages in addressing fluorescent macerals of micrometer dimensions (source rock and oil shale). Stasiuk (1999) employed LSCM to examine algae in oil shale.

LSCM can efficiently elucidate the cellular structure of algae. The author noted that algae exhibit a predominant and stable exterior cell wall. The outcome confirmed that kerogen has a specific preservation during its creation. LSCM was utilized to acquire images depicting the spatial distribution of organisms in mineral slides. Stasiuk (2001) analyzed the fluorescence spectra of lipoids and chlorophyll derived from diatoms. The mineral slides in the Saanich Inlet region contain abundant spore oil and reproductive spores, correlating with the seasonal spore blooming in spring. The outcomes of the LSCM may involve the preservation of spore oil at an appropriate depth.

Despite LSCM's superior spatial resolution and image clarity, the shortest wavelength of semiconductor lasers remains at 370 nm, and the fluorescence of single OGI is confined to the visible spectrum.

1.2 The UV-VIS spectra technique for single OGIs

As above mentioned, some organic composites in OGIs emit fluorescence at UV range. So it is necessary to develop UV-VIS spectra technique.

For overcoming above drawbacks, Kihle (Kihle, 1995) established an UV-VIS microspectroscopy for measuring excitation-emission spectra of single OGIs. The setup includes an upright tri-ocular deep ultraviolet (DUV) microscope, a condensed lens, a collimated objective, two fiber adaptors, two fiber cables, and an UV-VIS spectrometer. The DUV microscope is very expensive comparing with the common fluorescence microscope. But for measuring the UV-VIS spectra of single OGIs, it had to be used. The spectrometer was connected with the microscope by the two fiber cables, the condensed lens, the collimated objective and the two fiber adaptors. The spectrometer supplies the exciting light in the range of 220 – 900 nm and record the UV-VIS spectra.

The setup could successfully measure the UV-VIS spectra of single OGIs. But there are four points to be noted. The first is for obtaining a small focal point, one has to replace the original focusing mirror of the microscope itself with a condensed mirror which was settled down on a 3D adjustable adaptor. The second is the C-mount was connected with the fiber cable not a videograph head as origin. So one can only observe the sample by ocular and can't take photos of the samples or get the sizes of the single OGIs and focusing point precisely. The third is the expensive DUV microscope, its price is 9-20 times of a common fluorescence microscope. This will greatly limit it using. The fourth is that for effective collecting fluorescence an oil immersed objective had to be used, which is trouble to clean.

The aforementioned experimental setup is prohibitively costly, necessitating the reconstruction of the microscope. Can a conventional microscope be utilized in place of the DUV microscope to reduce costs while still detecting UV-VIS fluorescence spectra of individual OGIs? A micro-spectroscopy system designed to achieve this objective was established (Yang, 2009a, 2009b, 2011). This will be provided in the third portion of this chapter.

2. The geological context of the samples

The sand rocks came from the core drilling samples of five oil wells (Bai 95#, Hua5#, Jian 22#, Nong 29# and Fu 4#), Jinlin Oil Filed, Songliao Basin, northeast China. Bai 95# oil well localizes in the west slope of the basin, the other are in the south to the basin. The earth strata belong to middle to shallow, upper Cretaceous Period. The thickness of inclusion thin slide is about 150 nm .

Table 1 shows the information about the strata, depths, GOIs, average homogeneous temperatures ($\bar{T}_h s$) and salinities of the sand rocks.

There are two episodes OGIs in these samples. The first episode is in the earlier stage of quartz overgrowth, the second in the later stage of the quartz overgrowth.

Table 1. Information about the strata, depths, the measurement results of GOIs, $\bar{T}_h s$ and salinities of the sand rocks.

Oil well	Stratum	Depth	The first episode			The second episode		
			GOI (%)	$\bar{T}_h$(°C)	Salinity (%)	GOI (%)	$\bar{T}_h$(°C)	Salinity (%)
Bai 95#	K_2Y_{2+3}	419.3422.1	2 − 3 ±	75.3	9.93	5 − 6 ±	113.3	4.03
	K_2y_1	426.8436.5	3 ±	74.6	10.31	2 ±	95.4	2.98
	K_2qn_2	494.6	4 ±	72.3	10.95	2 ±	98.7	5.09
Hua 5#	K_2n_1	1449.8 1524.0	2 ±	}	}	0.5 ±	110.3	3.79
Jian 22#	K_2q_4	377.8 420.0	<1	81.0	6.77	20 ±	104.2	4.89
Nong 29#	K_2q_4	571.0595.9	2 ±	77.0	11.34	20 ±	116.8	7.86
Fu 4 #	K_2q_3	419.1426.6	2 − 3 ±	75.0	7.17	70 ±	135.2	4.35

The GOIs, $\bar{T}_h$ s and salinities are different in the two stages. In the first episode and for all strata, except for Jian 22# oil well, the GOIs belong to middle value (Liu, K. Y. & Eadington, P., 2005, as cited in Eadington, P, 1996). In the second episode, the GOIs of Bai 95# K_2y_1, K_2qn_2 and Hua 5# K_2n_1 decrease. But for Bai 95# K_2y_{2+3} stratum, the GOI doubles. The GOIs in the second episode increase from 10 to 23 times of Jian 22#, Nong 29# and Fu 4 # oil wells, which shows there are lots of oils charging these strata in the second episode.

In the first episode, $\bar{T}_h$ s are in the range of 72.3-81.0 °C. And in the second episode, $\bar{T}_h s$ increase. The increasing amplitude is in the range of 23 − 60°C

The salinities of all strata decrease in the second episode. The $\bar{T}_h$ s and salinities show that the deposition environment has a great change in the second episode. The second episode may be in the water expanding time.

Under the fluorescence microscope, the OGIs of the first episode are liquid phase and very small. They are dense and often distribute along a line or a stripe. Their colors are often deep brown or grey brown. The fluorescence intensities are very weak. It is very difficult to measure the fluorescence of such kind of oil inclusions. The second episode OGIs are liquid or liquid-gas phases. In a mineral grain, they distribute in scattering or in group. The liquids in the OGIs are light yellow, light brown yellow, brown and grey black. Under excited at UV, blue or green light, they often emit fluorescence. Fig. 3 shows the typical OGIs in the Bai 95# oil well. Fig. 3 a is a micro-photo of the first episode oil inclusions, and Fig. 3b is the second episode OGIs. Fig. 3 c is the fluorescence image of Fig. 3 b as excited at 365 nm (internal light source). The OGI emits light yellow fluorescence.

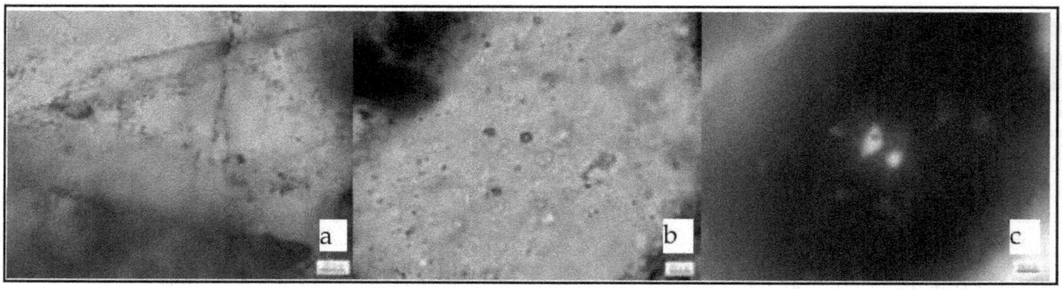

Fig. 3. The two episodes OGIs in Bai 95# oil well, in which, a is the first episode, the depth is 426.8 m ; b the second episode, the depth 420.1 m ; and c the fluorescence image of b under excited at 365 nm internal light source in the microscope. The scale in a is 20μ m, and in b and c 10μ m.

3. Utilizing a standard inverted microscope to assess UV-VIS spectra of individual OGIs and conduct color analysis

3.1 The experimental setup

Fig. 4 shows the schematic diagram of the experimental setup (Yang, 2009a, 2009b), which includes seven parts: (1) an inverted fluorescence microscope (IMF); (2) a reflecting microscope objective (RMO); (3) a 3D adaptor; (4) a micro lens; (5) a fiber cable; (6) a spectrometer and (7) a computer.

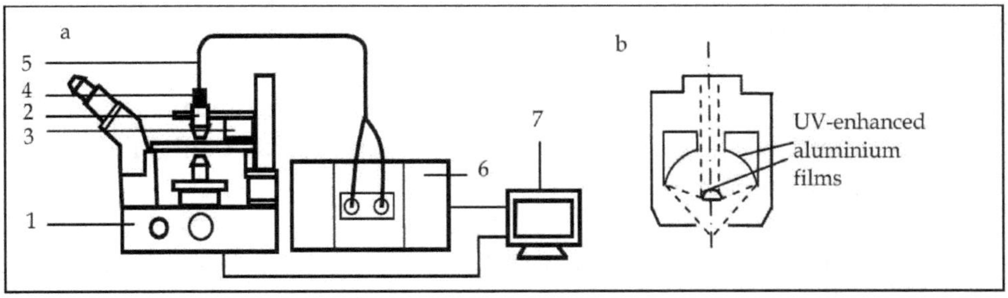

Fig. 4. Schematic diagram of the micro-spectroscopy system (a) based on common inverted microscope and the reflecting microscope objective (b). In a, 1 – IMF; **2—RMO; 3—3D adaptor; 4-micro-lens; 5-fiber cable; 6-UV-VIS spectrometer and 7-computer.**

The IFM includes an internal light source (mercury arc lamp), three (violet, blue and green) standard fluorescence "cubes" bandpass filters, a dichronic mirrors and beam splitters in the optical light path. The mercury arc lamp can be used as an internal excitation source. Using bandpass filters and fluorescence "cubes", a beam with tens of nm band width can be obtained. There are five differences between this system and Kihle's setup (Kihle, 1995). (1)

A common and cheap IFM was used but not an expensive upright DUV microscope. The 3D adjustable adaptor can conveniently connected to the main body of the IFM and not need any rebuilding to the microscope. (2) It is easy to adjust the RMO and the IMF coaxial. When the switch of the IMF is on and kept the least intensity, one can adjust the 3D adjustable adaptor while observe the reflected light intensity from the RMO. When the beam intensity is maximal, the two are coaxial. (3) The RMO has many advantages over refracting objective. It is an all-reflecting construction and free from chromatic aberration. The primary spherical aberration, primary coma and primary astigmatism have been corrected. The specific mirror coating is UV-enhanced aluminium film, which has as high as 89% average reflectivity in the range of 190 nm – 10μ m and is highly recommended for most UV use. Comparison with the refracting objective, the RMO has stronger focusing ability. (4) The RMO has a relative big numerical aperture (0.65), so it can be used as an excellent focusing element and also a good component to effectively collect fluorescence; (5) One can use the videohead to take photos for the samples and the focusing point in time.

The configuration includes a FluroMax-4 spectrometer (Horiba Jobin Yvon) equipped with a 150 W xenon lamp and a single-photon photomultiplier tube (PMT). The RMO comprises Ealing (52X). The micro-lens and fiber optic cable originate from Avertise. The fiber has a diameter of 400 μm and exhibits 90% transmission within the wavelength range of 240-800 nm. The fiber cable contains seven fibers. The fibers are organized in a club-like formation. The central fiber transmits the illuminating light from the xenon lamp in the spectrometer to the micro-lens. The peripheral six function as emitting fibers to direct the fluorescence into the spectrometer.

The system amalgamates micro-area localization, DUV light excitation, weak fluorescence detection, and real-time photography functionalities. A computer algorithm was developed to calculate the chromaticity coordinates of the OGIs based on spectra stimulated at 365 nm..

3.2 The primary measurement procedures

Several essential stages must be acknowledged in the experiment for monitoring, photographing, and measuring the fluorescence spectra of individual OGIs. The experiment was conducted in darkness to minimize environmental effect. The primary steps are as follows: Identify and position an OGI at the center of the visual field. Capture images when white light illuminates the OGI. Capture images using UV, blue, and green light from the internal light source for enhanced excitement. Exercise caution to avoid satiating the CCD. Adjust the 3D adaptor to ensure the RMO and the IMF goal are coaxial. When the two are coaxial, a prominent and round spot can be observed over the RMO. Observe the intensity of the green light emitted by the mercury arc lamp and wear protective goggles to safeguard your eyes. Acquire a minimal focus point. Attach the micro-lens to the 3D adaptor and connect the fiber wire to the micro-lens. Activate the spectrometer, allowing the excitation fiber to transmit the 550 nm green light from the spectrometer to the micro-lens and then to the RMO. The RMO concentrates the intense light into a small area and directs it onto the very thin slide. Upon satisfactory adjustment, a green focal spot will be seen on the slide. Based on the focal spot's position, one must consistently adjust the x, y, and z axes of the adjustable adaptor and reduce the focal spot's size. To get a minimal focal point, it is essential to modify the relative distance between the micro-lens and the RMO. Once a least focal spot appearance, this distance should be fixed and not change it in the whole experiment. (5) Capture images of individual OGI stimulated by an external light source (Xeon lamp) and document the fluorescence spectra.

The excited and emission slits measure 10 nm and 8 nm, respectively. A suitable filter must be positioned prior to the emission window of the spectrometer. Measure the microfluorescence spectra of OGI stimulated by the internal light source (UV, blue, and green). Ensure that the spectrometer's exciting slit is deactivated. The emission slit measures 5 nanometers. Measure the background fluorescence spectra when stimulated by an external light source. The wavelengths and the widths of the two slits are identical to those in step (5).

3.3 The least focal spot of the RMO

Fig. 5 is the least focal spot of the RMO. The background is a quartz grain in an inclusion thin slide. The size of focal spot is about 12μ m. The halo light around the focal spot is the result of the diffraction. It is difficult to be eliminated for a RMO. The focal spot includes most energy.

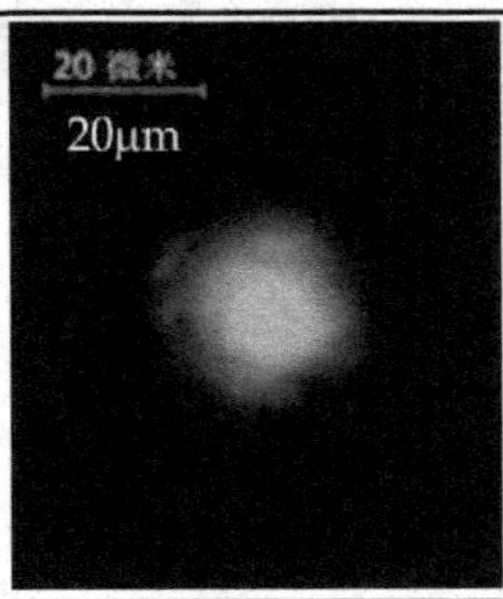

Fig. 5. The least light spot

3.4 The background fluorescence and the subtraction factor

The focal spot of the RMO is insufficiently small. When the focal spot size exceeds that of the OGI, both the OGI and the background emit fluorescence. The background fluorescence is intricate. Its roots may be as follows: The mineral grain may contain several tiny oil inclusions (Stasiuk & Snowdon, 1997; Meng, 2009), but these cannot be discerned under the IMF. The mineral grain's surface was polluted by oils present in the sandstone during the preparation of the inclusion thin slide. The mineral grain may contain thulium, which emits fluorescence when activated by light. The inclusion thin slide may be contaminated by airborne organisms or by human contact.

Fig. 6 shows the transmission micro-images of a single OGI (a, c, and e) and background (quartz grain)(b, d and f). In Fig. 6*a* and Fig. 6*b*, the exciting wavelengths are 250 nm and a violet fluorescence filter cube was used. In Fig. 6 c and Fig. 6 d, the exciting wavelength is 470 nm and a blue fluorescence filter cube was used. And in Fig. 6 e and Fig. 6 f, the 550 nm light was as exciting beam, the fluorescence filter cube was violet (The green fluorescence filter cube blocks the green light to get to the videohead, but the violet fluorescence filter cube permits the green light arrive to the videohead).

At the 550 nm light exciting, the borderline of the OGI and the focal spot are clear. This will advantage to determine the sizes of the OGI and the focal spot.

From Fig.6, one can see that the fluorescence of single OGI is the addition of the background and the OGI. For decreasing the influence of the background, a subtract factor was determined.

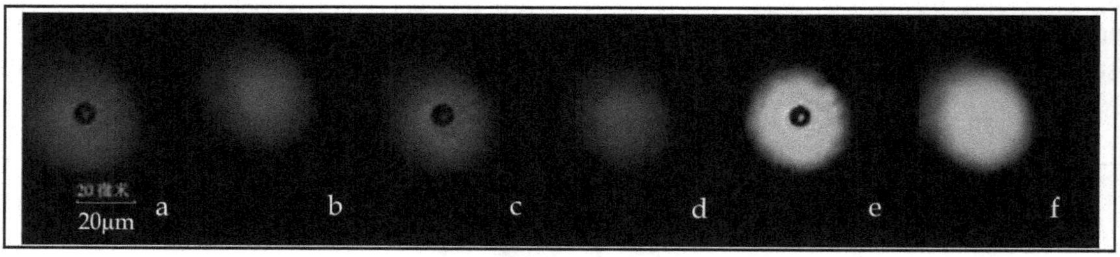

Fig. 6. Transmission micro-images of single OGI (a, c and e) and the background (b, d and f). In which, a **and** b, c **and** d, **and** e **and** f **are excited at** 250 nm, 470 nm **and 550 nm respectively. In** a, b, e, **and f , the violet fluorescence filter cube was used, which has a higher transmittance to the green light. In c and d, the blue fluorescence filter cube was used.**

The focal spot includes most exciting energy. The energy distribution is approximated to be homogeneous on the spot area. The OGI and the background are seen as surface light sources. Assuming the areas of the OGI and the light spot are S_1 and S_2. The two areas are easily obtained from Fig. 6 e and Fig. 6 f. The luminous intensities from unit surfaces of the OGI and the background are I_i and I_b. If the total fluorescence intensity of OGI and part background ($S_2 - S_1$) is $I_1(\lambda)$, the intensity of the background is $I_2(\lambda)$. Then

$$I_1 = S_1 I_i + (S_2 - S_1) I_b \tag{1}$$

$$I_2 = S_2 I_b . \tag{2}$$

From Eq. (1) and (2), the intensity of the OGI is

$$I_i S_1 = I_1 - \left(1 - \frac{S_1}{S_2}\right) I_2 = I_1 - F I_2 \tag{3}$$

$$F = 1 - \frac{S_1}{S_2} . \tag{4}$$

Here F is the subtract factor. It can be obtained from the micro-photos of the OGI and the background. By the experiment, one can obtain I_1, I_2, S_1, and S_2. According to Eq. (3) and

(4), one can further obtain the fluorescence spectra of single OGI. The following results based on above analysis.

3.5 The UV-VIS spectra of single OGIs

The first episode oil inclusions are very small and dense. The fluorescence is very weak. The background fluorescence greatly interferes the signal. The OGIs in the second episode are relative big and in scattering or in group in the mineral grains. It is possible to obtain the fluorescence spectra of single OGIs of the second episode, so these inclusions were chosen as samples.

In the experiment, we first observed dozens of OGIs in an inclusion thin slide then chose some typical OGIs to take photos and measure fluorescence spectra. The typical OGIs in an inclusion thin slide often have similar fluorescence colors. A relative clear area in the quartz grain was chosen as background. There are 38 typical OGIs to be measured. Table 2 shows the depths, strata, sizes, phases, F factors and corresponding oil wells of 38 OGIs.

Table 2. The depths, strata, sizes, phases (L is liquid, and G gas), F factor and corresponding oil wells of 38 OGIs. * means the OGI is in the cement.

Oil well	Stratum	Depth (m)	No.	Size (μm)	Phase	F (%)
Bai 95#	K_2Y_{2+3}	419.3	1	7.0, 12.3	L+G	86.0
		420.1	2	7.7, 10.1	L+G	79.0
	K_2y_1	426.8	3	4.9, 10.7	L+G	94.2
		431.0	4	8.3, 9.7	L+G	86.0
		434.5	5	7.0, 10.7	L+G	88.7
			6	7.7, 10.8	L+G	35.3
		435.0	7	12.5, 37.1	L	86.4
	K_2qn_2	494.6	8	14.1, 14.6	L+G	57.2
			9	6.5, 10.1	L	75.0
Hua 5#	K_2n_1	1449.8	10	4.5, 4.7	L+G	90.0
		1522.4	11	17.3, 53.7	L+G	42.7
		1523.0-1524.0	12*	11.1, 12.5	L+G	0.0
		1523.3	13	4.0, 4.0	L+G	96.0
Jian 22#	K_2q_4	377.8	14	16.6, 19.6	L+G	59.8
		396.6	15	9.42, 9.38	L+G	49.0
			16	15.0, 15.4	L+G	61.9
		401.9	17	9.6, 15.3	L+G	69.9
		404.7	18	5.7, 6.6	L+G	91.5
			19	21.6, 22.9	L+G	49.1

			20	10.8, 16.7	L+G	64.7
		420.0	21*	26.4, 27.1	L+G	38.3
			22	17.1, 17.1	L+G	36.6
Nong 29#	K_2q_4	571.0-572.0	23	11.0, 17.3	L+G	72.0
			24	27.2, 30.0	L+G	0.0
		573.0-574.0	25	7.4, 10.6	L+G	82.0
		595.9	26	8.7, 9.7	L+G	86.0
			27	12.7, 25.4	L+G	65.6
			28*	37.2, 37.4	L+G	0.0
			29	12.2, 25.3	L+G	50.0
Fu 4 #	K_2q_3	419.1	30	9.8, 11.7	L	81.6
			31	20.8, 24.2	L+G	39.0
		422.8	32	51.1, 71.3	L+G	0.0
			33	10.3, 15.7	L+G	62.2
			34	7.8, 13.6	L	78.0
		424.3	35	6.8, 7.0	L+G	89.2
			36*	22.6, 23.0	L+G	0.0
		425.2	37	15.9, 18.3	L+G	65.7
			38	15.5, 25.3	L+G	55.6

Fig.7-Fig. 22 are the transmission, reflective micro photos and fluorescence spectra of OGIs of No.6, No.7, No.12, No.15, No.20, No.25, No. 26 and No.34. No. 6 and No. 7 (Fig.7-Fig.10) are in Bai 95# oil well, the depths are 434.5 m and 435.0 m . No. 12 (Fig.11-Fig.12) is in Hua 5# oil well, the depth is 1523-1524 m. No. 15 (Fig. 13-Fig.14) and No. 20 (Fig.15-Fig.16) are in Jian 22# oil well, the depths are 396.6 m and 404.7 m . No. 25 (Fig.17-Fig.18) and No. 26 (Fig.19Fig.20) are in Nong 29# oil well, the depths are 573 – 574 m and 595.6 m . No. 34 (Fig.21-Fig.22) is in Fu 4# oil well, the depth is 422.8 m .

No.6, No.15, No.20, No. 25 and No. 26 are two phases (liquid and gas) inclusions. No. 7 and No. 34 are single phase (liquid) oil OGIs. No.12, No. 21 and No. 36 are in cements. The others are all in the quartz grains.

From the micro photos, one can see when the internal light source excites the aimed OGI, not only the OGI emits fluorescence, but also other OGIs and the cements around the mineral grains. The intensities of the cements are often stronger than that of the OGIs. Even the mineral grains also emit weak fluorescence. These are the reasons why the spectra look like each other even the OGIs with different phases and colours. For example, No. 6 and No. 7 OGIs are in the same oil well, the former is liquid-gas inclusion, the later is liquid phase. The spectra of the two excited by internal light source (see Fig. 8 a and Fi9.10 a) are similar. Obviously, the spectra were influenced by the background. According to the spectra excited by violet light of the microscope (see Fig. 8 a and Fig. 10 a, solid line), the peak positions are at 400 nm . The main aromatic hydrocarbons in the OGIs may be three or four cyclic hydrocarbons. But the other aromatic hydrocarbons in the OGIs are difficult to determine because except for the main peak, the other parts of the spectra are almost flat. The peak positions excited by the internal blue and green light are at 500-510 nm and 575 nm – 600 nm respectively.

When the external light source (250 nm, 365 nm, 440 nm, 470 nm and 546 nm) excited the aimed OGI, the focal spot size is small. The cements were not illuminated, so the fluorescence spectra of the single OGIs weren't interfered by the spectra of the cements. If the OGIs are not dense in the grain, only one OGI is excited, one can obtain the spectra of single OGI after subtracting the background.

From Fig. 8 b, Fig. 10 b, Fig. 12 b, Fig. 14 b, Fig. 16 b, Fig. 18 b, Fig. 20 b and Fig. 22 b, one can see that the spectra are in the range of UV-VIS at 250 nm exciting. These results show that the experimental setup can measure the UV-VIS spectra of single OGIs. Combined the spectra excited by different external lights with different wavelengths, the characteristics of the OGIs are as following:

1. For most OGIs, when excited at 250 nm , there are three main peaks in the range of 400 – 470 nm , which are corresponding to three and four condensed aromatic rings, the palaeo oils are the medium Oils (Liu, K. Y., & Eadington P., 2005; Abbas et al., 2006). There are also two secondary peaks near to 535 nm and 610 nm , the former is the fluorescence of four or more condensed aromatic rings, the later comes from resin and asphaltene, the palaeo oils are medium and heavy Oils (Liu, K. Y., & Eadington P., 2005; Abbas et al., 2006). The spectra in the range of 280 – 400 nm belong to two or three condensed aromatic rings, the palaeo oils are light Oils (Liu, K. Y., & Eadington P., 2005; Abbas et al., 2006). So the OGIs of No.6, No.7, No.12, No.15, No.20, No. 25 and No. 34 are all filled with light, medium and heavy oils. And medium oils are the major part. The aromatic hydrocarbons in these OGIs are mainly three, four and five cyclic hydrocarbons.

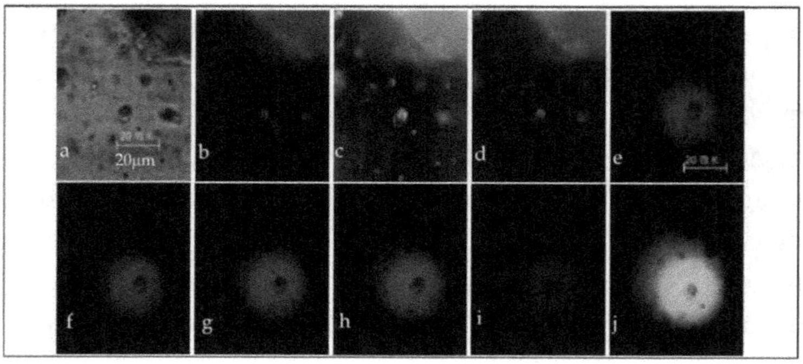

Fig. 7. The transmission (a , e, f, g, h, i and j) and reflective (b, c and d) microphotos of No. 6 OGI (Bai 95#, 434.5 m) when excited by white light (a), internal light source (c:violet; d:blue and e:green) and external light source (e:250 nm; f:365 nm; g:440 nm; h:470 nm; and i:546 nm), j is a transmission image when the 550 nm light exciting and violet fluorescence filter cube was used.

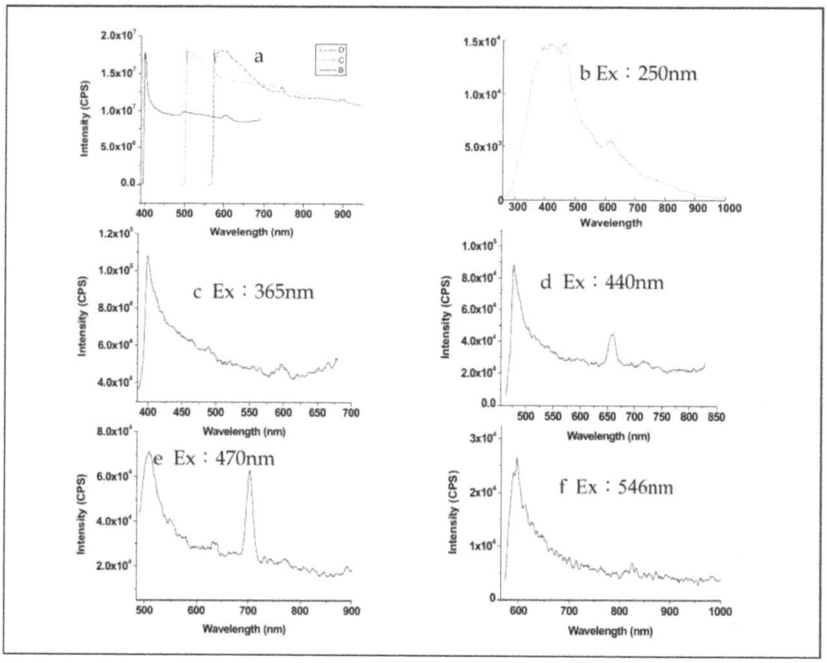

Fig. 8. The fluorescence spectra of No. 6 OGI excited by internal light source (in a, solid: violet, short dot: blue, and dash dot: green) and external light source (b-f).

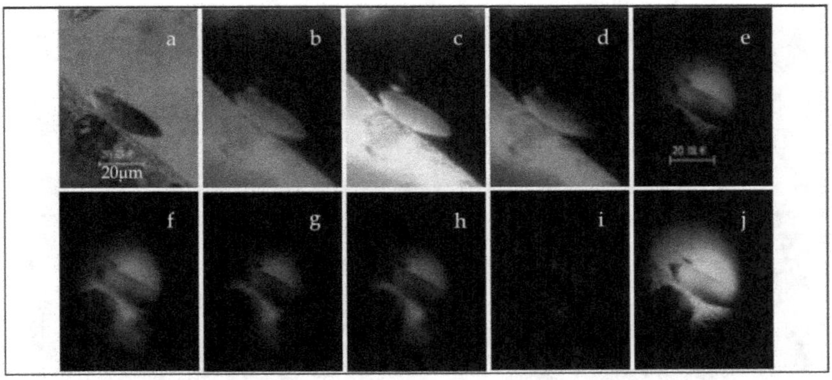

Fig. 9. The transmission (a, e, f, g, h, i and j) and reflective (b, c and d) microphotos of No. 7 OGI (Bai 95#, 435.0 m) when excited by white light (a and b), internal light source (b:violet; c:blue and d:green) and external light source (e: 250 nm ; f:365 nm; g:440 nm; h:470 nm and i:546 nm), j is a transmission image when the 550 nm light exciting and violet fluorescence filter cube was used.

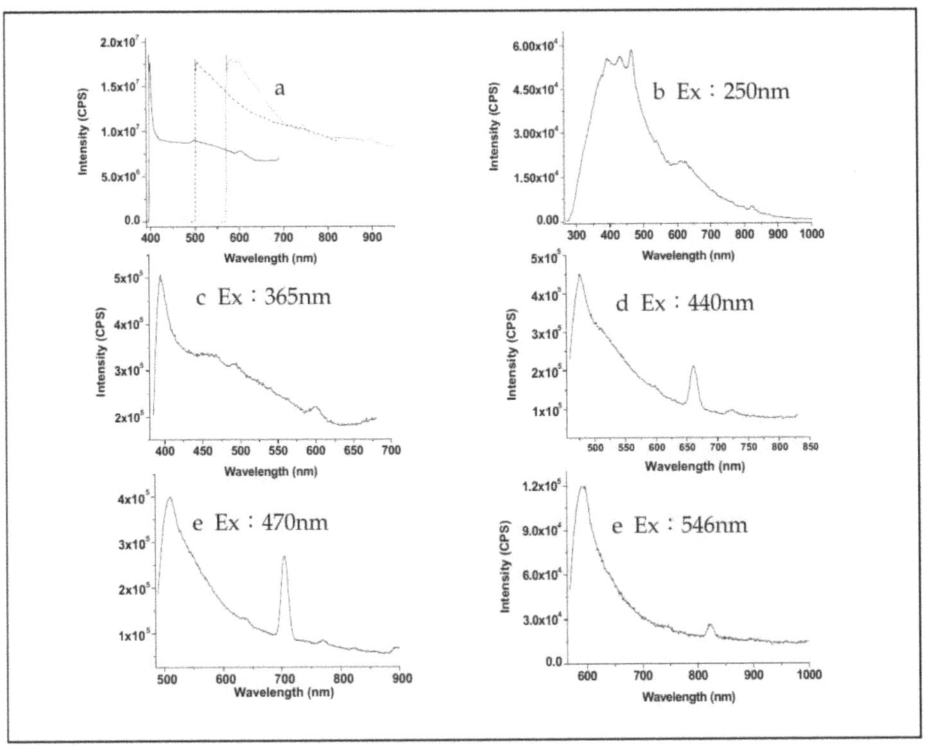

Fig. 10. The fluorescence spectra of No. 7 OGI excited by internal light source (in a, solid: violet, long dot: blue, and short dot: green) and external light source (b-f).

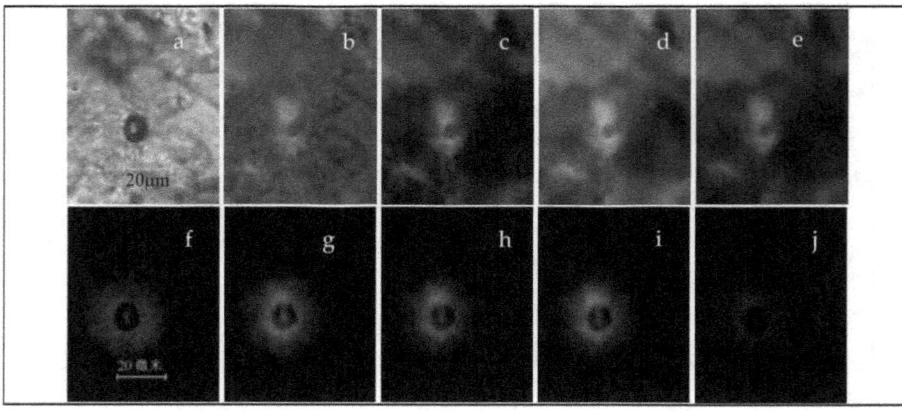

Fig. 11. The transmission (a, b, f, g, h, i and j) and reflective (b, c, d and e) microphotos of No.12 OGI (Hua 5#, 1523.4 m) in cement when excited by white light (a and b), internal light source (b and c:violet; d:blue and e:green) and external light source (f:250 nm; g:365 nm; h: 440 nm; i: 470 nm **and j:546 nm**), b **is a transmission and reflection image when white and violet light illuminating at the same time.**

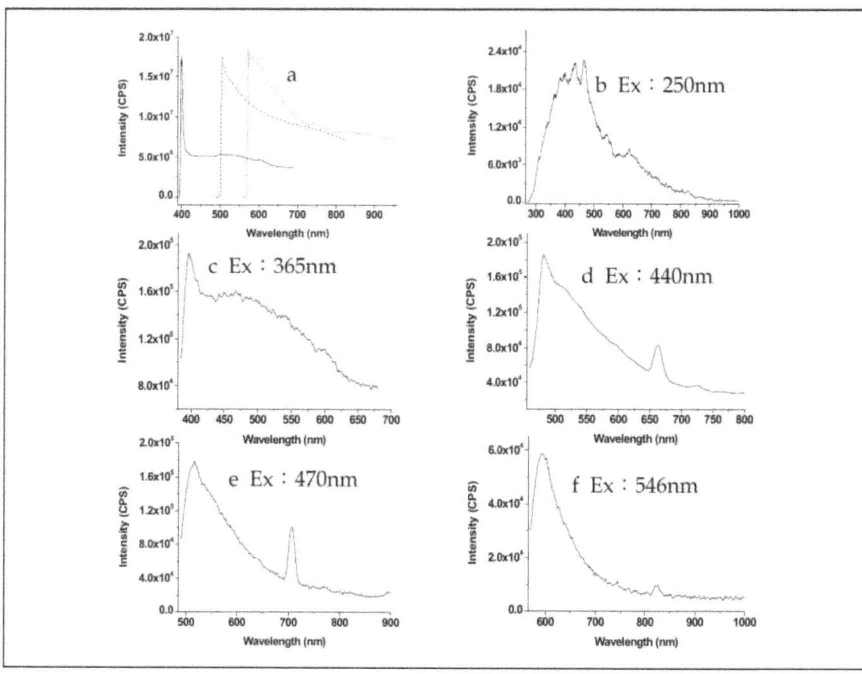

Fig. 12. The fluorescence spectra of No. 12 OGI excited by internal light source (in a, solid: violet, long dot: blue, and short dot: green) and external light source (b-f).

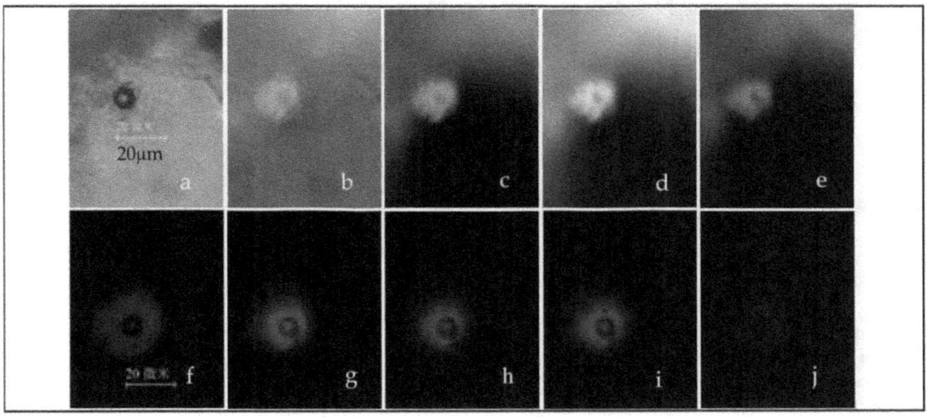

Fig. 13. The transmission (a, b, f, g, h, i and j) and reflective (b, c and d) microphotos of No. 15 OGI (Jian 22#, 396.6 m) when excited by white light (a and b), internal light source (b and c:violet, d:blue and e:green) and external light source (e:250 nm; f:365 nm; g:440 nm; h:470 nm and i: 546 nm **), b is a transmission and reflection image when white and violet light illuminating.**

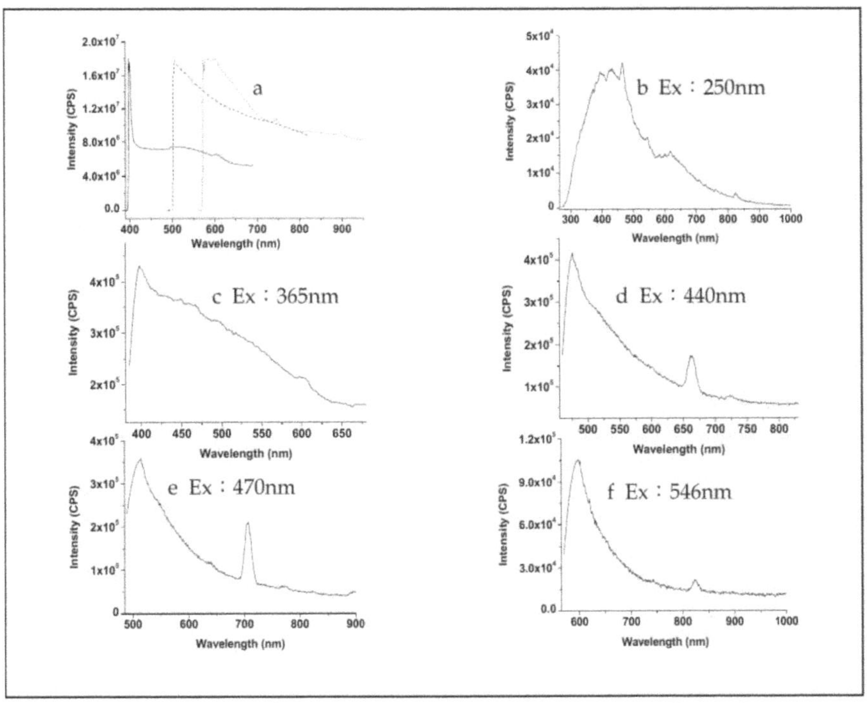

Fig. 14. The fluorescence spectra of No. 15 OGI excited by internal light source (in a, solid: violet, long dot: blue, and short dot: green) and external light source (b-f).

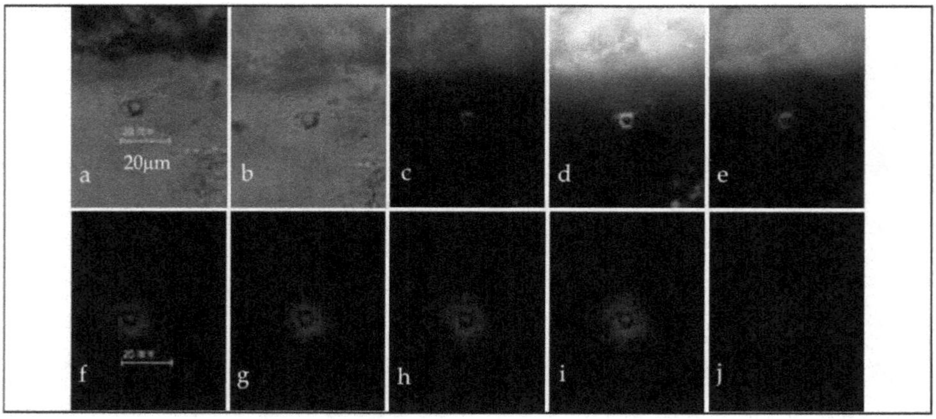

Fig. 15. The transmission (a, b, f, g, h, i and j) and reflective (b, c, d and e) microphotos of No. 20 OGI (Jian 22#, 404.7 m) when excited by white light (*a* and *b*), internal light source (*b* and c:violet; d:blue and e:green) and external light source (f:250 nm; g:365 nm; h:440 nm; i:470 nm and j:546 nm), b is a transmission and reflection image when white and violet light illuminating.

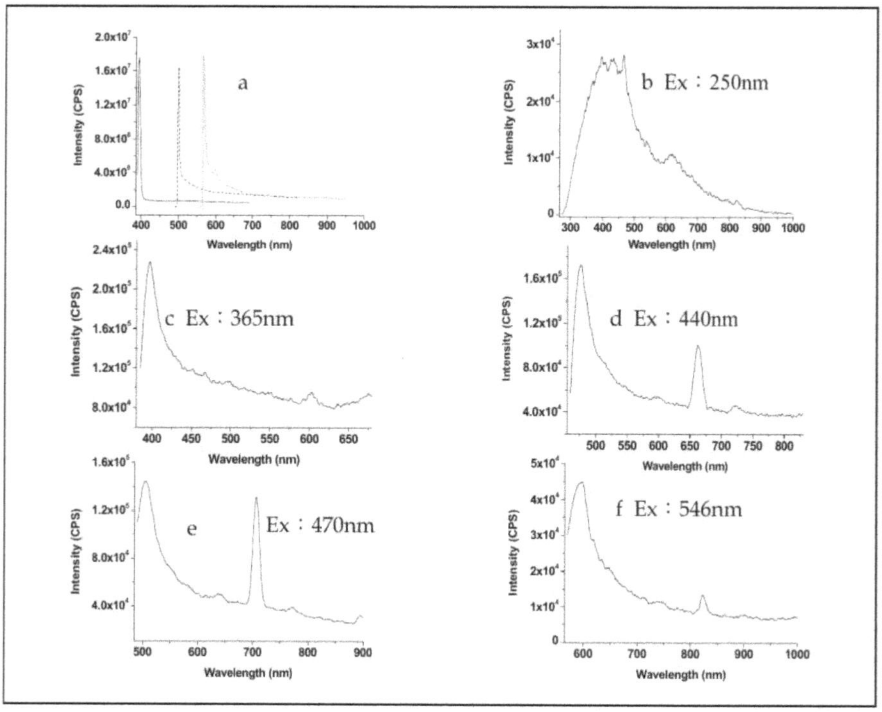

Fig. 16. The fluorescence spectra of No. 20 OGI excited by internal light source (in a, solid: violet, long dot: blue, and short dot: green) and external light source (b-f).

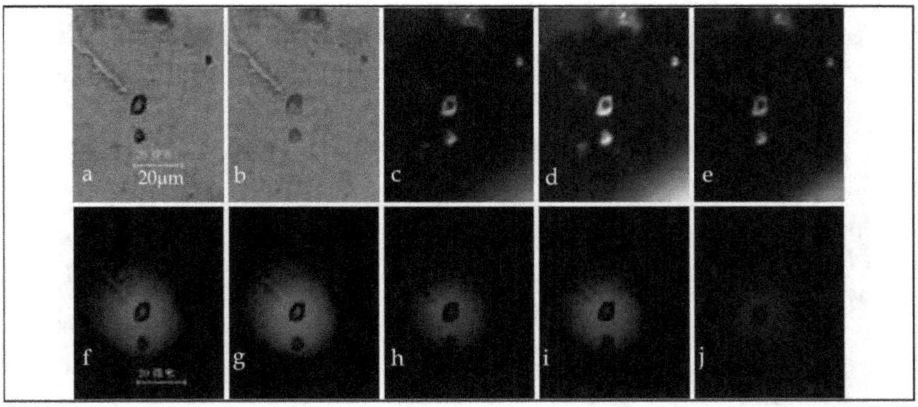

Fig. 17. The transmission (a, b, f, g, h, i and j) and reflective (b, c, d and e) microphotos of No. 25 OGI (Nong 29#, 573-574 m) when excited by white light (a and b), internal light source (b and c:violet; d:blue and e:green) and external light source (f:250 nm; g:365 nm; h: 440 nm; i: 470 nm **and j:546 nm). b is a transmission and reflection image when white and violet light illuminating.**

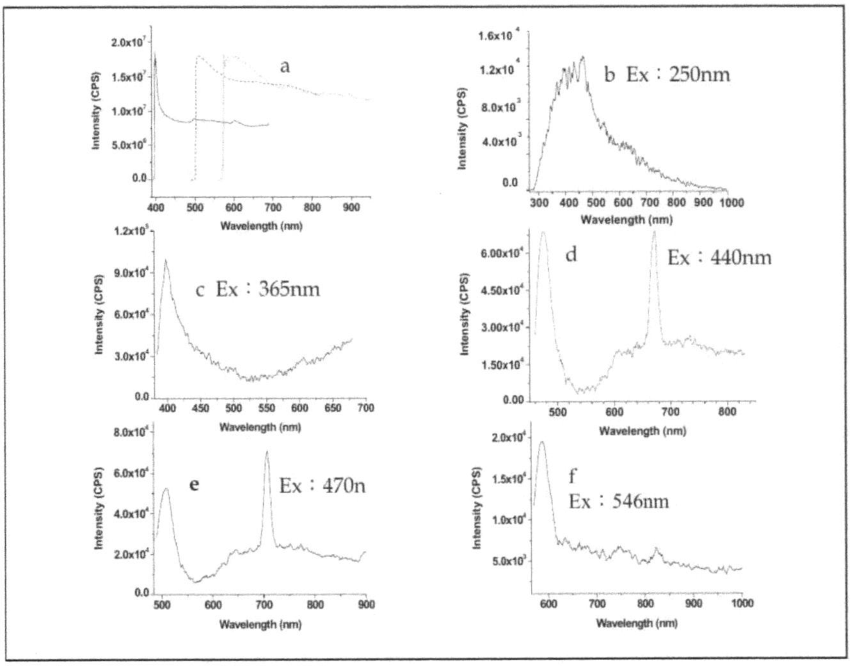

Fig. 18. The fluorescence spectra of No. 25 OGI excited by internal light source (in a, solid: violet, long dot: blue, and short dot: green) and external light source (b-f).

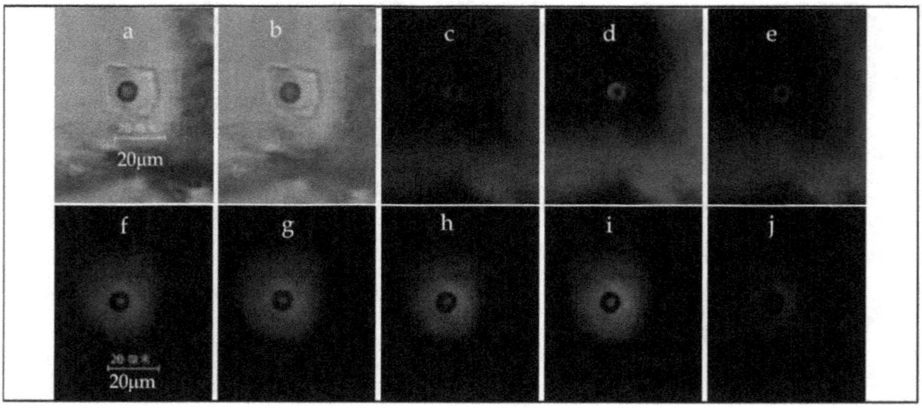

Fig. 19. The transmission (a, b, f, g, h, i and j) and reflective (b, c, d and e) micro photos of No. 26 OGI (Nong 29#,595.9 m) when excited by white light (a and b), internal light source(b and c : violet; d: blue and e: green) and external light source (f: 250 nm; g: 365 nm; h: 440 nm; i: 470 nm and j:546 nm), b is a transmission and reflection image when white and violet light illuminating.

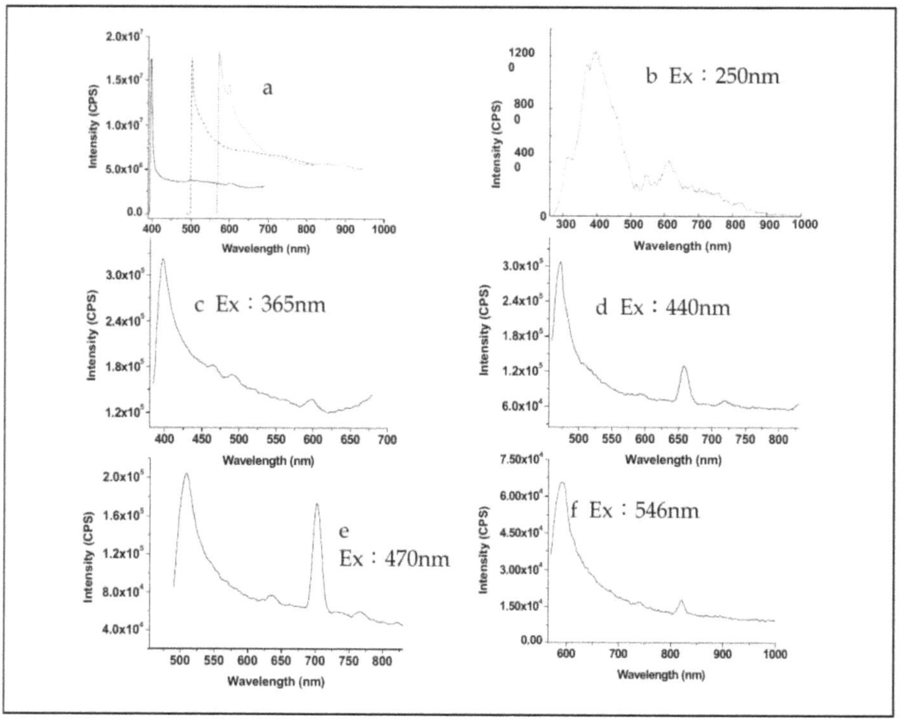

Fig. 20. The fluorescence spectra of No. 25 OGI excited by internal light source (in a, solid: violet, long dot: blue, and short dot: green) and external light source (b-f).

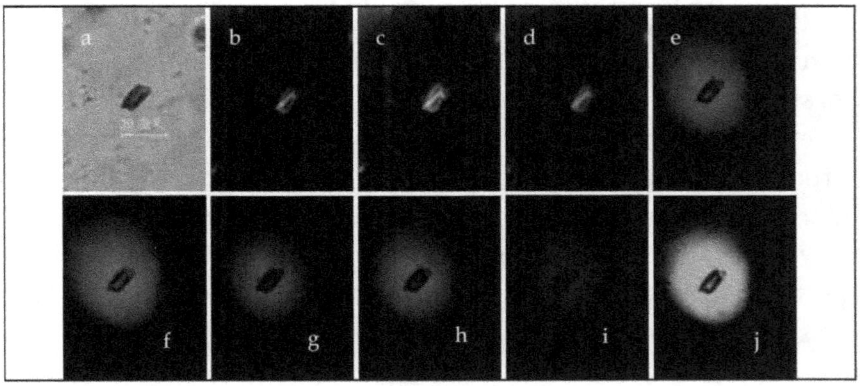

Fig. 21. The transmission (a, e, f, g, h, i and j) and reflective (b, c and d) micro-photos of No. 34 OGI (Fu4# 422.8 m) when excited by white light (a), internal light source (b: iolet; c:blue and d:green) and external light source (e:250 nm; f:365 nm; g:440 nm; h:470 nm and i:546nm), j is a transmission image when the 550 nm light exciting and violet fluorescence filter cube was used.

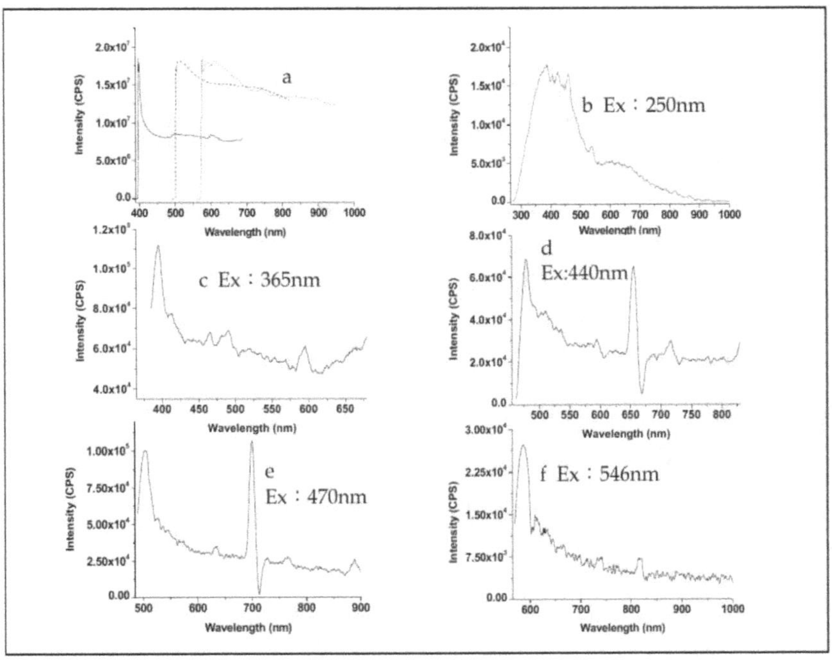

Fig. 22. The fluorescence spectra of No. 34 OGI excited by internal light source (in a, solid: violet, long dot: blue, and short dot: green) and external light source (b-f).

There are still some one and two cyclic aromatic hydrocarbons and heavy hydrocarbons. The spectrum (Fig. 20 b) of No. 26 OGI is different from others. There is one main peak at

400 nm . The full width of half maximum (FWHM) is obviously less than the others, which shows the main hydrocarbons in this OGI are light-medium oils. The secondary peaks in 575 – 605 nm show there are also some heavy oils in this OGI. The aromatic hydrocarbons in No. 26 OGI are mainly three and four cyclic hydrocarbons. There are still some two cyclic aromatic hydrocarbons and heavy hydrocarbons. For all 38 testing OGIs, the spectra of 90% are similar to No. 6 when exciting at 250 nm . This result shows that the palaeo-oils in the OGIs may come from two maternal sources with different maturities, one is high, another is low. So in the later time of quartz overgrowth, there may be two oil sources charging these strata at the same time.

2. In the same oil well, the FWHMs of liquid oil inclusions are wider than that of the liquid-gas OGIs at exciting 250 nm . For example, the FWHMs of No. 6 (liquid and gas) and No. 7 (liquid) inclusions are 175 nm and 185 nm respectively. But in different oil wells, this result is not always right. For example, the FWHMs of No. 12 and No. 15 (liquids and gas) are 189 nm and 190 nm , which are wider than that of No.7. So one should be cautious to say that the palaeo-oils in the liquid oil inclusions are more heavier than that of in the liquid and gas OGIs.

3. For all the 38 testing OGIs, when exciting at 365 nm (external light source), the spectra can be classified into two types. The first type has an obvious "bump" following the main peak. The wavelength range is in 425 – 650 nm. This range is relative to the medium and heavy hydrocarbons, which indicate that these inclusions including in more medium and heavy hydrocarbons (see Fig. 10 c, Fig. 12 c, Fig. 14 c). The second type has not an obvious "bump", which shows that these OGIs (see Fig. 8 c, Fig. 16 c, Fig. 18 c, Fig. 20 c and Fig. 22 c) don't include in enough medium and heavy hydrocarbons as in the first type. The palaeo-oil maturities of the second type are higher than the first one.

4. When exciting at 440 nm , except for the main peak near to 478 nm , there is another peak at 657 nm for all 38 OGIs . No. 25 OGI has another peak at 672 nm .

5. When exciting at 470 nm , except for the main peak near to 502 nm , there is another peak at 706 nm for all 38 OGIs.

6. The results of (4) and (5) show that a kind of characteristic matter exist in the OGIs. 440 nm and 470 nm are the effective exciting wavelengths for this kind of matter. This needs to make further analysis to determine what it is.

7. The characteristics of the spectra excited at 546 nm are not obvious. The reason is the absorption bands for the most hydrocarbons are in the range of UV and violet.

3.6 The hues of individual OGIs

To facilitate understanding, the colors of the OGIs stimulated at 365 nm are frequently employed to assess the maturities of palaeo-oils within the inclusions. However, visual assessments are inherently subjective. The outcomes will be affected by psychological, physiological, or environmental factors. It is necessary to calculate the chromaticity coordinates of the OGIs.

The calculation theory (Xu & Su,2004) is as following. Assume $P(\lambda)$ is the power spectrum distribution of an OGI, the tristimulus values are given by equation (5),

$$X = \int_{380}^{780} P(\lambda)\bar{x}(\lambda)d\lambda$$

$$Y = \int_{380}^{780} P(\lambda)\bar{y}(\lambda)d\lambda(5)$$

$$X = \int_{380}^{780} P(\lambda)\bar{z}(\lambda)d\lambda(5)$$

In which, $\bar{x}(\lambda), \bar{y}(\lambda)$ and $\bar{z}(\lambda)$ are the tristimulus values of standard light source. One can get them by looking up the CIE 1931 system. The chromaticity coordinates of an OGI are

$$x = \frac{X}{X+Y+Z}$$

$$y = \frac{Y}{X+Y+Z}(6)$$

$$z = \frac{Z}{X+Y+Z}(6)$$

In which, $x+y+z=1$ (x :red; y :green and z :blue). According to above theory, a computing program based on Matlab was established to calculate the chromaticity coordinates of OGIs.

The chromaticity coordinates show that for all OGIs, the chromaticity coordinates excited by internal light source (365 nm) are larger than that of external light source (365 nm). The main reason is that not only the OGI emitting fluorescence, but also the cements and the mineral grain itself when excited by internal light source, which influenced the spectra of the under studied OGI. The chromaticity coordinates deflect to light blue white or yellow white, and relative centralize for different OGIs. When the external light source excites the OGIs, the focal spot is small. The cements don't emit fluorescence and not influence the spectra of single OGIs. The fluorescence of the mineral grain around the OGI has been subtracted as the background. Such spectra are near to the real spectra of single OGIs.

For the OGIs in the cements in Jian 22# and Fu 4# oil wells, the chromaticity coordinates of these OGIs are less than the OGIs in the mineral grains whatever external or internal light source exciting (Fig. 23 c and Fig. 23 e), which indicates the hydrocarbons in the cements are lighter than those in mineral grains. The formation time of the OGIs in the cements may be later than that in the mineral grains. But for OGI in cements of Nong 29# oil well, the chromaticity coordinates basic overlap with the OGIs in the mineral grains. The OGIs in the cements and in the mineral grains may be formed at the same geology age.

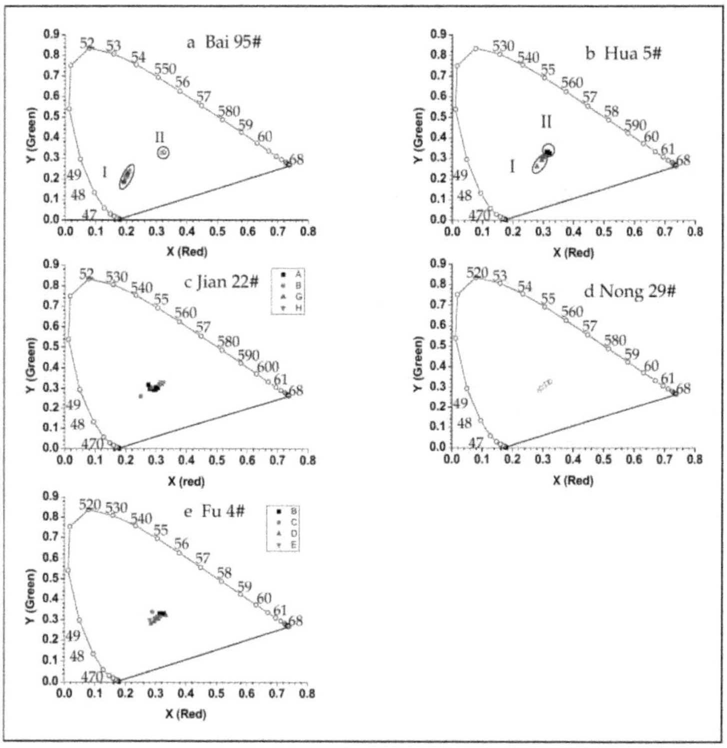

Fig. 23. The chromaticity coordinates of OGIs in the five oil wells excited by external and internal light sources, in which, a and b are Bai 95# , Hua 5# respectively, external (I) and internal (II); c—Jian 22#, external (A, B) and internal (G, H), B(G) is in the cement; d-Nong 29#, external (triangle) and internal (circle); e-Fu 4#, external (D, E) and internal (B, C), D(C) **is in the cement.**

The chromaticity coordinates are dispersive for single OGIs under external light source excited. This result shows that it is easier to distinguish OGIs with external light source exciting. For OGIs in Bai 95# oil well, the chromaticity coordinates are in the range of light blue white and light yellow white. For OGIs in Hua 5# oil well, the chromaticity coordinates are in the range of blue green and light yellow white. For OGIs in Jian 22# oil well, the chromaticity coordinates are in the range of green blue and light yellow white. For OGIs in Nong 29# oil well, the chromaticity coordinates are in the range of light blue white and light green blue. For OGIs in Fu 4# oil well, the chromaticity coordinates are in the range of light green blue, light blue white and white. Above results indicate that in the later time of quartz overgrowth, there may be two maternal source to charge these strata, one with high maturity, another with low maturity.

4. Concise Summary

The inclusion samples were derived from core drilling samples of five oil wells located in the Jinlin Oil Field, Songliao Basin, northeast China (Upper Cretaceous Period).The microscope testing shows:

- The experimental setup based on common inverted microscope, no need any rebuilding to the microscope and can measure the UV-VIS spectra of single OGIs. The cost was decreased greatly.
- By the spectra excited at 250 nm , the main aromatic hydrocarbons in an OGI can be qualitative determined. Almost all OGIs contain light, medium and heavy petroleum. The main aromatic hydrocarbons are three, four and five cyclic hydrocarbons. There are also some heavy hydrocarbons and non-hydrocarbons in these OGIs. In the same oil well, the FWHMs of liquid oil inclusions are wider than that of the liquid-gas OGIs at exciting 250 nm . But in different oil wells, this result is not always right.
- The peak positions of the spectra excited at 365 nm are in the range of 395 – 400 nm, the main aromatic hydrocarbons are three and four cyclic aromatics. The "bump" (425 650 nm) following the main peak shows there are more medium and heavy hydrocarbons in the OGIs. The OGIs without the "bump" don't have many medium and heavy hydrocarbons. The palaeo-oil maturities of the second type are higher than the first one.
- The special peaks near to 657 nm (440 nm exciting) and 706 nm (470 nm exciting) for all OGIs indicate there may be same special matters in these OGIs. This need further analysis combined with GC-MS and palaeo-biology.
- The chromaticity coordinates are more dispersive for single OGIs under external light source excited than internal light source. The calculating colours excited at external light source are more objective than judging by eyes under exciting at

internal light source. But one should note the different spectra may have same colour. So the spectra are important.

- The colours, the spectra excited at 250 nm and chromaticity coordinates show that there may be two maternal sources to charge these strata in the later time of quartz overgrowth. One is high maturity, another is low maturity.

This optical device is capable of measuring the UV-VIS spectra of individual OGIs. To further reduce the focal spot, one may consider utilizing a 74x RMO or decreasing the fiber diameter while enhancing the coupling efficiency between the fiber and the spectrometer. The UV-VIS spectra must be integrated with GC-MS analysis to identify the distinctive substances in these OGIs. This configuration shows potential for monitoring the fluorescence spectra of micro regions, including bioinclusions, micro fractures in mineral slides, and specific mineral grains.

5. References

Abbas, O., Rébufa, C., Dupuy, N., Permanyer, A., Kister, J. & Azevedo, D.A. (2006). Application of chemometric methods to synchronous UV fluorescence spectra of petroleum oils. Fuel, Vol. 85, No.17-18, (Dec. 2006), pp. 2653-2661, ISSN 0016-2361

Aplin, A.c., Macleod, G., Larter, S.R., Pedersen, K.S., Sorensen H. & Booth, T. (1999). Combined use of Confocal Laser Scanning Microscopyand PVT simulation for estimating the composition and physical properties of petroleum in fluid inclusions. Marine and Petroleum Geology, Vol.16, No.2, (Mar., 1999), pp. 97-110, ISSN 9153-7061

Caja, M. A., Permanyer, A., Munz, I. A. & Johansen, H. (2007). Preliminary data on oil and aqueous fluid inclusions of the fracture-fill in the Corones and Armancies Fms, Eocene, SEPyrenees. GEOGACETA, Vol.101, (n. d. 2007), pp. 127-130, ISSN 0213683X

Caja, M. A., Permanyer, A., Kihle, J., Munz, I. A. & Johansen, H. (2009). Fluorescence quantification of oil fluid inclusions and oil shows:Implications for oil migration (Armancies Fm, South-eastern Pyrenees, Spain). Journal of Geochemica Exploration, Vol.101, No.1, (April, 2009), pp.16, ISSN 0375-6742

Damaskinos, S., Dixon, A. E., Ellis, K. A., & Diehl-Jones, W. L. (1995). Imaging biological specimens with the confocal scanning laser microscope/macroscope. Micron, Vol.26, No.6, (n.d.,1995), pp. 493-502, ISSN 0968-4328

Kevin, B., Liesbeth, P., Niek, N. S., Stefaan, C. D. S. & Joseph, D. (2003). Three-Dimensional Fluorescence Recovery after Photobleaching with the Confocal Scanning Laser Microscope. Biophysical Journal, Vol.85, No.4, (Oct., 2003), pp. 2240-2252, ISSN 00063495

Kihle, J. (1995). Adaptation of fluorescence excitation-emission micro-spectroscopy for characterization of single hydrocarbon fluid inclusions. Organic Geochemistry, Vol. 23, No.11-12, (Nov.-Dec. 1995), pp. 1029-1042,ISSN 0146-6380

Liu, D. H., Xiao, X. M., Mi, J. K., Li, X. Q., Shen, J. K., Song, Z. G. & Peng, P. A. (2003). Determination of trapping pressure and temperature of petroleum inclusions using PVT simulation software-a case study of Lower Ordovician carbonates from the Lunnan Low Uplift, Tarim Basin. Marine and Petroleum Geology, Vol.20, No.1, (Jan., 2003), pp. 29-43, ISSN 9153-7061

Liu, K. Y. & Eadington, P. (2005). Quantitative fluorescence techniques for detecting residual oils and reconstructing hydrocarbon charge history. Organic Geochemistry, Vol. 36, No. 7, (July, 2005), pp. 1023-1036, ISSN 0146-6380

Liu, Y. R., Lu, X. B. & He, M. C. (2003). Progress on Application of Fluid Inclusion in Petroleum Exploration. Geology Bullet in of Mineralogy, Petrology and Geoch emistry, Vol.22, No.3, (July, 2003), pp. 245-250, ISSN 1007-2802

Lo, H. B., Wilkins, R. W. T., Ellacott, M. V. & Buckingham, C. P. (1997). Assessing the maturity of coals and other rocks from North America using the fluorescence alteration of multiple macerals (FAMM) technique. International Journal of Coal Geology, Vol.33, No.1, (Jan. , 1997), pp. 61-71, ISSN 0166-5162

Meng, D. W., Wu, X.L., Fan, X.Y., Meng, X. & Zheng, J.P. & Mason, R. (2009). Submicronsized fluid inclusions and distribution of hydrous components in jadeite, quartz and symplectite-forming minerals from UHP jadeite-quartzite in the Dabie Mountains, China: TEM and FTIR investigation. Applied Geochemistry, Vol.24, No.4, (April, 2009), pp. 517-526, ISSN 0883-2927

Sanches S. , Leitão C., Penetra A., Cardoso V.V. , Ferreira E., Benoliel M.J., Barreto Crespo M.T., Pereira V.J. (2011). Direct photolysis of polycyclic aromatic hydrocarbons in drinking water sources. Journal of Hazardous Materials, Vol. 192, (Jun. 2011), pp. 1458-1465, ISSN 0304-3894

Siegwart, R. (2001). Indirect Manipulation of a Sphere on a Flat Disk Using Force Information. International Journal of Advanced Robotic Systems, Vol.6, No.4, (Dec. 2009), pp. 12-16, ISSN 1729-8806

Sliney D. H. (1995). Risk assessment and laser safety. Optics & Laser Technology, Vol. 27, No. 5, (n. d. 1995), pp. 279-284, ISSN 0030-3092

Stasiuk, L. D. & Snowdon, L. R. (1997). Fluorescence micro-spectrometry of synthetic and natural hydrocarbon fluid inclusions:crude oil chemistry, density and application to petroleum. Applied Geochemistry, Vol.12, No.3, (May, 1997), pp. 229-241, ISSN 0883-2927
Stasiuk, L.D. (1999). Confocal laser scanning fluorescence microscopy of Botryococcus alginite from boghead oil shale, Boltysk, Ukraine: selective preservation of various micro-algal components. Organic Geochemistry, Vol.30, No.8, (Aug.,1999), pp. 10211026, ISSN 0146-6380

Stasiuk, L.D. & Sanei, H. (2001). Characterization of diatom-derived lipids and chlorophyll within Holocene laminites, Saanich Inlet, British Columbia, using conventional and laser scanning fluorescence microscopy. Organic Geochemistry, Vol.32, No. 12, (Dec., 2001), pp. 1417-1428, ISSN 0146-6380

Thiéry, R., Pironon, J., Walgenwitz, F. & Montel F. (2002). Individual characterization of petroleum fluid inclusions (composition and P-T trapping conditions) by microthermometry and confocal laser scanning microscopy: inferences from applied thermodynamics of oils. Marine and Petroleum Geology, Vol.19, No.7, (Aug., 2002), pp. 847-859, ISSN 9153-7061

Veld, H., Wilkins, R. W. T., Xiao, X. M. & Buckingham, C. P. (1997). A fluorescence alteration of multiple macerals (FAMM) study of Netherlands coals with "normal" and "deviating" vitrinite reflectance. Organic Geochemistry, Vol.26, No.3-4, (Feb., 1997), pp. 247-255, ISSN 0146-6380

Wilkins, R. W. T., Wilmshurst, J. R., Russell, N. J., Hladky, G., Ellacott, M.V. & Buckingham C. (1992). Fluorescence alteration and the suppression of vitrinite reflectance. Organic Geochemistry, Vol.18, No.5, (Sept., 1992), pp. 629-640, ISSN 0146-6380

Wilkins, R. W. T., Wilmshurst J.R., Hladky, G., Ellacott, M. V. & Buckingham, C. P. (1995). Should fluorescence alteration replace vitrinite reflectance as a major tool for thermal maturity determination in oil exploration? Organic Geochemistry, Vol.22, No.1, (Jan. , 1995), pp. 191-209, ISSN 0146-6380

Xiao, X.M., Wilkins R. W. T., Liu Z.F. & Fu, J.M. (1998). A preliminary investigation of the optical properties of asphaltene and their application to source rock evaluation. Organic Geochemistry, Vol.28, No. 11, (July,1998), pp. 669-676, ISSN 0146-6380

Xiao, X.M., Wilkins, R. W. T., Liu D.H. Liu, Z. F. & Fu, J. M. (2000). Investigation of thermal maturity of lower Palaeozoic hydrocarbon source rocks by means of vitrinite-like maceral reflectance - a Tarim Basin case study. Organic Geochemistry, Vol.31, No.10, (Oct., 2000), pp.1041-1052, ISSN 0146-6380

Xiao, X. M., Wilkins R. W. T., Liu, D. H. & Shen, J.Q. (2002). Laser-induced fluorescence microscopy-application to possible high rank and carbonate source rocks. International Journal of Coal Geology, Vol.5, No.2, (July, 2002), pp. 129-141, ISSN 0166-5162

Xu, X.R. & Su, Z. M. (Oct., 2004). Luminescence Theory & Luminescence Materials, Chemical Industry, ISBN: 7502554106, 9787502554101, Beijing

Yang, A.l., Zhang, J.L., Ren. W.W. & Tang. M.M. (2009a). Micro-spectroscopy system based on common inverted microscope to measure UV-VIS spectra of a micro-area. Proceedings of SPIE 7283, ISBN 0277-786X, Chengdu, China, Nov., 2008

Yang, A.l., Ren. W.W., Zhang, J.L., & Tang. M.M. (2009b). A micro-spectroscopy system to measure UV-VIS spectra of single hydrocarbon inclusions. Proceedings of SPIE 7384, ISBN 0277-786X, Beijing, China, April, 2009

Yang A.L.,Tang, M.M.,Ren. W.W.,Yang, Y. & Zhang, J.L. (2011). Investigation of the ultraviolet-visible micro-fluorescence-spectra and chromaticity of single oil inclusion. ACTA OPTICA SINICA, Vol.31, No.3, (Mar., 2011), pp. 0318002-1-6, ISSN 0253-2239.

Chapter-3

Validation of Hybrid Chemical Engineering Proton NMR for Food Analysis

1. Introduction: An Overview

In food engineering, chemical engineering and bioengineering frequently utilize data acquired via physicochemical approaches. Water species are significantly correlated with food quality, as evidenced by the International Symposium on Properties of Water (Eds. Rockland and Stewart, 1981). The water species contained in meals exhibit their multifunctional characteristics by constantly reacting to external factors such as temperature, pressure, dehydration processes, and water content. The scientific examination of multifunctional water species maintained in foods is particularly challenged by the impact of nonlinear dynamic changes on food quality. A multitude of scholars have quantitatively assessed the diversity of water species by examining numerous biological and physicochemical properties. Water activity (a_w) has frequently been utilized as a metric to assess food degradation (Fennema, 1976). The effective water diffusivity measure (De) in food has often been utilized to assess the dehydration rate (Jason, 1958). From a chemical engineering perspective, this is a standard approach for assessing water mobility inside the food tissue matrix. In contrast, magnetic resonance techniques have been utilized for the direct identification of water species at the molecular level to assess food quality, as evidenced in the International Conferences on Applications of Magnetic Resonance in Food Science (Eds. Belton et al., 2003).In our previous papers (Konishi et al., 2001, 2003), using separately chemical engineering and proton NMR techniques, water species retained in fish paste sausage and squid were roughly distinguished into two different species as a function of water content. The two species were water species A_1, to be recognized as the higher water content at $W_0 > 120\%$-d.b. accompanied by higher water diffusivity, De, and water species $A_2, W_0 < 120\%$-d.b., with lower De. This delineation, however, was unclear because of a vagueness of the boundary between the two water species. To clearly discriminate between the two water species, a new technique is needed for the design of food products. In the present study, responding to this need, a hybrid technique derived from joint research using chemical engineering techniques and proton NMR techniques was proposed. This study aims to: (1) illustrate the dynamism of water species affected by environmental changes and the dynamic transformation between the two water species, A_1 and A_2; (2) quantitatively assess the limitations of both the chemical engineering technique and the proton NMR technique; (3) employ a hybrid technique to select specific design parameters for distinguishing differences between the two water species and various food products; and (4) categorize twelve commercial jerky products as a representative example into several groups based on the characterization of design parameters utilizing the hybrid technique.

2. Experimental methodologies of the hybrid technique

2.1 Methods

Six commercially available beef and pork jerkies (BJ-1 to BJ-6 and PJ-1 to PJ-6, respectively), a reference pork meat (designated P_H) from Hokkaido, two beef samples (B_A from Australia and B_H from Hokkaido), as well as salmon (SM) and squid (SQ) were selected as food samples. All samples utilized in this study are enumerated in Table 1. The commercial jerkies utilized exhibited a variation of water content of 30 ~ 80%-d.b. (dry base, W_D), the pork and beef meats had the initial water contents of 230 ~ 320%-d.b. and 230 ~ 280%-d.b., respectively, and the initial water content of the salmon and the squid commonly ranged from 300 ~ 360%-d.b. To evaluate the effective diffusivity (De) of water species, each of the samples was placed in a stainless steel net tray (4 meshes) that was mechanically hung from a strain gage transducer in the dryer. The sample weight was, to evaluate the dehydration rate, continuously recorded by the output of strain-gage transducer using a data-logger. With molecular-level analysis, to evaluate the correlation time (τ_C) of the water species in the salmon, squid, 12 jerkies, B_H, B_A, and P_H, a nuclear magnetic resonance (NMR) technique was applied to measure the ^{1}H-NMR spectra and the spin-spin relaxation time (T_2) of water protons. All samples, cut into $2 \times 2 \times 10$ mm pieces, were inserted into an NMR sample tube (4 mm in inner diameter, 180 mm in length). ^{1}H NMR spectra were obtained using a JEOL A-500 FT-NMR spectrometer operating at 500 MHz for protons. The observed frequency width was 20 kHz . The 90° pulse width was 12.5μ s, and the number of pulse repetitions was 8 . The proton chemical shifts were measured by using a slight amount of water containing deuterium oxide as an external reference. All the NMR measurements were performed at 23.5 ± 0.5°C. The spin-spin relaxation times, T_2, were obtained by the spin locking method. The hardness (N/m^2) of the samples was measured using a creep tester equipped with a V-shaped plunger (30 mm wide, 1 mm thick) to press a 60% of the sample size of $2 \sim 8 \times 10 \times 50$ mm.

Table 1. Features of the pork meat (P_H), beef meat (B_A and B_H), squid (SQ), salmon (SM), and the commercial jerkies used for this study. PJ-1~-6: pork jerkies; BJ-1~-6: beef jerkies.

	Features	Group No.
PJ-1	soy-sauce base taste, about 1.0 mm thick	2
PJ-2	soy-sauce base taste, 1.1 ~ 1.4 mm thick	1
PJ-3	light taste, oily, smoked flavor, 0.8 ~ 5.9 mm thick	2
PJ-4	salt base taste, smoked flavor, strong taste, hard, 2.2 ~ 4.8 mm thick	2
PJ-5	salt base taste, smoked flavor, strong taste, soft, 1.3 ~ 2.7 mm thick	3
PJ-6	soy-sauce base taste, light taste, hard, 1.9 ~ 3.5 mm thick	3
BJ-1	salt base taste, light taste, hard, 1.4 ~ 2.2 mm thick	2
BJ-2	salt base taste, strong taste, soft, about 2.2 mm thick	3
BJ-3	miso-sauce base taste, strong taste, soft, 4.2 ~ 5.9 mm thick	3
BJ-4	salt base taste, light taste, soft, 1.6 ~ 3.1 mm thick	2
BJ-5	soy-sauce base taste, strong taste, hard, 3.1 mm thick	1
BJ-6	salt base taste, light taste, hard, 0.9 ~ 2.8 mm thick	2
P_H	pork meat produced in Hokkaido, soy-sauce base taste, hard, about 6 $\pm$ 1.0 mm thick	
B_H	beef meat produced in Hokkaido, about 8 $\pm$ 1.0 mm thick	
B_A	beef meat produced in Australia, about 8 $\pm$ 1.0 mm thick	
SQ	squid mantle, 50 $\times$ 50 $\times$ 7 $\pm$ 1.5 mm	
SM	salmon, 50 $\times$ 15 $\times$ 12 ~ 26 mm	

2.2 Evaluation of the parameters

2.2.1 Effective diffusion coefficient (De) of water species

As has been reported by a large number of researchers (Andreu and Stamatopoulos, 1986; Waananen and Okos, 1996; Litchfield and Okos, 1992; Kannan and Bandyopadhyay, 1995), diverse mathematical equations for evaluating the effective diffusion coefficient (De) of water in foods have been demonstrated in various forms. In the present study, De' s of all the samples were evaluated by applying Equation (1) to the dehydration response curves obtained. In the present experimental drying conditions, it was reconfirmed that the drying operations were within a falling-rate period. Through all the evaluation of De, the effect of the shrinkage of the samples (derived from the dehydration) on the values obtained was previously evaluated, and the sizes of the samples were corrected.

$$\frac{W - We}{W_D - We} = \left(\frac{8}{\Pi^2}\right)^3 \exp\left(\frac{-\Pi^2 \cdot De \cdot t}{4} \cdot \left(L_a^{-2} + L_b^{-2} + L_c^{-2}\right)\right) \tag{1}$$

Where W is the water content (%-d.b.) of the sample at $t = t(\text{ h})$, We is the equilibrium water content (%-d.b.), W_D is the water content (%-d.b.) at the given initial condition, De is the effective diffusion coefficient (m^2/h) of water species retained in the foods, L_a, L_b, and $L_c(m)$ are the half distances of the sample width, and t is dehydration time (h).

De is represented as a function of the structure parameters (porosity and labyrinth factor) of food and the activation energy, as shown in Equation (2).

$$De = \left(\frac{\varepsilon}{\chi}\right) \cdot D = \delta \cdot D_0 \cdot \exp\left[\frac{-E_D}{R \cdot (T_D + 273)}\right] \tag{2}$$

Where ε is porosity, x is labyrinth factor, D is diffusivity (m^2/h), D_0 is the frequency factor of D, δ is diffusibility $(= \varepsilon/\chi), E_D$ is activation energy of De, R is gas constant, and T_D is drying temperature.

The pre-exponential factor, $\delta \cdot D_0$, can be a useful tool to discriminate between the water species retained in different kinds of foods, as is described in upcoming sections.

2.2.2 Correlation time (τ_C) of water protons

The spin-locking pulse technique used was effective in detecting a very fast relaxation signal at low water contents. For evaluation of the relaxation time, T_2, the equation $M_t = M_0 \exp(-ts/T_2)$ was used, where M_t is the magnitude of the magnetization vector after the spin locking pulse, M_0 is magnitude of the macroscopic equilibrium magnetization vector, and ts is the spin locking pulse length. In the present study, the plot of ln $[M_t/M_0]$ vs. ts indicated good linearity (which demonstrates a correlation coefficient higher than 0.99) through all water content of the foods, suggesting that the evaluated T_2 value was reasonable. From T_2, the correlation time of a water proton, τ_C, was evaluated using Equation (3) as described by Abragam (1963):

$$\frac{1}{T_2} = \frac{r^4 \cdot \hbar^2 \cdot I(I+1)}{5r^6}\left(3\tau_c + \frac{5\tau_c}{1 + \omega_0^2 \cdot \tau_c^2} + \frac{2\tau_c}{1 + 4\omega_0^2 \cdot \tau_c^2}\right) \tag{3}$$

where γ is the gyromagnetic ratio of a proton ($= 2.675 \times 10^8 \text{rad} \cdot \text{T}^{-1}\,\text{s}^{-1}$), $\hbar$ is the modified Plank's constant (6.63×10^{-34} J · s), I is the nuclear-spin quantum number of a water proton (= 0.5), r is the proton-proton distance of a water molecule (0.16 nm), ω_0 is the resonance frequency of NMR ($= 3.14 \times 10^9\ \text{s}^{-1}$), and τ_C is the correlation time of a water proton (s).

3. Dynamic alteration of water species maintained in foodstuffs

3.1 Variety of water species in foods

3.1.1 Visualization from the proton NMR technique

The water species retained in foods are dynamically changed into multifunctional states according to environmental conditions such as temperature, atmospheric pressure, atmospheric relative humidity, seasoning components such as salt and sugar, concentration of seasoning components, etc. Figure 1 is a schematic explanation of the water species adsorbed on macromolecules such as proteins in foods. The water species are classified into three groups as: (1) monolayer adsorbed water localized in region-I, which is referred to as bond water, immobilized water, or non-freezing water; (2) multi-adsorbed water localized in region-II, which is referred to as strongly restricted water, glass-state water, etc.; and (3) free water localized in region-III, which is referred to as normal isotropic bulk water, capillary condensation water, etc. The proton NMR technique produced correlation times of the three regions, I, II, and III, as $\tau_C = 10^{-6\sim-7}$ s, $10^{-6\sim-11}$ s, and 10^{-12} s, respectively. This classification is interestingly correlated to the three regions of the adsorption isotherm evaluated by the chemical engineering technique, as shown in Figure 2 by curve 1.

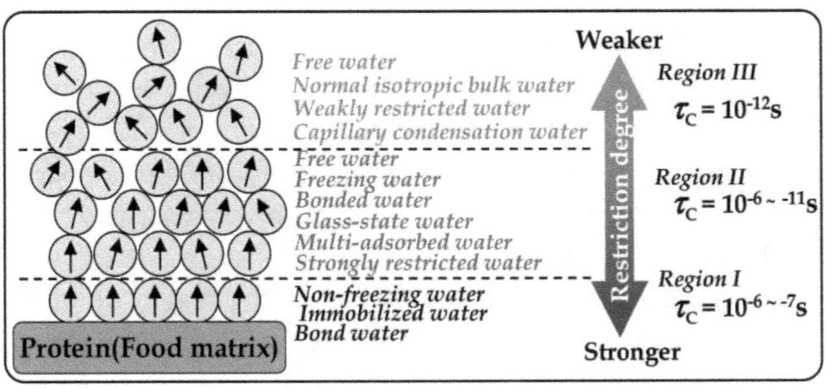

Fig. 1. Schematic explanation of the water species adsorbed on macromolecules in food.

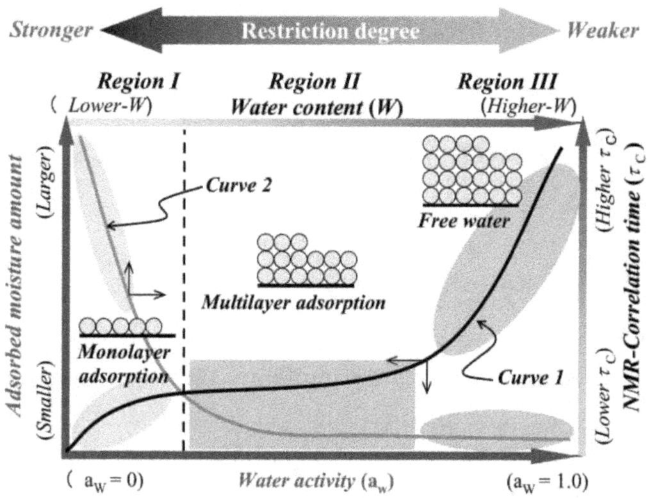

Fig. 2. Classification of the adsorbed water species in the food tissue matrix.

As has been reported in previous papers (Konishi et al., 2003, 2010), water species in the three regions are dynamically transformed by each other depending on environmental conditions in conjunction with the progress of dehydration. In atmospheric conditions, according to the progress of dehydration, the free water species in region III is preferentially removed rather than the two water species in regions I and II.

3.1.2 Visualization derived from chemical engineering methodology

Taking into account the water activity map presented by Labuza et al. (1970) and Schmidt (2004), one may visualize three regions on a water activity space. Fig. 2 shows a schematic explanation of the specified two curves: curve 1 for the water adsorption isotherm of food systems as a function of water activity (the isotherm was divided again into three regions, I, II, and III) and curve 2 for the NMR correlation time (τ_C) of water species retained in food as a function of water content (W_0). As can be seen from the comparison of the two curves, the water species adsorbed in region I with lower water activity contributes to the higher correlation time (τ_C). The higher τ_C means higher restriction strength derived from the food tissues. The water species adsorbed in the region III with higher water activity, contributing to lower τ_C, which means lower restriction strength. These results bring about an inverse relation between the two parameters, τ_C (curve 2), and the adsorbed water amount (curve 1) as shown in Fig. 2. This relation can be usefully employed in the hybrid technique as is described in upcoming sections.

The three water species distributed in regions I, II, and III presented in Figs. 1 and 2 can be further classified as species A_1 and species A_2. The water species in regions I and III belong to species A_2 and species A_1, respectively. The water species in region II belongs to an intermediate species between species A_1 and species A_2. Figure 3 schematically illustrates the water species A_1 and A_2 retained in foods as a model description. From the schematic visualization, one may recognize water species A_1 to be weakly restricted by macromolecules in foods, such as proteins, at greater water content and water species A_2 to be strongly restricted due to the progress of dehydration. The space formed due to the dehydration in region III in Fig. 1 would create micropores resulting in higher porosity (ε) and higher labyrinth factor (χ) in the food tissue matrix. These micropores would be filled again with water species when re-hydration occurred, indicating a reproducibility of the adsorbed water species as is discussed in upcoming sections (see Fig. 5).

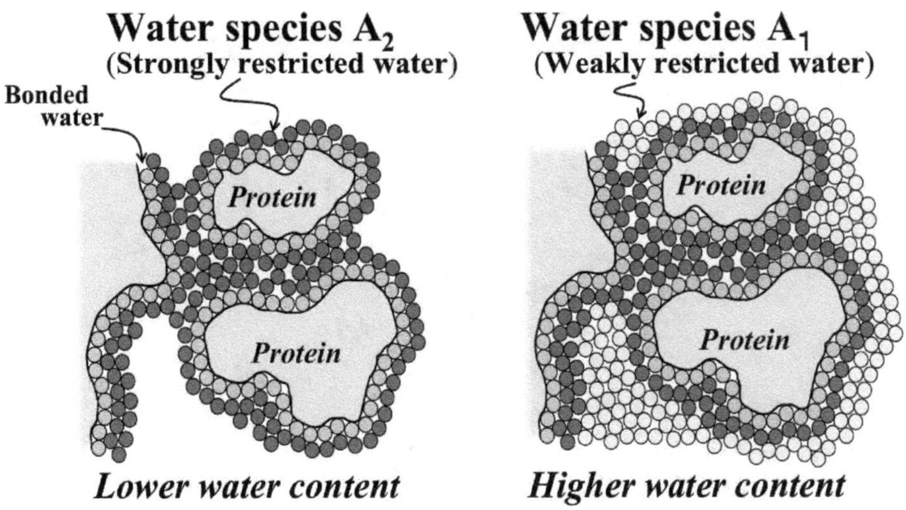

Fig. 3. Schematic explanations of the adsorbed states for water species in foods.

3.2 The dynamism and reversibility of water species in food analyzed using proton NMR approach

The proton NMR spectrum distinctly illustrates the variability of water species resulting from the sample's dehydration (Konishi et al., 2010). Figure 4 illustrates the dynamic decrease in water species (the dehydration of squid) as evidenced by a pronounced fall in the proton NMR spectra of peak-A. The three peeks, A, B, and C, were identified in our previous work (Konishi et al., 2010) as weakly restricted water, strongly restricted water, and lipids, respectively. One may recognize the reduction of only peak-A due to dehydration, whereas peaks-B and -C gave an identical value indicating that the two species are not influenced by dehydration.

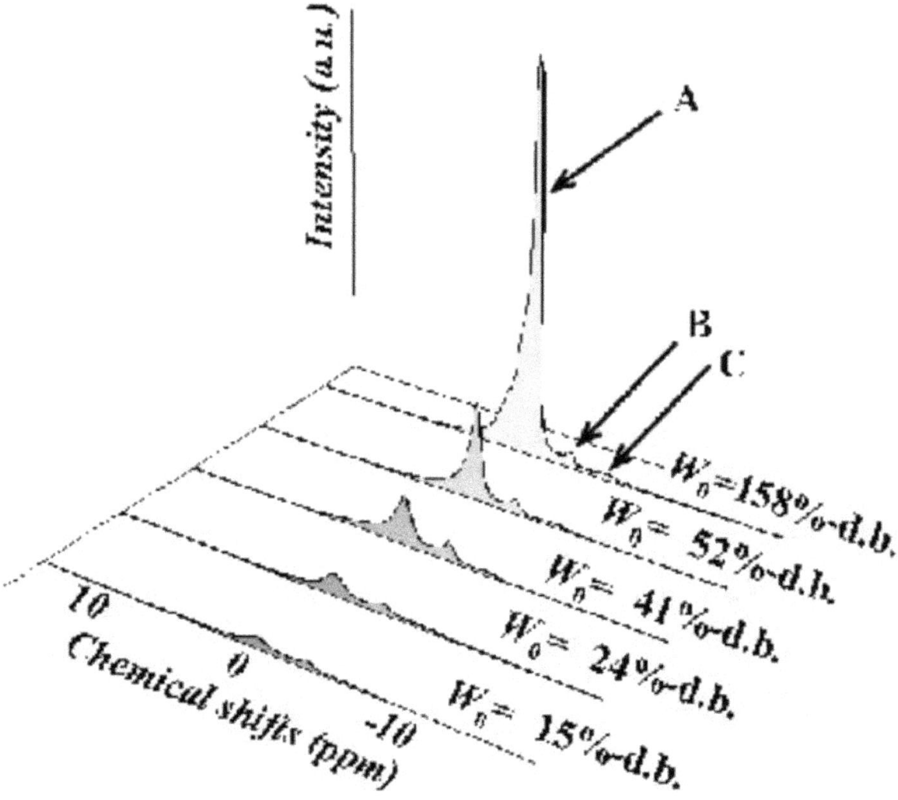

Fig. 4. Dynamic behavior of ^{1}H-NMR spectra for species A , B , and C in the squid muscle in the course of a drying operation.

By concentrating on the reversibility of peak-A, one can readily discern the dynamism of proton NMR spectra of water species in the squid, as illustrated in Figure 5. The intensity of peak-A significantly diminished due to dehydration (refer to curve (1)), and the peak height was reinstated upon rehydration (refer to curve (2)), while peaks-B and -C maintained consistent peak heights irrespective of dehydration and rehydration. The repeatability of peak-A clearly illustrates the capacity of water species A_1 and A_2 to be reconstituted inside the food tissue matrix by rehydration.

This reversibility is a very important factor for dried food products because, generally speaking, dried foods are frequently re-hydrated before being consumed (dried kelp, dried green onions, and dried noodles, etc.). Exploitation of reversible water species in dried foods could produce the same taste as when the food is freshly prepared. To evaluate the reproducibility of dried food taste, one should focus on τ_C because the τ_C value is closely related to the concentration of seasoning and the kind of food. The reproducibility of τ_C is, therefore, an important factor for taste reproducibility. Figure 6 demonstrates the reproducibility of τ_C as a function of W_0. The results show positive reproducibility of τ_C, indicating favorable taste reproducibility in the case of dried squid.

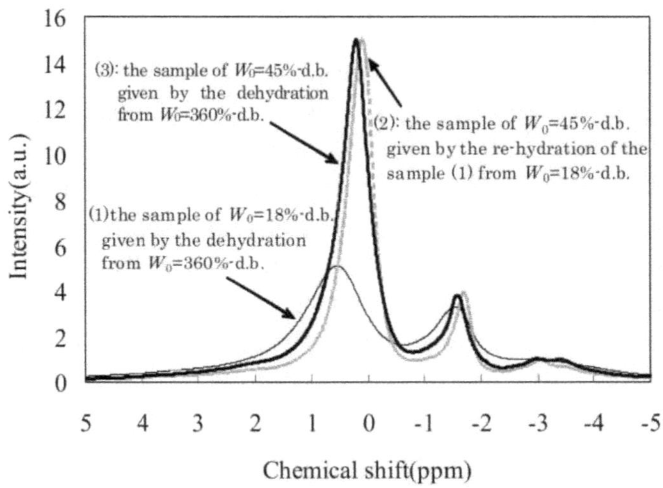

Fig. 5. Reproduced dynamic behavior of ^{1}H-NMR spectra for species-A, -B and -C in the squid muscle.

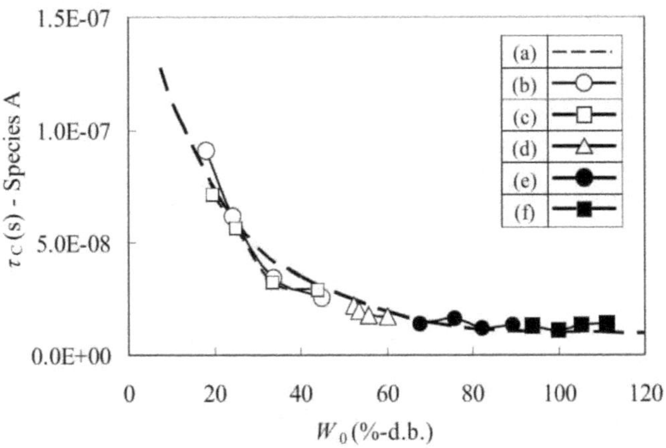

Fig. 6. Reproducibility of τ_C (species A) observed between the dehydration and the waterreadsorption operations for the squid muscle: (a) under the continuous drying operation from W_0 = 120%-d.b. to 10%-d.b., (b) after the gradual

readsorption of water from W_0 = 18% d.b. to W_0 = 45%-d.b., (c) from W_0 = 20%-d.b. to W_0 = 44%-d.b., (d) from W_0 = 50%-d.b. to W_0 = 60%-d.b., (e) from W_0 = 68%-d.b. to W_0 = 89%-d.b., and (f) from W_0 = 94%-d.b. to W_0 = 111% – d.b.

4. Constraints of the proton NMR methodology

To develop the food products demanded by the markets, the selected design criteria must effectively differentiate between various types of foods based on specific requirements. The water species kept in foods should be described by the types of foods and environmental factors, including temperature, atmospheric pressure, and water content. While the proton NMR water species evaluation methodology is a valuable tool, it is important to acknowledge that the method encounters challenges in differentiating between water species maintained in foods. Figure 7 depicts the τ_C values in relation to W_0 for B_A, B_H, and P_H. The τ_C~W_0 curves generated were unable to differentiate between the three meats, B_A, B_H, and P_H (Konishi and Kobayashi, 2009), despite the water species A_1 and A_2 being approximately discriminated at the critical point of τ_C=Cτ_C=10^(-8) " " s. The ambiguous distinction among the three curves can be perceived as a constraint of the proton NMR technique.

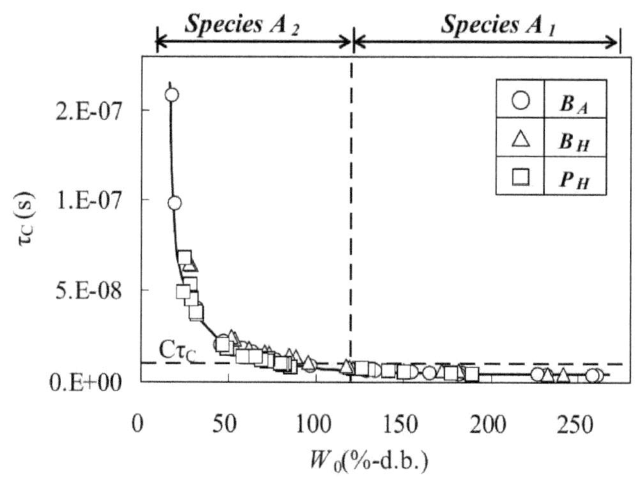

Fig. 7. τ_C **as a function of** W_0 **for** B_A, B_H, **and** P_H.

Figure 8 illustrates τ_C as a function of water content (W_0) of the 12 jerkies and the P_H. The solid line for P_H indicated a steep increase with decreasing W_0, between the values of $\tau_C = 1.0 \times 10^{-8}$ s and $\tau_C = 1.1 \times 10^{-7}$ s. Since τ_C indicates rotation time of the water species, the steep increase of τ_C demonstrates that the restriction strength of the water species in this region increases markedly upon dehydration from W_0 = 120 to 20%-d.b. Similar results were also obtained for the beef (Konishi and Kobayashi, 2009). The τ_c for the jerkies, on the other hand, fell around the solid line of P_H. From these results, one can recognize that it is difficult to distinguish among the three $\tau_C - W_0$ curves for the pork jerkies, beef jerkies, and P_H, indicating again a limitation of the proton NMR technique.

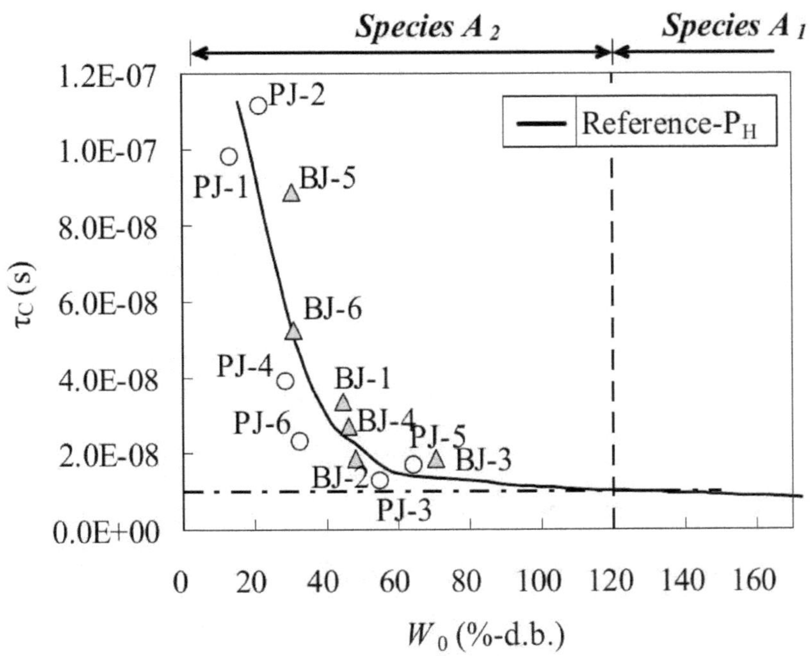

Fig. 8. τc as a function of W_0 for the pork and beef jerkies (open circle and closed triangle) and the pork (solid line).

5. Constraints of the chemical engineering methodology

5.1 Ambiguous differentiation of water activity in food within chemical engineering methodologies

Water activity (a_w) has conveniently been used to design food products because of the ease of evaluating the variable using the chemical engineering technique (Fennema, 1976). Figure 9 illustrates a_w as a function of W_0 for the squid and the salmon. As can be seen from the comparison between the two curves, an identical curve was obtained without detecting any difference between the squid and the salmon. This result strongly indicates a limitation of the chemical engineering technique. To develop a new food product, one should choose another parameter for discriminating among products. To address this issue, a computer simulation method for developing new food design parameters is proposed in the next section.

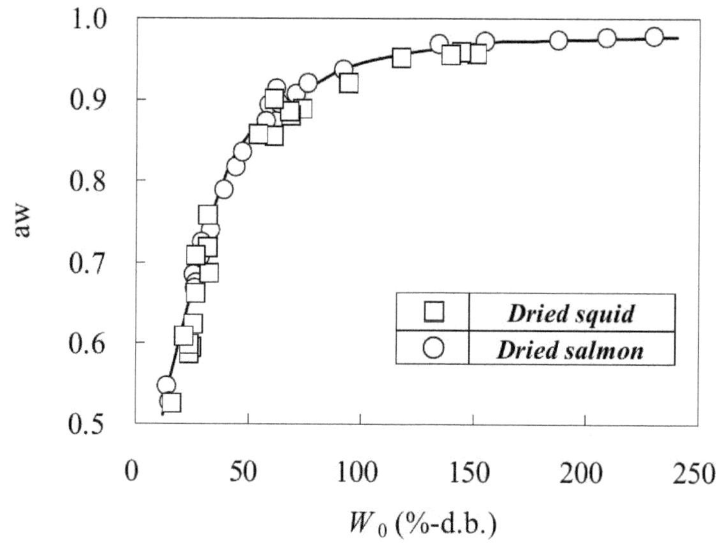

Fig. 9. a_w **as a function of** W_0 **for the salmon and squid.**

5.2 Unclear physical interpretation of the chemical engineering parameters obtained from a computer simulation of the dehydration dynamics of salmon

5.2.1 The dehydration dynamics of aquatic species in salmon

To characterize the water species retained in the salmon, dehydration response curves could effectively be employed. Figure 10 illustrates the dehydration response curves as a function of the water content of the salmon. The five samples of $W_0 = 54,99,144,200$, and 252%-d.b. were prepared from a sample of $W_0 = 342$%-d.b., and each of the samples was treated by a PUP-operation (poultice up: the samples were stored in a dark room at $T =$ 2°C for 24 h before use; this operation was conducted to provide homogeneity of water distribution in the sample). The PUP operation contributes to the stepwise increase of the dehydration rate, as shown in Figure 11, by the curves of $b' \sim f'$. Focusing on the five points of $b \sim f$ on the dehydration curve of $W_0 = 342$%-d.b. in Fig. 11(A), the dehydration rates of the five samples clearly demonstrate a stepwise increase, as shown in Fig. 11(B), by the curves of $b' \sim f'$ due to PUP operation. One can recognize that curves $b' \sim f'$ demonstrate a monotonous decay curve reaching finally to the curve of a' with an increase in the elapsed time. Using the dehydration rates obtained, the effective diffusivity of the water species, De, can be evaluated as a function of W_0 by applying Equation (1).

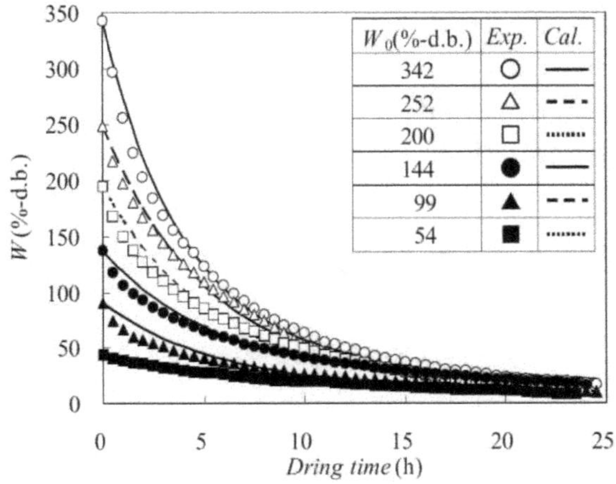

Fig. 10. Dehydration response curves of the salmon ($T_D = 45°C$).

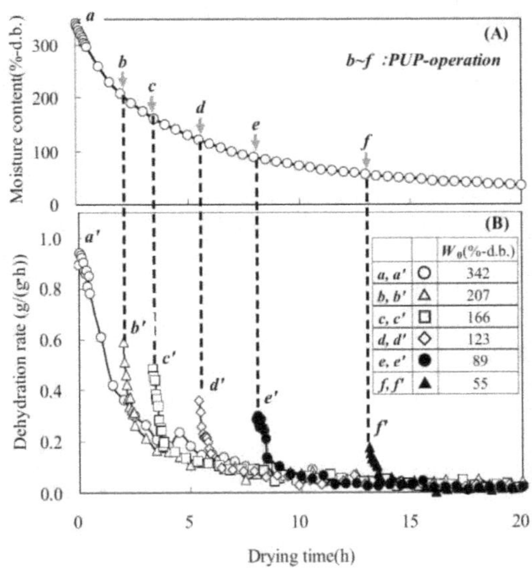

Fig. 11. Acceleration behavior of dehydration rate derived from the PUP operation for the salmon ($T_D = 45°C$).

5.2.2 Physicochemical mobility (De) of the aqueous species

The De value can be identified as the physicochemical mobility of the water species retained in food. Figure 12 depicts De for the salmon as a function of W_0 and temperature. Despite the data's extensive dispersion, one can approximately identify two distinct water species (species A_1 and A_2) generated in the salmon, contingent upon the values of W_0. The two water species were categorized at about W_0=120%-d.b., signifying a uniform De for the water species A_1 region between W_0=120 and 360%-d.b., alongside a steady decline in De for the water species A_2 region as W_0 decreased at temperatures ranging from 35 to 55°C. By employing a preliminary assessment of De acquired at the three temperatures, one would approximately create the Arrhenius plots depicted in Figure 13. The gradients of the linear equations appear to yield varying activation energies based on the moisture content of the samples.

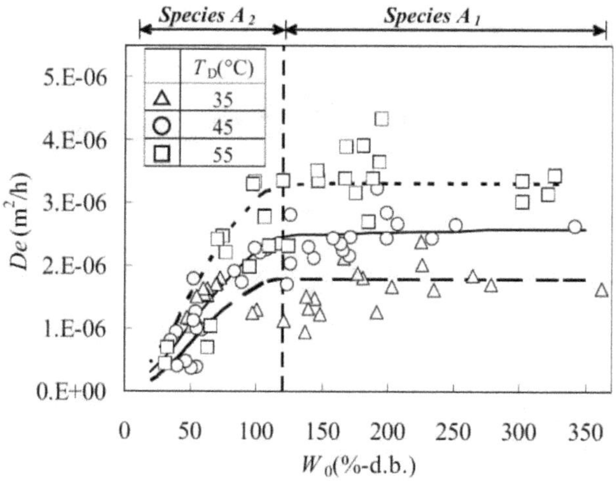

Fig. 12. De as a function of W_0 in the salmon.

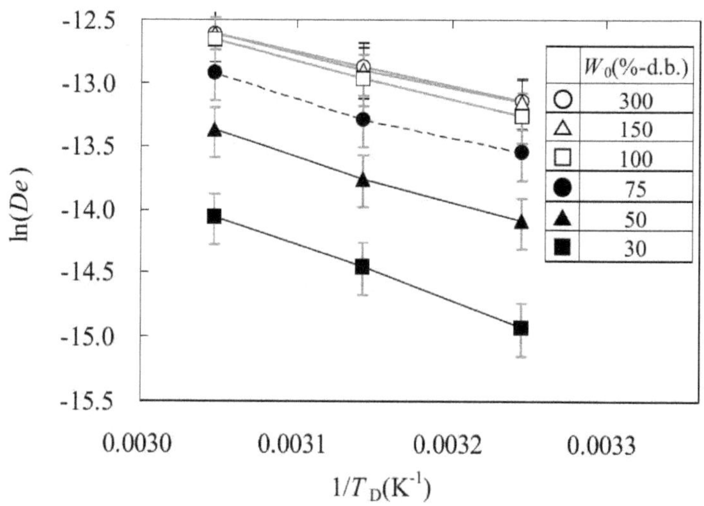

Fig. 13. Arrhenius plots of De for the salmon.

Figure 14 demonstrates E_D as a function of W_0. The activation energy obtained, the E_D values, demonstrates again two water species, A_1 and A_2, divided at $W_0 = 120\%$-d.b. E_D in the water species A_1 region gave an identical value of $22(\pm4)$kJ/mol and a gradual increase from 20 to 38(±8)kJ/mol with decreasing W_0. Taking into account the restriction of water species growth that accompanies increased dehydration, one can observe evidence of the activation energy growth (indicating a difficulty of the molecular diffusion of water species in the food tissue matrix due to the growth of the restriction strength) as a result of the decrease of water species mobility (De).

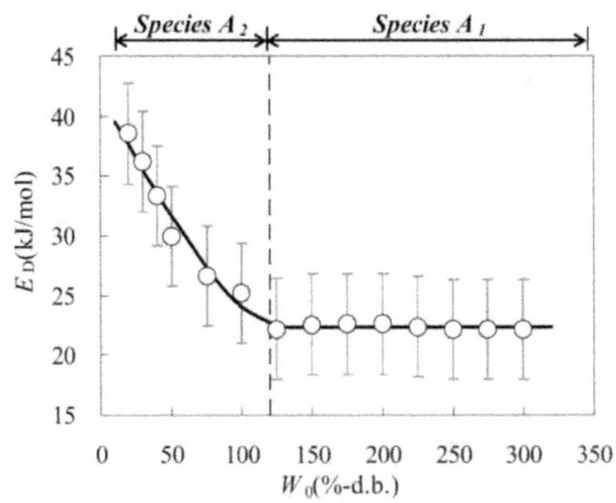

Fig. 14. E_D as a function of W_0 for the salmon.

As for a reason that the activation energy was drastically changed at $W_0 = 120$%-d.b., it can be attributed to a change in the physical structure of the food tissue matrix due to progressive dehydration. The value of $W_0 = 120$%-d.b. corresponds exactly to the critical value of $\tau_C (C\tau_C = 10^{-8}$ s), at which water species A_1 and A_2 were divided according to their restriction strength, and the pre-exponential factor ($PF = \delta \cdot D_0$) of De showed steep growth with decreasing W_0 (as will be discussed in the upcoming section, see Fig. 16). The steep growth of PF resulted from the drastic change in the porosity and labyrinth factor of the salmon at $W_0 = 120$%-d.b. accompanied by the gradual reduction of De decreasing W_0.

5.2.3 Model of a water reservoir for simulating dehydration reaction profiles

The transient response curves of dehydration during food drying operations have been modeled using several mathematical approaches (Chhinnan, 1984; Madamba et al., 1996; Thompson et al., 1968; Noomhorm and Verma, 1986). The dehydration response curves of the salmon depicted in Fig. 10 are simulated using a water tank model as an initial approximation in the current investigation. The water species contained in the sample are allocated to two distinct water tanks, each filled with a separate water species: A_1 (weakly restricted water species) and A_2 (highly restricted water species), as illustrated in Figure 15. (Konishi et al., 2001). During the drying operation, the two water species in the tanks are simultaneously drained by regulating the two valves as mass transfer coefficient, $k_W(h^{-1})$ and $k_S(h^{-1})$.

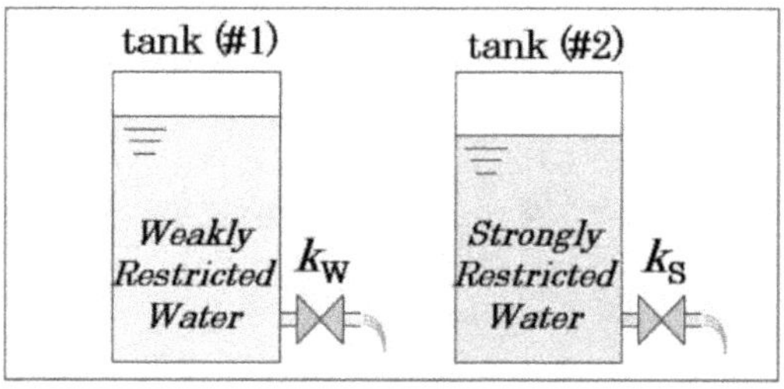

Fig. 15. Schematic explanation of the water tank model proposed as a dehydration of food.

On the material balance equation in the course of the dehydration operation presented in Fig. 10, the water content (W), presented as a perpendicular axis of Fig. 10, can be replaced by the water ratio, $W_R = (W - We)/(W_D - We)$, where We is the equilibrium water content (%d.b.) and W_D is the initial water content of the sample. The dynamism of W_R for the two tanks is expressed by Equation (4).

$$\left(-\frac{dW_{RI}}{dt}\right) = \left(-\frac{dW_{RW}}{dt}\right) + \left(-\frac{dW_{RS}}{dt}\right) \tag{4}$$

$$-\frac{dW_{RW}}{dt} = k_w \cdot W_{RW} \tag{5}$$

$$-\frac{dW_{RS}}{dt} = k_s \cdot W_{RS} \tag{6}$$

where W_{RW} is the ratio of the weakly restricted water species ($= (W_W - We)/(W - W_D)$), W_{RS} is the ratio of the strongly restricted water species $\left(= (W_S - We)/(W - W_D)\right)$, W_W is the amount of weakly restricted water species (%-d.b.), and W_S is the amount of strongly restricted water species (%-d.b.).

The differential equation (4) is easily solved and expressed by Equation (7).

$$W_R = \frac{W - We}{W_D - We} = W_{R1} \exp\left[k_s \exp\left(\frac{-E_D}{R \cdot (T_D + 273)}\right) \cdot t\right] + W_{R2} \exp\left[k_w \exp\left(\frac{-E_D}{R \cdot (T_D + 273)}\right) \cdot t\right] \tag{7}$$

$$k_s = \beta_1 \cdot De^0 \tag{8}$$

$$k_w = \beta_2 \cdot De^0 , \tag{9}$$

where W_{R1} and W_{R2} are a proportion of the strongly restricted water amount and the weakly restricted water amount, respectively, and $k_S(h^{-1})$ and $k_W(h^{-1})$ are the mass transfer coefficient of the strongly restricted water species and of the weakly restricted water species, respectively. $De^0(= \delta \cdot D_0)$ is the pre-exponential factor of De, and β_1 and β_2 are the constants evaluated by a curve fitting simulation between the experimental and the calculated curves as shown in Fig. 10. The curve fitting obtained showed favorable agreement with the experimental dehydration curves as shown in Fig. 10 by solid curves.

5.2.4 Discrimination of water species A_1 and A_2 due to the pre-exponential factor (De ^{0}) of De

The pre-exponential factor, De^0, can easily be evaluated by the extrapolation of the Arrhenius plot lines against the perpendicular axes in Fig. 13. Figure 16(A) illustrates De^0 as a function of W. The De^0 values obtained were of identical value regardless of W in the water species A_1 region, whereas in the water species A_2 region, there was a steep increase with decreasing W_0.

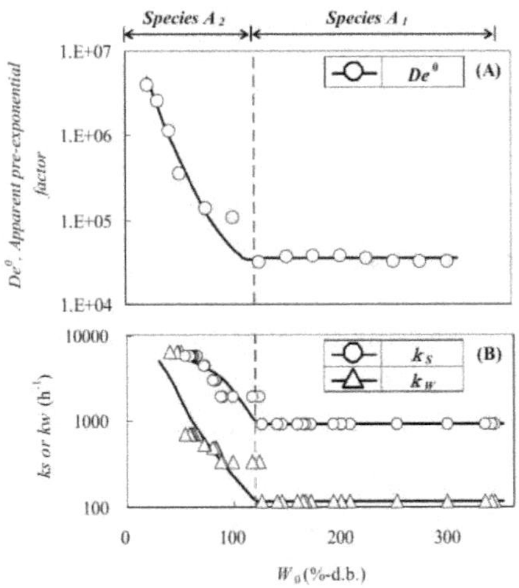

Fig. 16. Pre-exponential factor (De^0) and the dehydration parameters for the computer simulation (k_S, k_W) as a function of $W_0(T_D = 45°C)$ for the salmon.

Figure 16(B) illustrates k_S and k_W as a function of W_0. The two values show tendencies similar to De^0 even though they can be evaluated by the computer simulation. Comparing the two values between k_S (or k_W) and De^0, one can generate a linear relation between the two values as shown in Figure 17. The parameters of k_S and k_W roughly demonstrate the equations of $k_S = 0.025(\pm 0.008)De^0$ and $k_W = 0.007(\pm 0.004)De^0$. Using Equations (8) and (9), one may evaluate $\beta_1 = 0.025(\pm 0.008)$ and $\beta_2 = 0.007(\pm 0.004)$. Concerning the physical meaning of β_1 and β_2, since the two parameters were evaluated by the computer fitting simulation, they seem to be meaningless. This evidence indicates a disadvantage of the chemical engineering technique, even though the linear relation between k_s (or k_w) and De^0 would cause the presumption of a relationship between the two parameters.

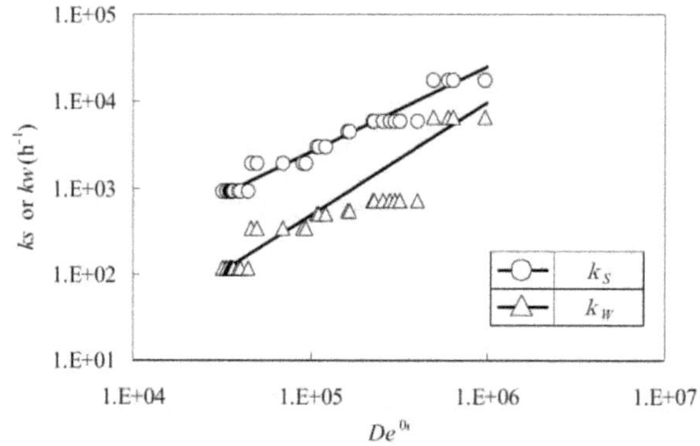

Fig. 17. Dehydration parameters (k s and kw) as a function of pre-exponential factor (De^0)($T_D = 45°C$) for the salmon.

6. Benefits obtained from the hybrid method

6.1 Visualized correlation between molecular mobility (τ_c) and physicochemical mobility (De)

Physicochemical mobility, De, which is evaluated by the chemical engineering technique (Fig. 12), and molecular mobility, τ_C, which is evaluated by the proton NMR technique (Fig. 8), are important parameters for characterizing the water species retained in foods as the diffusion rate in the three dimensional space of the food tissue matrix and as the rotation time of water molecules, respectively. The two parameters (De and τ_C) for the pork meat (P_H) and 12 jerkies are interestingly related, as shown in Figure 18. This clearly shows an advantage of the hybrid technique. De for P_H demonstrates a drastic change at $\tau_C = 10^{-8}$ s, indicating an identical $De = 2.2(\pm 0.2) \times 10^{-6}$ m^2/h in the species A_1 region and $De = 3.5(\pm 1.0) \times 10^{-7}$ m^2/h in the water species A_2 region.

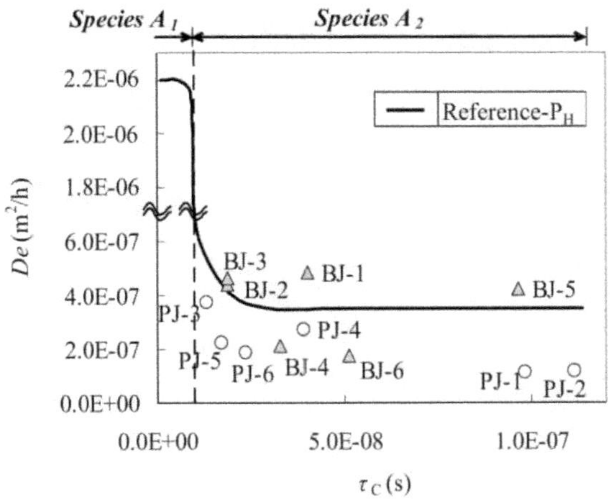

Fig. 18. De as a function of τ_C for jerkies and P_H.

All the De values for the commercial jerkies are distributed in the water species A_2 region, indicating an identical $De = 3.0(\pm 1.8) \times 10^{-7}\ m^2/h$, which, even though the data are widely scattered, is the same as P_H. Regarding the jerky data, although all the samples distributed in the market commonly use the water species A_2 not A_1 and the effective diffusivity, De, falling in the range of $1.0 \sim 5.0 \times 10^{-7}\ m^2/h$, τ_C is widely distributed in the range of $10^{-8} \sim 1.1 \times 10^{-7}$ s. Since each of the jerkies used in this study has own taste different from other, the wide distribution of τ_C has an important meaning in developing different tastes in jerky. This idea is also supported in an upcoming section.

6.2 Distinct differentiation of the two water species based on diffusivity (De)

As has been demonstrated in Fig. 12, the $De \sim W_0$ plots could not clearly discriminate the water species A_1 and A_2. The $De \sim \tau_C$ plots derived from the hybrid technique, on the other hand, are conveniently able to visualize the discrimination between beef and pork and the two water species, A_1 and A_2, as shown in Figure 19. In particular, the boundary of the two water species is clearly recognized as $C_{\tau C} = 10^{-8}$ s, indicating an advantage of the hybrid technique. Focusing on the behavior of B_A, B_H, and P_H, the De values gave a constant value in the two water species regions: in the A_1 region, as $3.1(\pm 0.5) \times 10^{-6}$, $4.7(\pm 0.1) \times 10^{-6}$, and $2.2(\pm 0.3) \times 10^{-6}\ m^2/h$, respectively; in the A_2 region, as $1.8(\pm 0.5) \times 10^{-6}$, $2.0(\pm 0.4) \times 10^{-6}$, and $7.0(\pm 1.0) \times 10^{-7}\ m^2/h$, respectively, even though τ_C is significantly changed from 10^{-8} to 4.5×10^{-8} s.

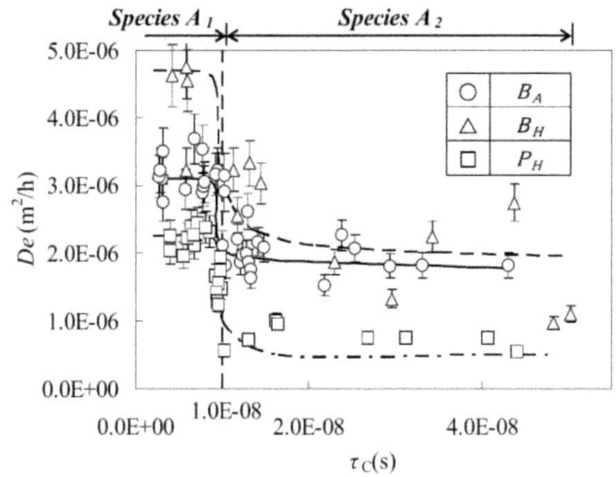

Fig. 19. Comparing behavior of De among B_A, B_H, and P_H as a function of τ_C at 70°C.

These results suggest the water species A_2 has the interesting characteristic of being able to make a wide shift of the molecular mobility, τ_c, without changing in the De values. This result indicates that, by changing the value of τ_C as a design parameter for food products, the character of various meats can be significantly altered. The hardness of various meats and salmon, for example, can also be controlled by a change in τ_C, as shown in the next section.

6.3 Distinct differentiation between the two water types based on hardness

Figure 20 demonstrates the hardness of B_A, B_H, P_H, and salmon as a function of τ_C. The hardness of all the samples shows a drastic increase at $C_{\tau_C} = 10^{-8}$ s with increasing τ_C in the water species A_2 region. This clear discrimination between the water species A_1 and A_2 is an advantage of the hybrid technique. One may recognize that the N_P value in the water species A_1 region commonly showed $1.0(\pm 0.5) \times 10^6$ N/m^2 regardless of τ_C in all the samples, whereas in the water species A_2 region and at the range of $\tau_C = 3 \times 10^{-8} \sim 6 \times 10^{-8}$ s, it varied widely depending on the samples, as with $N_P = 1.5 \times 10^7$ for B_A, 1.3×10^7 for B_H, and 1×10^7N/m^2 for P_H at $\tau_C = 5.0 \times 10^{-8}$ s. From these results, one can recognize that it is possible to design food products with different N_P values without changing the τ_C values, and it's possible to design products with different τ_C values without changing the N_P values. Keeping $N_P = 1.0 \times 10^7$ N/m^2, for example, one can have different τ_C values such as $\tau_C = 2.9 \times 10^{-8}$ s for B_A, $\tau_C = 3.5 \times 10^{-8}$ s for B_H, and $\tau_C = 4.8 \times 10^{-8}$ s for P_H. Although the physical meaning of this τ_C value shift is still unclear, it can be attributed to the taste difference between B_A, B_H, and P_H because concrete evidence that the τ_C value changes depending on the kind of food, the concentration of seasonings, and the kind of seasonings has been obtained. Details of these results will be reported in an upcoming work (Konishi and Kobayashi, 2011).

The visualization of $N_P \sim \tau_C$ exhibited in Fig. 20 clearly demonstrates the hybrid technique as a useful tool for discriminating between the two water species and among the kinds of food.

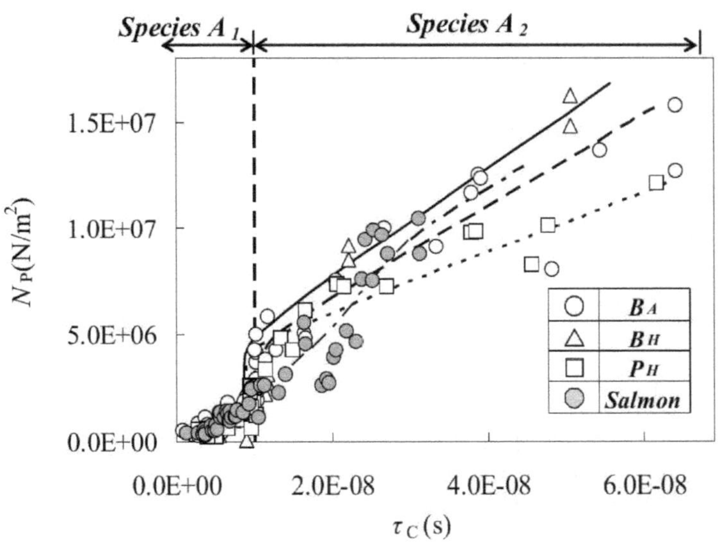

Fig. 20. Comparing behavior of N_P as a function of τ_C between B_A, B_H, and P_H at 50°C.

6.4 Physical meaning of $C_C = 10^{-8}$ s derived from the pre-exponential factor (De ^{0}) of De

Focusing on the boundary ($C_{\tau_C} = 10^{-8}$ s) between water species A_1 and A_2, one may be interested in what is happening in the food tissue matrix at the $C_\tau = 10^{-8}$ s. The information for the food tissue matrix can be obtained from Equation (2), and the equation is rewritten by Equation (10). The pre-exponential factor (δD_0) of Equation (10) consists of a function of ε, χ, and ΔS. Since the δD_0 value can be evaluated from the extrapolation of the Arrhenius plots against the perpendicular axes in Fig. 13, those evaluated for B_A, B_H, and P_H can be plotted as a function of τ_c, as shown in Figure 21. Three δD_0 curves obtained demonstrate a steep decay at the $C_{\tau_C} = 10^{-8}$ s. Based on the rough evaluation of the steep decay for the three curves to be $1/30 \sim 1/150$, the reduction of the δD_0 values should be attributed to a simultaneous change in the ε value, which becomes smaller while the χ value becomes larger and the ΔS value becomes smaller because of the steep change in the physical structure of the B_A, B_H, and P_H 's tissue matrix at the $C_{\tau_C} = 10^{-8}$ s derived from dehydration.

$$De = \delta D_0 \cdot exp\left(\frac{-E_D}{R\cdot(T_D+273)}\right) = \left(\frac{\varepsilon}{\chi}\right)\cdot\left(\frac{k\cdot(T_D+273)}{h}\right)\left[exp\left(\frac{\Delta S}{R}\right)\right]\cdot exp\left(\frac{-E_D}{R\cdot(T_D+273)}\right) \tag{10}$$

where k is Boltzmann constant, h is Planck constant, $\delta(=\varepsilon/X)$ is diffusibility, ε is porosity, X is labyrinth factor, and ΔS is activation entropy.

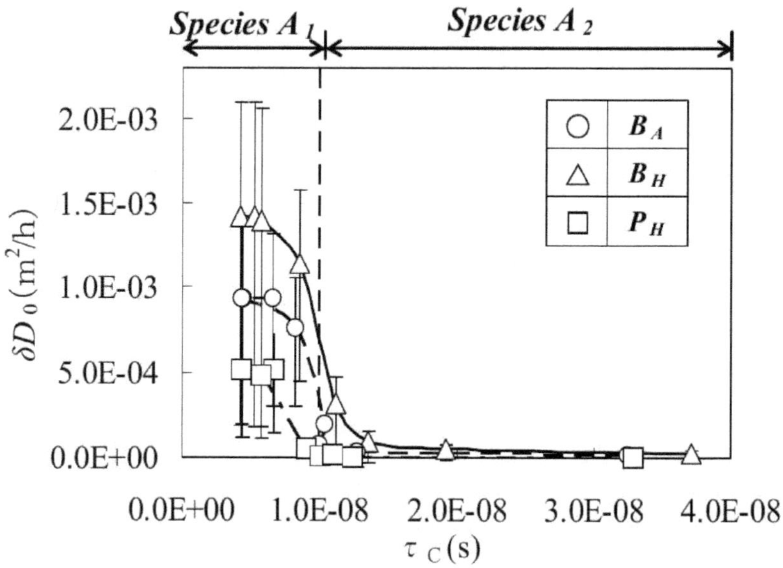

Fig. 21. Comparing behavior of the pre-exponential factor between B_A, B_H, and P_H at 50°C as a function of τ_C.

Figure 22 illustrates the optical microscope photographs of the cross-section for B_H. Comparing Figs. 22(A) and (B), one can clearly recognize that a large number of pores (complicated structures) appeared in the sample of $W_0 = 54\%$-d.b. rather than that of $W_0 = 127\%$-d.b., indicating a growth of the labyrinth factor (χ). This growth of χ contributes to the reduction of δ and is supported by the experimental evidences of the steep reductions of both δD_0 in Fig. 21 and De in Fig. 19 in the water species A_2 region.

The $\delta D_0 \sim \tau_C$ plots again demonstrate the hybrid technique to be a useful tool for discriminating between the two water species and among the kind of foods. The drastic change of δD_0 should derive the change of water species from A_1 to A_2, suggesting the diffusion mechanism change between the two regions. This idea is strongly supported by experimental evidence as shown in Figure 23. All E_D 's of water species A_2 for B_A, B_H, and P_H demonstrated a drastic reduction at $C\tau_C$ in the water species A_2 region. This reduction of E_D strongly suggests a change in the diffusion mechanism from the water species A_1 region.

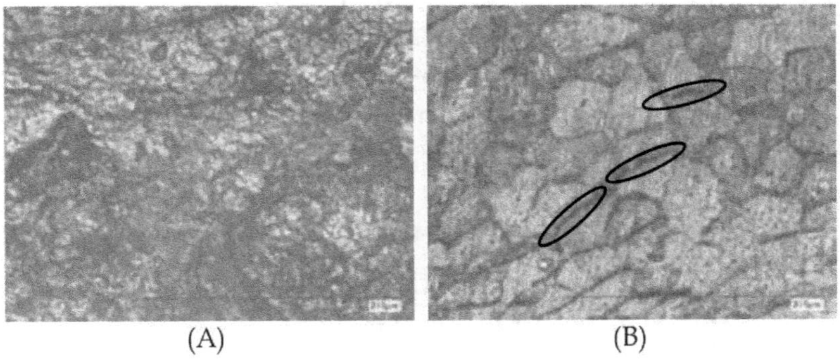

Fig. 22. Optical microscope photographs ($\times$ 1000) **of the cross-section for** P_H(A): 127% – d. b. , **(b): 54%-d.b.**

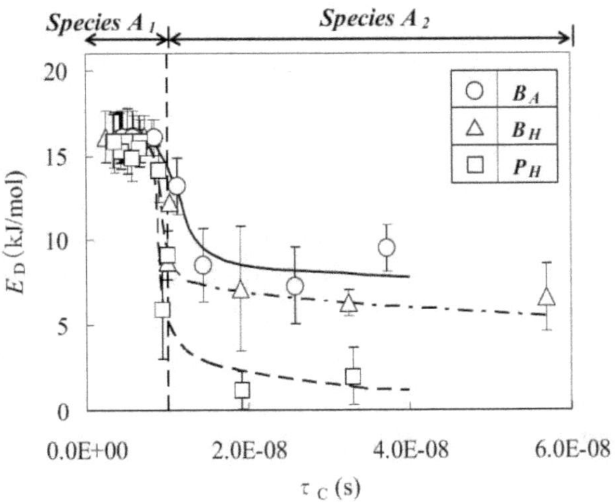

Fig. 23. Comparing behavior of E_D among B_A, B_H, and P_H as a function of τ_C at 50 ~ 70°C.

7. A novel approach to jerky product design utilizing factors derived from the hybrid process

Our interest is focused on a classification of the 12 jerky products influenced by multifunctional water species. Hills et al. (1999) empirically demonstrated a linear relation between the NMR relaxation rates ($1/T_2$) and a_w. In addition, Hills (1999) derived a theoretical equation to explain the linear relation. Since the relaxation time, T_2, can be replaced by τ_C using Equation (3), it is proposed that a_w should be related to τ_C. Figure 24 demonstrates a_w as a function of τ_C for the 12 jerky samples. All the samples fell roughly around three straight lines, indicating the existing of three groups. The three lines started from $a_w = 0.92$, which is an identical value independent of the samples and empirically evaluated, meaning a specified value for all kinds of foods (Konishi and Kobayashi, 2010). The $a_w = 0.92$ value always gave $\tau_C = 10^{-8}$ s independent of the kinds of foods and seasoning species. As has been discussed in the previous section, one can recognize both $a_w = 0.92$ and $\tau_C = 10^{-8}$ s to be critical values for foods in which the water species drastically changes from A_1 to A_2 and the pore structure of the food tissue matrix provides regulation of ε and χ values accompanied by drastic increases or decreases.

Focusing on the three straight lines in Fig. 24, line-1 generates Group-1 (BJ-5 and PJ-2), line2, Group-2 (BJ-4, BJ-1, PJ-4, BJ-6, and PJ-1), and line-3, Group-3 (PJ-5, BJ-3, BJ-2, and PJ-6). Based on the experimental data presented here, although one cannot demonstrate an exact physical meaning of the three groups' difference, it might be presumed that each group should have the specified seasonings different from others. Concerning the seasonings in the three groups, Group-1 is characterized by soy sauce, Group-2, by a mixed seasoning such as soy sauce, smoke flavoring, and salt, and Group-3, by a mixed seasoning such as salt, soysauce, and miso sauce. Although each of the three groups could not be distinguished by the chemical engineering (the aw $\sim W_0$ curves and the $De \sim W_0$ curves) and proton NMR techniques (the $\tau_C \sim W_0$ curves), the hybrid technique (the aw $\sim \tau_C$ linear relation) clearly distinguished the characteristics of the seasonings for the 12 different food products. This evidence strongly demonstrates the hybrid technique to be a useful tool to design various food products requested by the commercial fields.

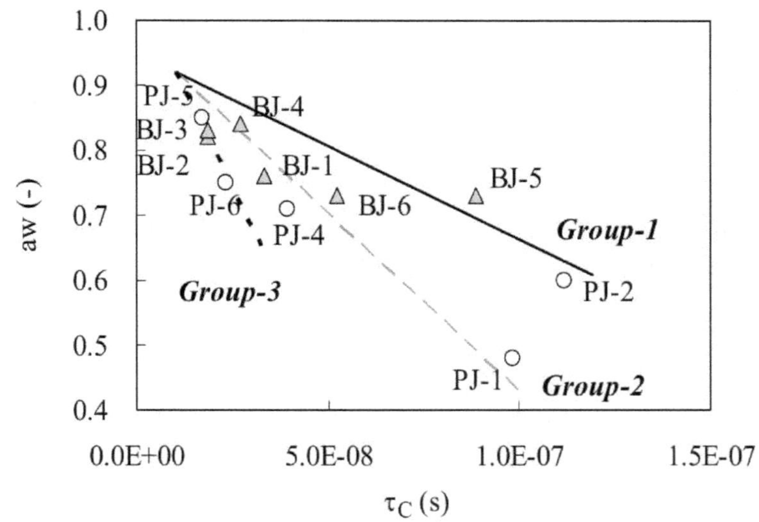

Fig. 24. a_w **as a function of** τ_C **for the beef and pork jerkies.**

8. Concise Summary

The integration of proton NMR and chemical engineering methodologies was utilized to differentiate 12 commercially available jerkies, namely salmon, B_A, B_H, and P_H, as well as to distinguish between the two water species, A_1 and A_2, employing four parameters: De, τ_C, N_P, and W_0.

1. The dehydration and adsorption of the water species, A_1 and A_2, retained in the foods were reversibly repeated and the transformation between the two water species was reproducible.

2. The proton NMR technique and the chemical engineering technique could respectively not discriminate among the $\tau_C \sim W_0$ curves and among the $a_w \sim W_0$ curves obtained for the foods. Although the computer simulation model proposed by the chemical engineering technique generated a good fitting against the dehydration response curves, the unknown parameters evaluated by the curve fitting, β_1 and β_2, were recognized as meaningless.

3. The hybrid technique proposed demonstrates a clear discrimination between water species A_1 and A_2, divided at the critical value of $C_\tau = 10^{-8}$ s, where the drastic change in the values of De, E_D, and N_P appeared. The physical meaning of C_{τ_C} was understood as a drastic change in the pre-exponential factor of De derived from the change in the food tissue matrix due to dehydration.

4. Using the hybrid technique, a variety of food products could easily be designed by changing the value of τ_C even though the values of a_w are identical. The meat jerkies commercially distributed in Japan were reasonably classified into three groups characterized by the three $a_w \sim \tau_C$ straight lines. The advantage of the hybrid technique brought a conclusion that the $a_w \sim \tau_C$ straight lines give characteristic slopes depending on the seasonings of the 12 jerkies. This could be useful in designing a variety of food products.

9. References

Abragam, A., (1963). The Principles of Nuclear Magnetism. Oxford at the Clarend Press, p347349.

Andreu, J. & Stamatopoulos, A.A. (1986). Durum wheat pasta drying kinetics. Lebensm. Wiss. u.-Technol., 19, 448-456.

Belton, P.S.; Gil, A.M.; Webb, G.A., & Rutledge, D., Eds.(2003). Magnetic resonance in food science-latest developments, The Royal Society of Chemistry, Cambridge.

Chhinnan, M.S. (1984). Evaluation of selected mathematical models for describing thin-layer drying of in-shell pecans. Trans. ASAE, 27(2), 610-615.

Fennema, O.(1976). In principles of food science, Part 1, Marcel Dekker, New York.

Hills, B.P. (1999). NMR studies of water mobility in foods. In"Water Management in the Design and Distribution of Quality Foods" ISOPOW 7(Y.H. Roos, R.B. Leslie, and P.J.Lillford, eds) Technomic Publishing, Lancaster, PA.

Hills, B. P., Manning, C. E. & Godward, J. (1999). A multistate theory of water relations in biopolymer systems. In: Advances in Magnetic Resonance in Food Science (edited by P. S. Belton, B. P. Hill and G. A. Webb). Pp.45-62. Cambridge, UK: Royal Society of Chemistry.

Jason, A.C.(1958). A Study of evaporation and diffusion processes in the drying of fish muscle in fundamental aspects of dehydration of food stuffs, ed. Society of Chemical Industry. McMillan, London, pp.103-134.

Kannan, D. and Bandyopadhyay, S. (1995). Drying characteristics of a tropical marine fish slab. J.Food Sci. Technol., 32(1), 13-16.

Konishi, Y.; Horiuchi, J., & Kobayashi, M.(2001). Dynamic evaluation of the dehydration response curves of foods characterized by a poultice-up process using a fish-paste sausage-I. Determination of the mechanisms for moisture transfer, Drying Technology, 19(7), 1253-1269

Konishi, Y.; Horiuchi, J., & Kobayashi, M.(2001). Dynamic evaluation of the dehydration response curves of foods characterized by a poultice-up process using a fish-paste sausage-II, A new tank model for a computer simulation, Drying Technology, 19(7), 1271-1285.

Konishi, Y.;Miura, K. & Kobayashi, M.(2003). Drying efficiency design using multifunctional dynamics of water molecules in foods-H-NMR analysis of a fish paste sausage and squid, AIDIC Conference series, vol.6, 183-190.

Konishi, Y. & Kobayashi, M.(2009). Quantitative evaluation of the design parameters requested in beef and pork drying operation, AIDIC Conference Series, 9, 177-186.

Konishi, Y., & Kobayashi, M., (2010). Food product design using the water species as a probe molecule. Annual meeting of Japanese Society of Food Engineering.
Konishi, Y., Kobayashi, M., & Miura, K. (2010). Characeterization of water species revealed in the drying operation of Todarodes pacificus Steenstrup using water proton NMR analysis, International Journal of Food Science and Technology., 45.1889-1894.

Konishi, Y. & Kobayashi, M.(2011). Dynamism of the water species as a probe molecule in food, Chemical Engineering Transactions, Ed. Sauro Pierucci, 24, 475-480.

Labuza, T.P., Tannenbaum, S.R., & Karel,M. (1970), Water Content and Stability of Low moisture and intermediate moisture foods. Food Technol.24, 543-550.

Litchfield, J.B. & Okos, M.R. ,1992. Moisture diffusivity in pasta during drying. J. Food Engng., 17, 117-142.

Madamba, P. S., Driscoll, R. H. & Buckle, K. A.(1996). The thin-layer drying characteristics of garlic slices. J. Food Engng., 29, 75-97.

Noomhorm, A. & Verma, L.R.(1986). Generalized single layer rice drying models. Trans. ASAE., 29(2), 587-591.

Rockland, L.B. and Stewart, G.F., Eds.(1981). International symposium on properties of water, water activity: Influences on food quality, Academic Press Inc., London.

Schmidt, S.J., Water and Solids Mobility in Foods, (2004). Advances in Food and Nutrition research, volume 48, 2004, Edited by S.L. Taylor, Elsevier

Thompson, T.L., Peart, R.M. & Foster, G.H. (1968). Mathematical simulation of corn drying (A new model). Trans.ASAE.,24(3),582.

Waananen, K.M. & Okos, M.R. (1996). Effect of porosity on moisture diffusion during drying of pasta. J. Food Engng., 28, 121-137.

Chapter-4

Modeling Methodology for the Redesign of Technical Processes

1. Introduction: An Overview

Currently, the design and development of new goods or the modification of existing ones (redesign) is a crucial aspect for enhancing innovation and competitiveness in industrial organizations. Design is becoming increasingly crucial for distinguishing one product from another.

Design is the process of delineating a product description that fulfills a certain set of requirements (Umeda 90). Redesign is the process of altering the specifications of an existing product to meet a new set of requirements (Brown 98). Design engineering encompasses both the creation and modification of designs. The literature presents various terms to delineate design and redesign, including preliminary, conceptual, functional, creative, routine, non-routine, personified, parametric, and innovative. However, the characteristic activities of global design engineering can be categorized as follows [Subba-Rao 99], as illustrated in Figure 1:

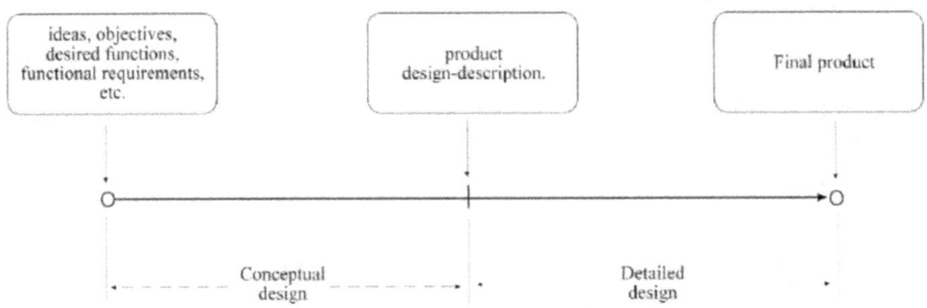

Fig. 1. Product design path.

- Conceptual (re)design, the phase where the global goals, requirements and operation of the product are established based on abstract concepts. The research presented in this chapter deals with this aspect.

- Detailed (re)design, the phase where the results of the conceptual design are used to physically implement a product.

Design engineering involves a wide range of activities. Design engineering can appear in a broad variety of domains, from the assembly of brakes to complex industrial plants and from simple chips to the most advanced super computers. Both design and redesign consist of two main components: the (re)design process and the (re)design object. The (re)design process involves all the (re)design activities performed over the (re)design object, which is the subject entity to be (re)designed. In engineering domains is common to refer to the (re)design object as the artefact. An artefact is a type of product to denote physical and technical devices. The (re)design process over an artefact is characterised by the map between functional requirements to structural requirements. Design and redesign, both can be considered as a dialectic process between goals (what it is desired) and possibilities (real constraints), directed to the satisfaction of functional specifications and performance (Stephanopoulos 90b).

The industry deals with complex technical processes where their behaviour is only predicted by means of complex numerical simulators. The redesign of a process is sometimes necessary when certain time has passed from its implantation or when they must adapt to economical, technological, or environmental requirements. The redesign is not part of the maintenance stage but must be considered into the process life cycle.

From a general point of view, the redesign is done typically in three steps: design-description acquisition (modelling), problem analysis (diagnosis) and proposal of modifications (generation of alternatives). In real redesign situations, human designers intuitively create mental abstract models by removing superfluous information about the process. Such models are based on functions of the components of the process and its context.

Since the early 1960s, Artificial Intelligence methodologies have been employed in design, including constraint-based systems, case-based reasoning, model-based reasoning, planning, neural networks, and genetic algorithms. While these methodologies have effectively addressed the modeling and simulation of processes, they have introduced a new issue: the employed knowledge representations necessitate such intricate details that they can be challenging to comprehend.

The primary aim of this study is to establish a support framework to aid the human designer in the reconfiguration of intricate technological procedures. The framework's structure must be grounded in the typical redesign activities executed by human designers in actual redesign scenarios. Consequently, the framework must effectively diminish the complexity of the processes slated for redesign, thereby facilitating the redesign operations.

The framework was developed by mixing model-based reasoning with case-based reasoning methodologies. The original process can be hierarchically modeled using model-based reasoning. Case-based reasoning allows for the acquisition of alternative process components from other processes, which must then be fitted to the original process. The framework was applied in the Chemical Engineering sector owing to the intricacy of the processes and the collaboration with domain specialists.

1. The complexity of the process must be high.
2. Complex numerical simulators can be used to model the behaviour of the process.
3. The process is already implanted, which means there is a design solution that satisfies the original requirements of such process.
4. Human designers can understand the process intuitively identifying its functional sections.
5. The process can be represented by functional abstract concepts.

2. Background of the research

2.1 The process of redesign

The body of studies about the redesign process is extensive. The many research methodologies are delineated from a broad perspective to a more particular one. A selection of the most compelling definitions of design identified in the literature is summarized as follows:

- Design is formally a search problem in a large space of objects that satisfy multiple constraints (Chandrasekaran 90).
- Design is the task of devising courses of action to change or create better ones (Simon 96).
- Design starts with an intended activity or use (Maher 97b) and uses available knowledge to arrive at a description of an artefact which will produce those results (Gero 90b).

Defining design is challenging as the phrase encompasses both a product (the object to be designed) and a process (the act of designing). The design reasoning process facilitates the transition from a functional notion to a product solution. Consequently, the design process can be regarded as a synthesis activity, significantly shaped by the designer's talents and cognitive frameworks. Regarding the process of redesign we have identified the following definitions in the literature:

- Redesign is considered as design in which there is a priori knowledge on the general and specialised functions to be performed and on the working principles to be selected [Salomons 95].
- Redesign is an inherent part of most design processes; in which new requirements or new domain knowledge influence the original design process (Brazier 96); but can also be seen as a family of design methods in itself (Pos 97).
- Redesign is part of design, which proposes suitable modifications free for inconvenience of existent artefacts (Kitamura 99).

Independently of the point of view of redesign, three types of redesign can be identified (Dixon 89): parametric redesign, component redesign, and structural redesign. In order to perform any of these redesign types, it is essential that some form of knowledge is available that allows the adaptation of existing designs. According to Pos (Pos 97) and based on the previous mentioned definitions of design/redesign, is possible distinguish two general points of view about the relationship between design and redesign, these are:

1. Viewing the design as total set which contains redesign as a subset. In order to satisfy this relationship, all the components of the design reasoning process should be satisfied for redesign. However redesign as a specialised subset would not be applicable in the same contexts as the more general notion of design. Here, design is viewed as an iterative process that uses intermediate results to get a final design description that fulfills the requirements. The task of redesign on the basis of a design created earlier produces a new temporary design description that is closer to the specification than the former design description.
2. Viewing both design and redesign as independent sets joined by a small common subset. For this relationship to be satisfied there is an expectation that some crossover or overlap will occur, thus only some of the components of design reasoning will be applicable in the redesign context and viceversa. Here, redesign starts with a previously constructed design description and a new set of requirements. The previously constructed design description must now be modified to fulfill the new set of requirements. Adopting any of the above points of view, basically minimal differences can be distinguished. In both cases the

important issue is to bridge the gap between a set of requirements and an existing design description. We can see that design starts from scratch, however, redesign starts with an existing design description, which is modified until it fulfills the current requirements. Both points of view can be captured by a single spectrum of problem-solving methods for redesign. Several authors (Akin 82, Chandrasekaran 93, Eldonk 96, Brazier 96, Bridge 97, Pos 97) state that the required knowledge for redesign is based on the following two principles:

- Minimise changes in the current design, and
- Maximise existing properties and benefits of the current design.

Many systems that solve redesign problems have been described in literature (Mitchell 83, Howe 86, Fischer 87, Daube 89, Goel 91, Smyth 96, Eldonk 96, Kitamura 99). However when one takes a closer look at the different variants of the redesign task, subtle differences exist that have an impact on how the task can be performed and what kinds of knowledge are involved.

There are a variety of research works referring to design or redesign; from (re)design of abstract (for example, components in software engineering) to physical entities (for example, a reactor in chemical engineering), for a general review see (Brown 97), for some details see (Akin 82, Mitchell 83, Howe 86, Fischer 87, Mostow 89, Goel 91, Bras 92, Stroulia 92a, Chandrasekaran 93, French 93, Brazier 96, Eldonk 96, Pos 97, Price 97, Umeda 97, Gero 98, Culley 99, Culley 99, Kitamura 99, Kraslawski 00, Grossmann 00, Arana 01, Maher 01).

Models in design and redesign are particularly important to guarantee they represent the intentions by which they were created. In general, the models are abstractions of the reality that guarantees communication of ideas by joining concepts, aggregations and relations (Bridge 97). Akin (Akin 82) outlines that the representational aspects to determine the utility of a model in design are:

- The represented information must be in a level of abstraction suitable for its intention.
- The contents must be in such a way that they are compatible with the expected results according to the mental representations of the designer.
- The model must be consistent with the reality that tries to reflect.

A substantial amount of research has focused on defining models of design (French 85, Tomiyama 87, Treur 89, Brown 89, Chandrasekaran 90, Gero 90a, Takeda 90a, Alberts 92,

Vescovi 93, Ohsuga 97, Brown 97). Most of this research highlights that the modelling of the functionality (or properties) of the design object description is an important aspect of the overall design process.

It is possible to represent explicit knowledge in (re)design by means of modelling functions of artefacts. This facilitates the systematisation of the reasoning and some tasks of (re)design. The reasoning based on functions allows abstracting information of the design on the same way as it is made in the reasoning of the initial stages of the design. The process of design of an artefact starts with the conceptual or functional design followed by the basic design and the detailed design (Stephanopoulos 90a). Within these, the functional design plays the central role since it guarantees the quality of the design and the innovation of the product (Umeda 97, Culley 99). The idea of function is fundamental in design since the work of the designers is to design artefacts that must achieve explicit functions (Chandrasekaran 00). Functional modelling is useful to model the object of (re)design, this modelling of objects enhances the formulation of (re)design strategies and the overall (re)design process. Functional modelling "hides" sections of the artefact structure at a lower abstraction level facilitating the manipulation of the artefact description.

Most of the research work on (re)design considers redesign as a knowledge-intensive field; wherein the processes (e.g., tasks) performed, descriptions of sequencing of processes, descriptions of the information within the system, and knowledge employed to perform a task are explicitly modelled most of the times by means of knowledge-based systems. These modelling frameworks try to model the (re)design so the (re)design object as well as the (re)design process are understandable by humans. To do this, human designers use the object specifications to propose a reasonable (re)design approach need to be understood (Leveson 00). Reasoning strategies employed in (re)design are derived or extensions of the commonly named problem-solving strategies. Examples of strategies are hypothesis and test (Hempel 66, White 05), pattern recognition (Doyle 62, Kirsch 64, Mitchell 97), skeletal plan refinement (Friedland 85, Tu 89), heuristic classification (Clancey 85), propose and revise (Goel 89), propose critique modify (Chandrasekaran 90), decision tree search (Raiffa 68, Qi 92), means-ends analysis (Newell 63, Rasmussen 86), and reasoning by analogy (Gick 80, Gentner 83).

In the above strategies, the human designer needs to formulate an explicit model of expertise as an integration of two types of models: a domain model and problem solving method. The domain model corresponds to the (re)design object and the problem solving method model corresponds to the (re)design process. Work on domain modelling has only recently attracted the attention of knowledge based system researchers (Stephanopoulos

90a, Schoen 91, Gruber 93, Skuce 93, Sowa 95, Kitamura 98, Fensel 01b, Gomez-Perez 04). The problem-solving methodology dictates the utilization of entities inside the model during the actual problem-solving process. Domain-specific concepts, relationships, and associated information are encapsulated in the domain model via ontologies (Chittaro 93, Kitamura 99, Fensel 01b, Kuraoka 03). Regardless of the models and methodologies utilized in the (re)design, it is crucial that facts and knowledge are documented consistently for future comprehension of the (re)design; this is referred to as (re)design justification.

2.2 The significance of function in the design process

Functions in design are crucial since they ensure the quality of the design and the inventiveness of the product (Umeda 97, Culley 99). Function is defined as the intended purpose of a design item; it serves as a manageable representation of the entity's overall behavior (Price 98). Certain authors characterize function as an abstraction of its intended behavior, closely linked to its context (Gero 90a, Goel 92, Stroulia 92a, Chittaro 93, Brown 97, Chandrasekaran 00). The human designers prioritize functions before addressing specific features. Functions may exist at varying levels of abstraction, contingent upon the design phase and the prevailing focus of design interest. During the initial design phases, functions often remain independent of the working principle; but, in subsequent phases, when functions are elaborated, they increasingly rely on the chosen working principle. A distinction is drawn among three levels or kinds of functions:

- General functions. (Keuneke 91, Lind 94, Kitamura 98, Bo 99) proposed a restricted list of general functions dealing with the transformation of matter, energy and/or information, which are independent of the working principle.
- Specialised functions or subfunctions. Act on flows, forces, moments etc., independent of the working principle.
- Working principle dependent function. Salomons (Salomons 95) defines it as the realisation of a specialised function (by means of physical phenomena). Several alternative solutions for fulfilling working principle dependent functions can exist without changing the working principle itself.

2.3 The design methodology

The design process is a multifaceted and not yet fully comprehended cognitive endeavor undertaken by humans (Salomons 95). The design process pertains to the sequence of actions and decisions undertaken to achieve a finalized product design. Design process models offer a systematic delineation of the design procedure. The models vary in their foundational formalizations and have been depicted in forms such as:

- blackboard architectures (Ball 92),
- algorithms (Alberts 93b),
- SOAR (cognitive architecture for developing systems with intelligent behaviour) (Steier 91),
- task models or problem solving methods (Brown 89, Brazier 94, Wielinga 97), or
- agent architectures (Dunskus 95, Berker 96, Lander 97).

The following models of the design process can be distinguished:

- Prescriptive models (Salomons 95)
- Descriptive models (Stephanopoulos 90b, Ohsuga 97, Sumi 97)
- Opportunistic design (French 93)
- Decision support problem (Bras 92, Thornton 93, Ullman 91)
- Theorem problem solving process (Takeda 94a)
- Human learning process (Gero 04)
- Multiagent design (Wood 01, DeLoach 04).

2.4 The design object

The design object serves as the primary focus throughout the entire design process. This may represent a model of a component, artifact, process, or system. Historically, design objects were produced through technical drawings; but, with the emergence of computers, design objects have transformed into digital models that can be quickly displayed, altered, and removed. Consequently, various models of artifacts have been employed in design.

Certain authors (Rasmussen 86, Douglas 88, Hoover 91, Lind 94, Turton 98, Leveson 00) have noted that abstractions of the design object are crucial in the design process for the manipulation of design objects. In this regard, Hoover (Hoover 91) has noted that:

- the design object evolves through abstractions and refinements.
- abstractions and refinements are selected opportunistically and are characterised by the designer focusing on a few aspects of the design object at a time.
- refinements are made within the framework of abstractions.

- conceptual, layout, and detailed stages are not distinct steps in the design process.

Numerous research studies have been conducted on the (re)design of object manipulation. The most pertinent methodologies on this matter are model-based design and case-based design.

2.4.1 Model-based design

One of the most used approaches in the manipulation of the (re)design object is model-based design which really is a branch of model-based reasoning (MBR) applied to (re)design. Model-based reasoning constitutes a set of techniques applied in several domains and it is used to create models and reasoning about the domain. Mainly the most used technique from MBR has been compositional modelling (Falkenhainer 91, Falkenhainer 92, Nayak and Joskowicz 96), which is an approach to construct a model of an artefact (components, devices, processes, systems, etc.) on the basis of a description of the artefact and a query about the composition of the artefact. The modelling of functions (functional representation) (Sembugamoorthy 86, Chittaro 98, Chandrasekaran 00) is crucial in compositional modelling. Functional Representation is a top-down approach to describe functions on devices (function), its structure (structure) and its causal processes (behaviour) of the device that culminate with the achievement of the function. Functional modelling reduces drastically the amount of information if simulation is required (Price 98). The approaches of functional modelling can be classified in two groups:

- State-based representations. It uses units of function representation, which are abstractions of behaviour states. Behaviour states and hence functions may be associated even with static objects which do not cause any state change.
- Flow-based representations. Flow-based representations are based on the concepts of flow and effort. In this approach exists a predefined set of functions, and functions of all existent components are expressed in terms of these primitive functions. This approach is based on the System Theory (Bertalanffy 50) and its derivatives (Abstraction Hierarchy (Rasmussen 86), Qualitative Process Theory (Forbus 84), and Multilevel Flow Modelling (Lind 90, Lind 94)).

2.4.2 Case-based design

Case-based design is a branch of case-based reasoning (CBR). CBR is a general paradigm to solve problems based on the recovery, reuse, revision and retention of specific experiences (cases) (Aamodt 94). CBR has been applied to component-based systems (Maher 97a) which is, however, mostly concerned with the manipulation of design object descriptions. This paradigm is particularly attractive in domains where explicit models do not exist or its understanding is difficult (Kolodner 93). In CBR, similarities between formal methods implemented in computer programs and informal observations from designers are taken into account (Maher 97a). The applications of CBR can be for classification or synthesis tasks. (Re)design problems are within the synthesis tasks. The direct or analogical use of previous designs or plans of design can reduce and improve the quality of design because take advantages of previous experiences (Maher 95). CBR is viewed as a redesign process for the adaptation of a case where a new artefact (named goal) is designed to achieve certain function, its physical structure can be inferred in analogical way from some physical, chemical or biological object (named source) whose function is similar to the required function. CBR has been applied to solve problems of real world; there are several works about casebased design for example (Goel 89, Qian 92, Sycara 92, Bhatta 94, Borner 96, de Silva Garza 96, Maher 01, Price 97).

2.5 Redesign approaches

Here the division between research on design and redesign is remarked, only research from the redesign perspective is presented.

2.5.1 Generic approaches in engineering

Goel et. al. [Stroulia 92a, Stroulia 92b, Goel 94b, Goel 97a] presented a control architecture for model-based redesign in the context of case-based redesign. They state that the redesign task is characterised by small differences in the functions desired of and delivered by an existent known design. The redesign is divided in three subtasks: a) generation of modifications to the structure of the old design, b) realisation of the modifications on the structure, and c) evaluation of the new design. Eldonk et. al. [Alberts 93a, Bakker 94, Eldonk 96] presented a redesign approach based on techniques developed in model-based diagnosis. Eldonk et. al. state that redesign activities are diagnosis and respecification. The objective of this approach is to find the part of the system, which causes the discrepancy between a formal specification of the system to be redesigned and the description of the existing technical system. Kitamura and Mizoguchi [Sasajima 95, Kitamura 99] proposed a redesign approach based on ontologies of functional concepts. They concentrate on elucidating the justifications for the design of an artifact and on structuring overarching methods for redesign. Initially, they employ an ontology of functional ideas that facilitates the identification of functional structures and the automatic representation of a portion of the design rationale. Regarding the second argument, they employ an ontology of redesign strategies. This methodology comprises the following stages: functional comprehension, requirements analysis, alternative proposal, and evaluation.

2.5.2 Mechanical engineering

Aranna et. al. [Fothergill 95, Forster 96, Forster 97b, Arana 00] proposed a redesign environment called DEKLARE, which supports acquisition, representation and reuse of redesign knowledge. It allows the designer to use design techniques to suggest alternative designs that fulfill specific requirements. Gupta et. al. [Das 94] proposed a methodology that automatically provides suggestions of redesign for reducing setup costs for mechanical parts. This approach is based on the interpretation of the design as a collection of mechanical features. The objective is to generate alternative mechanical features by means of geometric changes of the original parts and adding them to the feature set of the original part. Kim [Kim 93] proposed an approach for redesign of assemblies by means of planning techniques. Kim deals with the absence of required design information using the replay and modify principle. He employs a reverse engineering model to infer information about the process executed when creating a given design, and using the inferred information for design recreation or redesign. The propose model consists of the three stages: knowledge acquisition, construction of the default design plan, and redesign based on cases.

2.5.3 Electrical and electronic engineering

Steinberg and Mitchell developed a system to redesign VLSI circuits [Steinberg 85]. This redesign approach is based on planning techniques and causal and teleological reasoning [de Kleer 79]. The subtasks of this approach are: a) focus on an appropriate section of the circuit, b) generate redesign options to the level of proposed specifications for individual modules, c) rank the generated redesign options, d) implement the selected redesign option, and e) detect and repair of side effects resulting from the redesign. Maulik et. al. [Maulik 92] proposed the use of optimisation techniques to redesign CMOS analog circuits. The optimisation approach is guided by three principles. First, equations that describe device characteristics are encapsulated and separated from equations that describe the performance of the circuit topologies. Secondly, constrained optimisation techniques are employed to synthesise the redesigned-scaled CMOS circuit. Finally, constrained optimisation allows the solution of some final constraints over specific variables. Based on the approach of Umeda et. al. [Umeda 92], Tomiyama et. al. [Umeda 94] describe an extension of their approach taking into account the potential functions of the components of an artefact to redesign it. The architecture consists of sensors, which monitor the machine, and a model-based reasoner diagnoses faults and plans repairs. The system generates a FBS model (Function-Behaviour-State) based on the design object, and then searches the model for candidate redundant function. The FBS model consists of a function hierarchy that represents the designer's intentions, and a behaviour network that describes how the function hierarchy is carried out. The system first tries a control type strategy that adjusts various machine parameters. If the strategy fails the system applies a strategy based on functional redundancy, it uses the potential functions of existing parts in a slightly different way from the original design. Heo et. al. [Heo 98] presented a redesign approach of digital electronic systems by means of evolutive programming. They use directed acyclic graphs known as task flow graphs (TFG) to represent the redesign object. Each node of the graph represents computational tasks; an edge represents a transfer of data. The design process consists of five tiers: a) system-level design, b) architectural design, c) logic design, d) circuit design, and e) physical design.

2.5.4 Chemical engineering

The (re)design of chemical processes is made with the purpose of adapting existing processes to changes in economic, technological or environmental requirements. In the eighties, there were mainly significant advances on saving energy by means of two constraint-based approaches: a) pinch methodology (Tjoe 86, Smith 87, Linnho. 88) and b) mathematical programming on synthesis and design of processes [Papoulias 83, Pistikopoulos 87, Vaselenak 87]. In the nineties, Gundersen [Gundersen 90] made a revision of systematic methods of redesign of processes, which were broadly tackled. In such revision, he emphasised two important observations:

- Most of the projects in the industry of processes were redesign projects.
- The systematic methods of redesign of processes are based on methods of design of processes.

Doherty et. al. [Fischer 87] developed a systematic procedure of redesign by means of opportunistic searches; the procedure considers modifications in the structure of the flowsheet and in the dimension of equipments. Kirkwood et. al. [Kirkwood 88] implemented a methodology of redesign by means of an expert system by using heuristic rules to construct hierarchical structures. Nelson and Douglas [Nelson 90] developed a systematic procedure considering alternative reaction routes; the procedure is hierarchical and provides guides to identify viable processes. Rapoport et. al. [Rapoport 94] proposed an algorithm to design units of process by means of the redesign of already existing ones. The algorithm consists on hierarchical levels and heuristic rules; this approach is similar to synthesis of processes. Stephanopoulos et. al. [Han 95] developed an approach based on agents to synthesis of processes; they modeled the process of design like a set of tasks that can be executed by agents. Systems have also been developed to satisfy economic, environmental and safety constraints. Kraslawski et. al. [Kraslawski 00] developed a methodology centred on the identification and elimination of bottlenecks in reaction and separation sections. Sylvester et. al. [Sylvester 00] optimised processes within the concept of Greener Process. Hertwig et. al. [Hertwig 01] applied techniques of MINLP (Mixed-Integer Non-Linear Programming) to optimise configuration of processes. Pasanen [Pasanen 01] developed a tool for conceptual design of processes, this is called Phenomenon Driven Process Design (PDPD). This methodology emphasizes the systematization of design concepts for chemical processes. Uerdingen et al. [Uerdingen 01] introduced a screening method predicated on the examination of flow path patterns. Performance indicators are employed to evaluate the economic impact of each element within the flowsheet along the flow path.

2.6 Modelling

2.6.1 Modelling the redesign process

The comprehensive redesign process is contingent upon the employed problem-solving method. To initiate the redesign process, the problem must be articulated in terms of the objectives that the original artifact must fulfill and the criteria that can be employed to evaluate the alternative designs. A synthesis process occurs, resulting in a collection of potential designs. Each alternative is analyzed and assessed according to the established objectives and design criteria. Ultimately, one alternative is chosen for implementation. The method is extremely iterative; outcomes from subsequent phases are sent to earlier stages to adjust objectives, criteria, design options, and so forth. Design alternatives are produced by analyzing the composition of the system. The designer deconstructs the system (artefact) into a collection of subsystems (components), together with the functions and limitations imposed on the design of each individual subsystem. These issues are examined concerning the targeted performance characteristics and limitations of the system. The approach is iterative until a satisfactory design alternative is attained. Upon completion of this phase, all components must be delineated with sufficient precision to facilitate the execution of the entire object.

2.6.2 Modelling the redesign object

The understanding of the redesign object depends strongly on the mental models of the human designer. Usually designers communicate their ideas more easily in terms of abstract, high-level descriptions to describe complex concepts (Price 03). The description of the redesign object can be done in many different ways, depending on the context and purpose for which the description is to be used. In the early phases of redesign, highly abstract descriptions (e.g. qualitative or causal) might be helpful, whereas in later phases, more detailed and quantitative descriptions provide more suitable information.

Considering the notion of function, some researchers (Sembugamoorthy 86, Goel 89, Franke 92, Keuneke 91, Chittaro 93, Iwasaki 93) organise the knowledge in a domain by means of functional concepts. The main claim of these approaches is that functions and intentions can provide important additional information for understanding and reasoning about the structure and behaviour of physical systems. In addition, other researchers have directed their extentions to hierarchical modelling by means of different aggregation levels (Liu 91, Rajamoney 91) or different approximations (Weld 86, Kuipers 87, Struss 91, Falkenhainer 91) to organise the knowledge. Independently of the tools and representations employed, several authors (Fischo. 78, Checkland 81, Jaffe 91, Vicente 92) suggest that two important aspects must be addressed if computer tools are used to tackle activities of complex systems:

- Content, the semantic information that should be contained in the representation given the goals and tasks of the users. The content gives the basic issues to understand the information about the redesign object. Independently of the amount and complexity of the information, the designer can conceive, in general terms, the objectives of the redesign object.
- Structure, how to design the representation to facilitate that the user can extract the required information. The structure concerns to the organisation of the process components.

Rasmussen (Rasmussen 85) observed that the complexity of a system depends on the level of resolution in which the system is considered. The complexity can only be measured comparing with other systems observed at the same level of abstraction. The complexity can be manageable with more or less detail in the representations: then, hierarchical modelling can be seen as a way to handle complex systems. Models of complex artefacts can be expressed in terms of a hierarchy of levels of organisation, each one more complex than the previous. This modelling approach is named Hierarchy Theory (Rasmussen 86). Rasmussen studied the protocols developed by people working on complex systems and found that human users structure the system along two dimensions: a part-whole abstraction and a means-ends abstraction.

Some authors (Umeda 90, Franke 92, Lind 94) propose representation approaches for physical systems which maintain a clear separation between knowledge of structure and behaviour on one side and knowledge of function or purposes on the other side. This feature makes them useful in redesign of technical complex systems. The hierarchical functional modelling approaches employed in the present work are Multilevel Flow Modelling (MFM) (Lind 90, Lind 94, Lind 96, Lind 99) and Multimodelling (Ziegler 79, Praehofer 91, Chittaro 93), which are following described. Both approaches provide a more intuitive vision of reasoning on each task to be performed, and thus the redesign activities are enhanced.

2.6.2.1 Multilevel flow modelling

MFM offers a visual and methodical foundation for employing means-end and whole-part hierarchical decompositions in the modeling of intricate systems, such as industrial facilities. A system is characterized by its aims, functions, and the physical components involved, as delineated by the separation between means and ends. Simultaneously, each of these descriptions can be articulated at various levels of whole-part decompositions. Figure 2 illustrates the primary types of decomposition. These are functional models characterized by a significant degree of abstraction, integrated with a teleological depiction of the objectives or purposes of the modeled system. Lind has proposed a syntax for a formal language and outlined the fundamental concepts for utilizing the MFM representation.

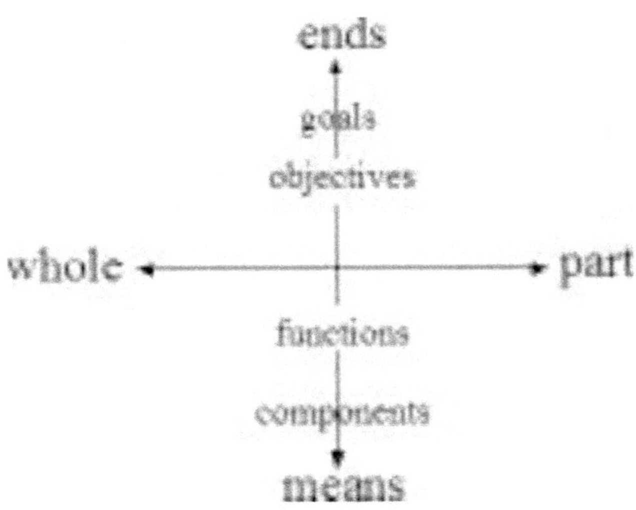

Fig. 2. Means-ends and part-whole dimensions in MFM.

An MFM model is a prescriptive description of a system, a representation of what it has been designed to do, how it should do it, and with which information it should do it. Thus, the three basic concept types of MFM are:

- goals, which are the objectives or purposes of the system, i.e., the ends that the designers and operators want that the system reaches.
- functions, which are the means by which the goals are obtained, i.e., the powers or capabilities of the system.
- physical components, which are the different elements of the system, the equipment of which it consists.

The concept of goal is central to MFM, as it is the "descriptor object" for teleological information. It is important to be able to recognise and describe goals, as they play an important role in every activity using means-end information. Without knowing the goals, it is virtually impossible to know the available functions. Three different types of goals can be recognised:

- production goals, which are used to express what enables production. For example, a specific process variable should be kept within a given interval.
- safety goals, which are used to express reasons of safe operation. For instance, a particular process variable should be kept above or below some value, or inside or outside an interval.
- economy goals, which are used to express considerations of overall process optimisation.

The function is the second important concept on MFM. A function is always associated with a goal, and correspondingly, goals are always associated with functions. MFM describes the functional structure of a system as a set of interrelated flow structures on different abstraction levels. The levels are connected via achievement and condition relations; the flow structures consist of connected flow functions. Thus, the following types of flow structures can be:

- mass flows
- energy flows
- information flows

These flows are of completely different types, but they have many properties in common. Most flow functions can appear in each type of flow structure, thus, there are three flow types of flow functions. In MFM plant functions are represented by a set of mass, energy, activity and information flow structures on several levels of abstraction. The levels are interdependent and form means-end structures. Mass and energy flow structures are used to model the functions of the plant and activity and information flow structures are used to model the functions of the operator and the control systems. The mass and energy flow functions are:

- source, the capability of a physical system to act as an infinite reservoir of mass, energy, or information.

- transport, the capability of a system to transfer mass, energy, or information from one part of the system to another (from one medium to another).
- barrier, the capability of a system to prevent the transfer of mass, energy or information from one part of the system to another (from one medium to another).
- storage, the capability of a system to accumulate mass, energy, or information.
- balance, the capability of a system to provide a balance between the total rates of incoming and outgoing flows.
- sink, represents the capability of a system to act as an infinite drain of mass, energy or information.

These functions can be used to describe information flows. There are also some specific information flow functions:

- observer, the capability of a system to translate physical observations to information.
- decision maker, represents the decision-making capabilities of a system.
- actor, represents the capability of a system to turn information into physical consequences.

In addition to the flow functions, some organisational functions are used. They are concerned with expressing support and control:

- network, which is used to group a flow structure and connect it to a goal.
- manager, which describes control and supervisory systems, including human operators.

2.6.2.2 Multimodelling

The Multimodelling approach represents many diverse and explicit models of a system, which are used in a cooperative way in specific problem solving tasks. The fundamental assumptions about knowledge modelling and reasoning mechanisms do not identify a unique way of representing a physical system and reasoning about it. On the contrary, the Multimodelling approach is an abstract and general framework that allows for a variety of specific implementations. The fundamental concepts in Multimodelling are:

1. Ontologies. An ontology contains the descriptions of entities in the real system. Two types of ontologies can be distinguished:

- Object-centred ontology. The real world is made up of individual objects whose properties can be stated in an objective, context independent and general way.
- System-centred ontology. The real world is made up of systems, intended as organised units, whose elements cannot be defined in isolation.

2. Representational assumptions. This issue concerns about what to represent of the real system in the model. This involves two basic aspects:

- The scope of the model, i.e., the aspects of the real system which are considered relevant to the purpose of the model.
- The precision of the model, i.e., the degree of accuracy of the representation

3. Epistemological types. The type of knowledge represented in the model. These types can be:

- Structural. The knowledge about system topology, i.e., the components that constitute the system and how they are linked.
- Behavioural. The knowledge that describes how components work and interact in terms of the physical quantities (variables and parameters).
- Functional. The knowledge about the role components play in the physical processes in which they take part. This knowledge relates the behaviour of the system to its goals, and deals with functional roles, processes, and phenomena.
- Teleological. The knowledge about the goals assigned to the system by its designer and about the operational conditions that allow their achievement through correct operation.
- Empirical. The knowledge concerning the explicit representation of the system properties through empirical associations (such as observation, experimentation, and experience). This knowledge may include subjective competence that usually human experts acquire through direct interaction with the system.

4. Aggregation levels. The degree of granularity of the represented knowledge.

3. The framework for redesign

The Multilevel Flow Modelling (MFM) and Multimodelling methodologies effectively depict the behavior of human designers throughout the redesign process. MFM is employed for high levels of abstraction, while Multimodelling is better appropriate for intermediate and lower levels. The structure and behavior of the components (equipment) are abstracted through the Multimodelling approach, which is subsequently transferred to the MFM approach. The connection between the two approaches lies in the functions of the equipment inside the domain. This functional modeling serves as the foundation for managing the project throughout the whole redesign phase. The primary objective is to hierarchically describe the process and reason utilizing functional abstract ideas. Thus, the designer can "navigate" in both top-down and bottom-up directions within the representation, akin to the manner in which the designer formulates mental models of the process. From an abstract perspective, three entities operate independently within this framework.:

1. The simulator. The commercial software used to obtain the design description of the process and to implement and evaluate the generated alternative process designs.
2. The reasoner. The software modules required to model the process, identify the suitable equipment/section to be modified and obtain similar equipments/sections based on the selected equipment/section.
3. The human designer. The human user interpreting the results.

The stages of the proposed redesign framework (Figure 3) are:

1. Identification of objectives and criteria. This stage covers the design-description acquisition and the identification of candidates stages of the framework.
2. Generation of design alternatives. This stage is similar to the obtaining alternatives and adaptation stages in the framework.
3. Evaluation of alternatives. This stage is carried out manually by the human designer.
4. Implementation of alternatives. This stage is also carried out manually by the human designer.

This redesign framework may oversee complex technological processes (the subject of redesign). The modeling technique was chosen to emulate the actions of human designers in real-world redesign situations of such processes. The primary aim is to support human designers rather than to independently undertake the redesign without human participation.

3.1 Acquisition of design descriptions

The initial phase of the framework involves acquiring the design specifications of the process slated for redesign. This description suffices to execute the redesign activities, and only a few adjustments are necessary to achieve the redesign target. This phase consists of two subtasks: data collecting and functional identification.

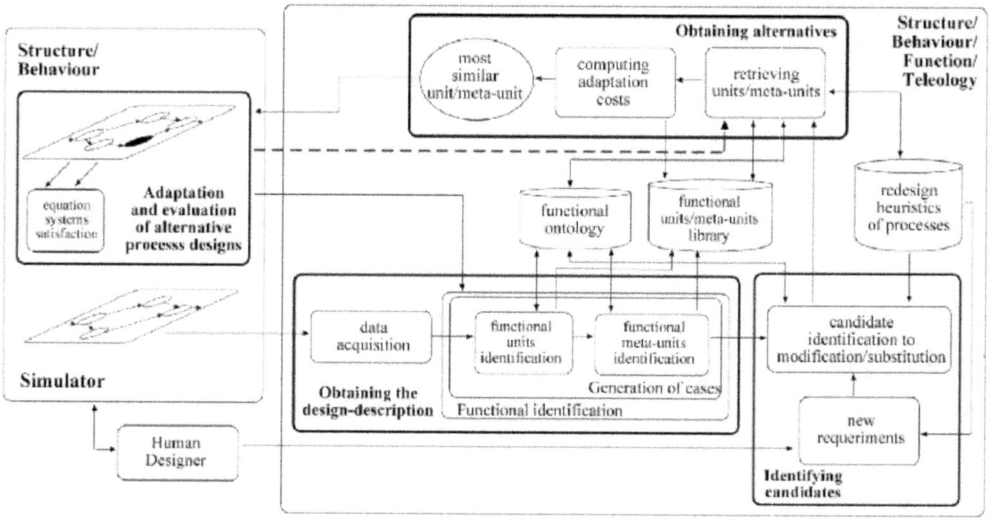

Fig. 3. The proposed redesign framework.

3.1.1 Data acquisition

It deals with data extracted from the specialised simulator used to implement the process to be redesigned. It was conceived as an appropriate step to reduce human intervention on the introduction of data to the reasoner module. The aim of the data acquisition is to obtain only the most useful data to generate the appropriate knowledge useful for the redesign of the process; thus irrelevant or superfluous data is ignored. Based on this data, the following types of knowledge are generated:

- Structural. Knowledge related to the topology of the process, i.e., the equipments conforming the process and the connection between them.
- Behavioural. Knowledge related to the values of variables and parameters that characterise the behaviour of each equipment.

3.1.2 Functional identification

The data obtained from the simulator is used to model hierarchically the process. To do this, the functions of each equipment in the process must be identified. Based on the identified functions, the functional sections of the process can incrementally be identified.

In the rest of the chapter any equipment will be named unit and a functional section will be named metaunit. This stage is divided into functional unit identification and functional meta-units identification. Here it is necessary to specify an ontology about the functional issues of the existing equipments in the process. By using this ontology, it is also required to specify a priority order of functions and the process variables related to each one. The grouping of functions depends strongly on such priority order.

Functional unit identification

The function of each unit is inferred by analysing their inputs (preconditions) and their outputs (postconditions), the variables involved, and the neighbour units. This process involves the analysis of the behaviour of the unit and its consequences in the surrounding units (the units connected to it). The next classification of functions was obtained based on MFM and Multimodelling:

- Broad function. Denotes a process-independent function that can be achieved considering only flows of mass, energy, or information.
- General function. Denotes a function that can be achieved by several equipments in a domain. These functions deal with the transformation of mass and energy and are independent of the physical phenomena.
- Specific function. Denotes the abstract function as it is known in the domain of the process. These functions relate flow variables with a specific physical process. According to the domain, these are functions to denote functional sections into the process.
- Working function. Denotes a function that can be achieved by a specific single unit. These functions relate specific flow variables with a specific physical phenomenon.

A unit can have several functions but only one goal (the objective, intention or purpose of the artefact). Several units can have a common goal. Therefore, based on the data extracted from the simulator it is possible to infer the following knowledge:

- Functional. Knowledge about the roles of each unit. The functional knowledge connects the behaviour (physical phenomena and processes) of the unit to its goal.
- Teleological. Knowledge related to the goals of each unit by considering the required input operational conditions and the output operational conditions that were meant to be produced.

The functional knowledge is independent of the process (the same functional knowledge can be found in others processes), while the teleological knowledge are the goals assigned to the units by the designer).

As result of this stage, each unit is represented by structural, behavioural, functional, and teleological models. The aim of this stage is to model the process in a higher level of abstraction (respect to the simulator).

Functional meta-unit identification

Based on the functions inferred in the functional unit identification stage, it is possible to identify the functional sections of the process (named meta-units). The incremental identification of these functional sections denotes the most important sections of the process. This incremental identification is carried out by generating different representations of the process at different levels of abstraction. The function of a unit is a working function because the unit (representing real equipment) was designed to perform only such function. The functional sections of the process denote specific and general functions by means of meta-units. Meta-units representing general functions are composed by meta-units with specific functions, not necessarily of the same type. A meta-unit represents a functional section at an abstract level. Thus a meta-unit at a higher abstract level can contain several units and other meta-units. Two or more meta-units can generate a more abstract meta-unit. Units/meta-units with lower priority functions are "absorbed" by units/meta-units with higher priority functions, as shows Figure 4.

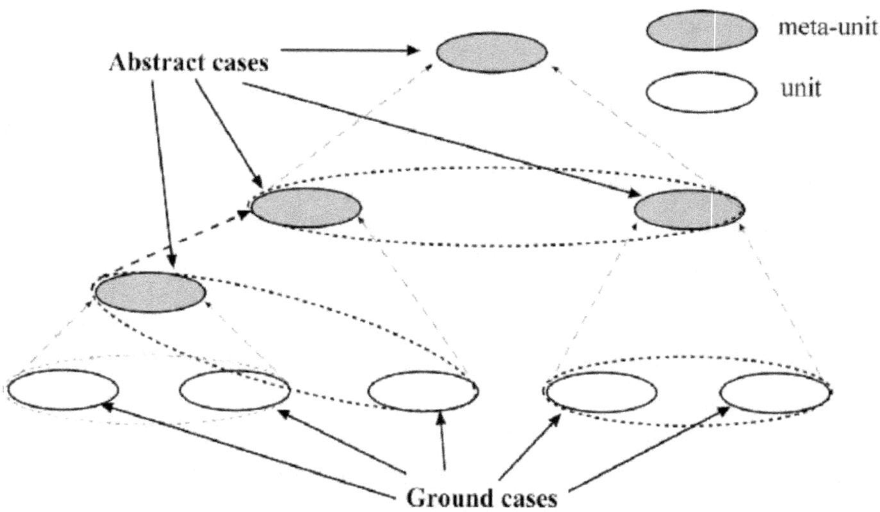

Fig. 4. Grouping of units/meta-units.

Every functional section forms a hierarchy of meta-units and units where meta-models are connected in a same level (intralevel), as shows Figure 5, and at different levels (interlevel), as shown in Figure 6.

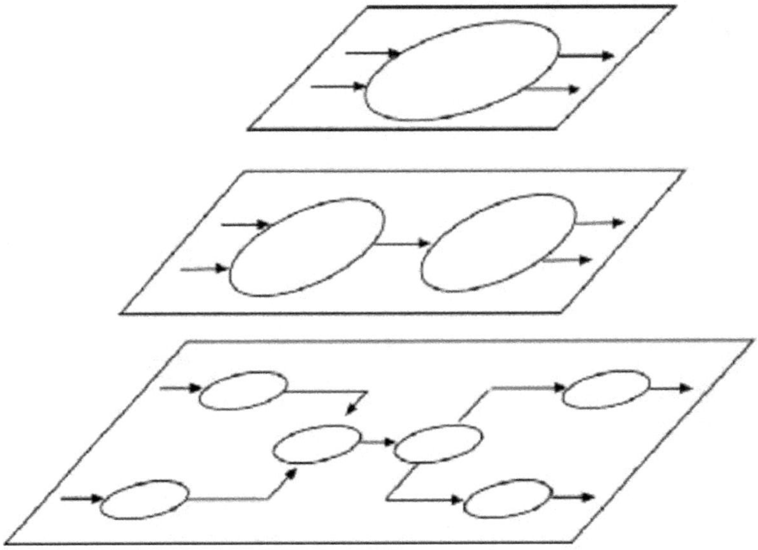

Fig. 5. Intralevel meta-models.

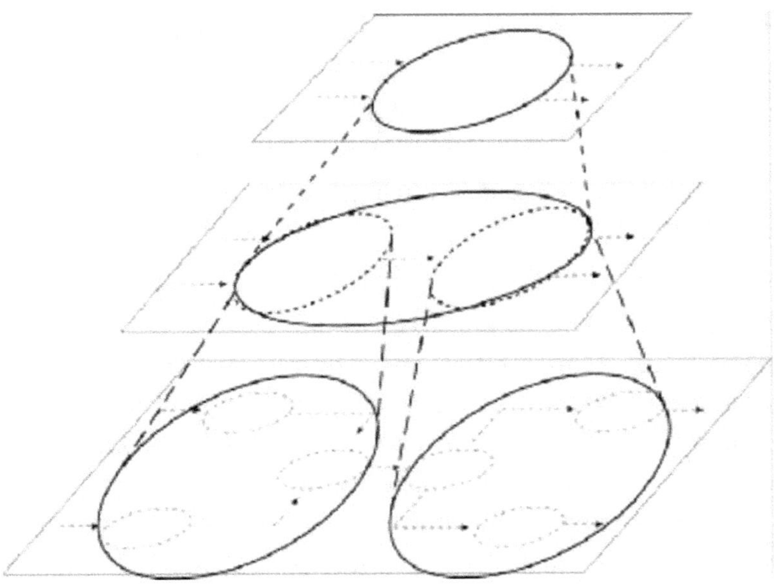

Fig. 6. Interlevel meta-models.

The entire process is depicted through various functional parts that indicate flow structures. The gradual identification of these functional parts signifies the most critical components of the process. This process concludes once the most critical functional portions are determined. This pertains to the "blackbox" from which the initial design may commence. (Figure 7).

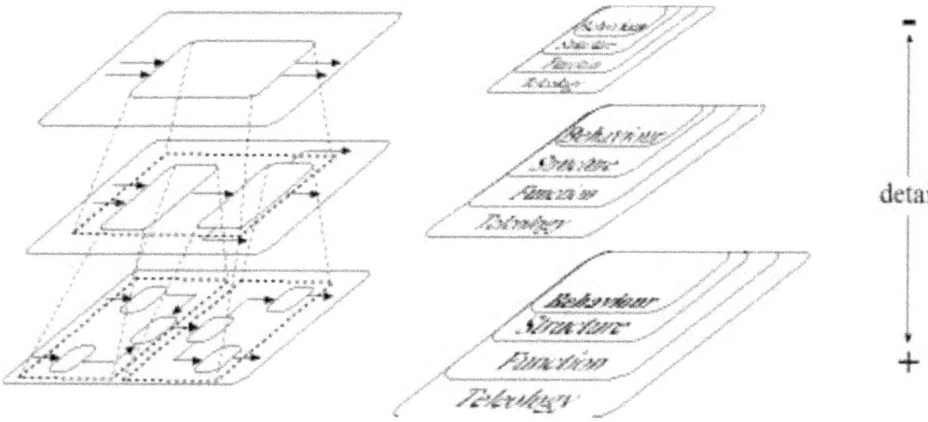

Fig. 7. Abstraction of a process.

3.2 Identification of candidates

The objective of this phase is to identify the appropriate unit or meta-unit for modification to achieve the redesign goals. The initial redesign should concentrate on a process variable. Upon identification of the variable, a diagnostic algorithm is employed to ascertain the units/metaunits influencing the process variable. This reasoning process is predicated on the functions delineated during the functional analysis phase. This step consists of two substages: the specification of redesign requirements and the identification of the appropriate unit or meta-unit for change or substitution.

3.2.1 Specification of redesign criteria

The human designer must delineate the new requirements that the process must fulfill. Redesign needs can be categorized into two types: functional requirements and physical requirements. A design specification invariably includes a singular functional need and may also encompass a collection of physical criteria. A functional requirement embodies an abstraction of the anticipated behavior of the artifact. It may serve as a broad, particular, or operational function. There is no direct correlation between the requisite function and the physical mechanism that delivers it. A physical requirement embodies an abstraction of the physical process variables that fulfill the functional need given in the design specification. It shows preferences of the designer's aims related to a specific area of the process. The redesign specification may be delineated by either the functional and physical requirements or solely by the physical requirements. A redesign specification is a criterion defined by the functionalities that must be integrated into a process to achieve enhanced functionality.

3.2.2 Candidate identification

Here At this stage, a diagnostic algorithm is employed, which operates on the functional concepts delineated in the functional analysis. Diagnosis facilitates the identification of defective components that fail to meet the overall performance standards of the process. The design description and new specifications are utilized to identify potential candidates for change or substitution. The diagnostic algorithm produces a sequenced list of units or meta-units. The diagnostic algorithm functions on abstract functional notions, therefore eliminating the need for simulation. The diagnostic method does not identify the single unit or meta-unit accountable for the "faulty" behavior; instead, it provides a list of units/meta-units that fail to meet the overall performance criteria given in the redesign specification. The human designer is tasked with selecting the appropriate unit or meta-unit from the resultant list for modification or substitution in the process. As this unit/meta-unit is linked to others via a flow path, the "cause" and "consequence" units/meta-units must also be delineated. A cause unit/meta-unit refers to the unit(s)/meta-unit(s) positioned before to the current unit/meta-unit in the flow route. They are responsible for establishing the suitable operational settings for the relevant process variables pertaining to the unit/meta-unit of interest. A consequence unit/meta-unit refers to the unit(s)/meta-unit(s) positioned subsequent to the current unit/meta-unit in the flow route. They are the unit(s)/meta-unit(s) influenced by the operational conditions specified by the unit/meta-unit of interest. The cause and consequence units/meta-units are not always proximate neighbors.

3.3 Development of alternatives

The objective of this phase is to acquire analogous units (equipment) or meta-units (sections) for integration into the existing process, guided by the appropriate unit/meta-unit specified by the human designer in the preceding step. The optimal method to acquire analogous units/meta-units is through comparable processes. By incorporating any acquired unit or meta-unit into the relevant process, an alternative process design is achieved, which constitutes the ultimate objective of the redesign framework. A suitable method for executing this phase is case-based reasoning (CBR), which facilitates the application of previous experiences to novel circumstances.

Beginning with the chosen unit/meta-unit during the candidate identification phase, analogous units/meta-units can be extracted from alternative processes. The adapted unit/meta-unit corresponds most closely to the functional and teleological models of the unit/meta-unit in question. This method necessitates that the performance and operational conditions of the cause and consequence units/meta-units related to the retrieved unit must closely resemble those of the original case.

According to the later stage, the overall target process is modelled as a graph denoting a hierarchy of functions. Therefore hierarchical case-based reasoning (Smyth 01) is required. Thus, based on the levels of abstraction, two kinds of cases are distinguished:

- ground cases. Cases located at the lowest level of abstraction, units (real equipments)
- abstract cases. Cases represented at higher levels of abstraction, meta-units (non-existent "meta-equipments") The organisation of cases into the library is performed according to the type of functions of the unit/meta-unit. In this way, several groups can be distinguished according to the general function type: source, transport, barrier, storage, balance, and sink. Within each functional group, units and meta-units are grouped based on their specific functions. Again, within specific functions groups, the units/meta-units are grouped based on the working function achieved. There are not distinctions between units and meta-units with the same specific function. The case library is organised by an abstract hierarchy based on function groups. This structure denotes the organisation of the functional ontology used in the framework.

To retrieve cases from the case library, a similarity engine is used. Only units/meta-units of the same specific functional group are considered. The similarity engine uses functional and teleological targets to search into the library of cases. Functional and teleological models denote strongly the relationship between the units/meta-units and its neighbours. Two types of similarity are computed, local and global, which are defined as follows:

- Local similarity. Similarity between two cases is based on the local similarity between each feature of such cases. The computation depends on the type of the feature and the value that it could take.
- Global similarity. Once a set of local similarities has been computed for each known feature-value pair, the CBR system computes the global similarity of the candidate cases based on such set.

As final result, a set of cases is obtained which contains meta-units (with its corresponding units/meta-units) or units. The set is ranked according to the global similarity between the target case (the unit/meta-unit of interest) and the source cases (the retrieved units/metaunits).

3.4 Adaptation and evaluation

For this stage the human designer intervention is required since the use of the specialised simulator employed in the data acquisition is mandatory to test the proposed candidates. The adaptation is highly domain-dependent and requires online simulation of the process to verify its correct performance. Since information of abstract cases can not be used directly, the adaptation and revision of equipment must use information of ground cases (real equipments on the simulator). The human designer must fit the ground cases with the process variables involved.

To facilitate the adaptation, an adaptation cost is computed to suggest the human designer the adaptability of the chosen unit/meta-unit. The adaptation cost is based on the differences of the selected unit (source case) and the cause and consequence units/metaunits identified with the diagnostic algorithm. Thus, the cost is a normalised numerical value denoting the difference on the values of the process variables involved in the performance of the unit and the values of the process variables involved in the performance of the neighbour units/meta-units.

The adaptation cost has a value between 0 and 1. Values close to zero mean the adaptation is difficult. The designer experience is determinant because modifications on equipments may affect the overall performance of the process. The modification of the original process, based on the adaptation of a retrieved case, generates an alternative of the process for every unit/meta-unit adapted.

4. Evaluation and results

The framework has been implemented in Chemical Engineering due to the compatibility of chemical processes with the proposed methodology. A chemical plant may consist of one or more chemical processes.

The software modules of the redesign framework have been executed in Java (Sun 05). Supplementary libraries utilized include JESS (JESS 08), Ozone (Ozone 08), and The Selection Engine (Wetzel 00). User contact occurs via a graphical interface.

The implementation of the concepts of the proposed redesign framework is based on the following ontological commitments.

- The chemical processes typically operate at steady-state. That means that values of variables do not change with respect to time.
- A chemical process is constituted of real and abstract units. The abstract units are the sections of the process that appear as atomic elements in conceptual models. All real equipments can be viewed as descendants of the generic real equipment.
- A generic real equipment can be modelled as an object having four attributes: structure, behaviour, function, and teleology to describe all the properties of any real equipment.

Since the framework requires functional concepts, a crucial point is to define the type of functions by means of an ontology. The functional ontology obtained is formed by highlevel and low-level concepts in a similar way to the SUMO (Suggested Upper Merged Ontology) ontology structure (Niles 01). SUMO structures the concepts using metaconcepts, where terminology of general purpose is situated at higher levels, while terminology to specific domains is situated at lower levels. The ontology developed has extended generic concepts of SUMO such as process, objects and mereological and topological concepts. Additional specific concepts have been defined: physico-chemical processes, thermodynamic processes, substances (mass and energy), substance roles (of chemical compounds), functional roles, devices (equipments and connections), measure units, tasks, operations, and relations.

Most of the concepts in the ontology correspond to physical entities. The high-level concepts denote very abstract concepts, which can be found in several domains. The middle-level includes the functional concepts proposed in the Multilevel Flow Modelling and Multimodelling approaches, which are: source, transport, barrier, storage, balance, and sink. The low-level functional concepts come from the well-known chemical process design methodologies developed by Douglas (Douglas 88) and Turton (Turton 98). The low-level functional concepts can be grouped as: reaction, separation, temperature change, pressure change, and flow change. These concepts are called general functions. Each specific function is divided into more specific ones named specific functions, which denote the function of the equipment into the process. Also each specific function is divided in more specific ones, called working functions. A working function can be associated with one or more units and a unit can be related to more than one function. But from the several working functions, only one is the main function in the process.

Over the identified functions an importance functional order and the variables involved in such functions (carried out by the equipment's of the process) have been defined. This order was defined with the aim of forming groups of functions where more important functions "absorb" functions with minor importance, see Figure 8.

The framework was tested over 50 chemical processes (Lopez-Arevalo 05):

1. Acetaldehide from ethanol
2. Acetaldehide from ethylene and oxygen
3. Ethyl acetate
4. Vinyl acetate
5. Acetone
6. Acetic acid
7. Acrylic acid
8. Cyanhydric acid
9. Nitric acid
10. Acrolein
11. Ammonia from natural gas and pure N2
12. Ammonia from pure N 2 and H 2
13. Phthalic anhydride from naphtalene
14. Phthalic anhydride from o-Xylene
15. Maleic anhydride
16. Bencene and methane
17. Bencene and o-Xylene
18. Bencene, Toluene and Styrene
19. Separation of Chlorine-Bencene and Bencene
20. Ethyl-Bencene
21. Cumene
22. 1,3-Butadiene
23. Cyclohexane
24. Allyl Chloride

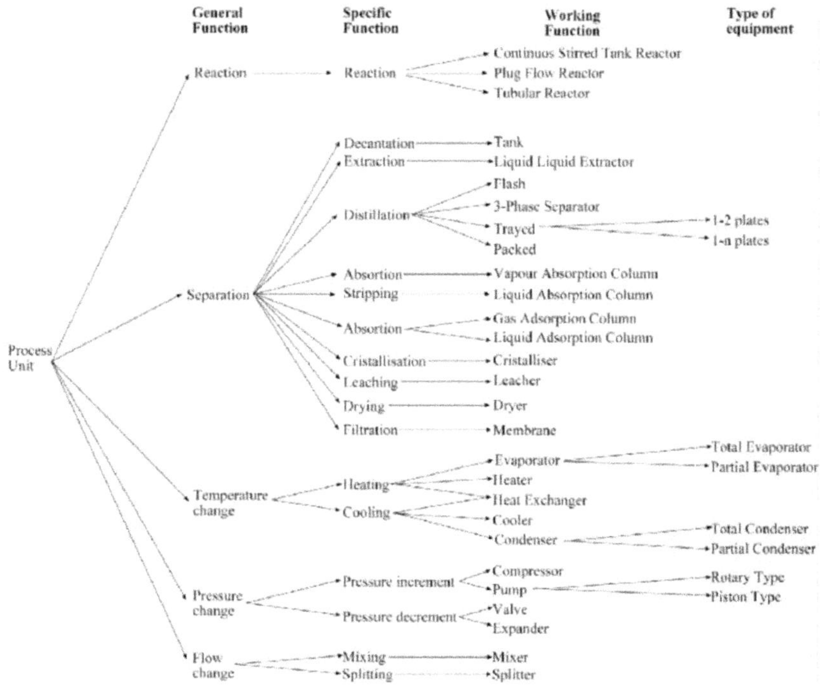

Fig. 8. The hierarchy of functions

25. Separation of Ciclohexane
26. Chloroform
27. Ethanol
28. Purification of Ethanol
29. Dimethyl ether
30. Ethyl tert-butylic ether (ETBE)
31. Methyl tert-butylic ether (MTBE)
32. Tert-amyl Methyl ether (TAME)
33. Styrene
34. Separation of ethane, n-heptane y n-octane
35. Ethylene
36. Ethylene oxide
37. Formaldehyde
38. Formaline

39. Methyl formate
40. HP gas
41. Heptane
42. Hydrogen
43. Separation of metane
44. Separation of metane and ethane
45. Methanol from natural gas
46. Methanol from carbon monoxide
47. Oxygen and nitrogen
48. Purification of parafins
49. Propyleneglycol and dipropylene glycol
50. Vinyl chloride

4.1 Modeling the ammonia synthesis process

The ammonia production process (Figure 9) has been selected as case study because it is one of the most relevant chemical processes in the industry. Ammonia is one of the most important chemicals commodities because of its role in the production of fertiliser and hence of food. It is produced in over 80 countries worldwide with a volume of 130 million tonnes annually (GIA 04).

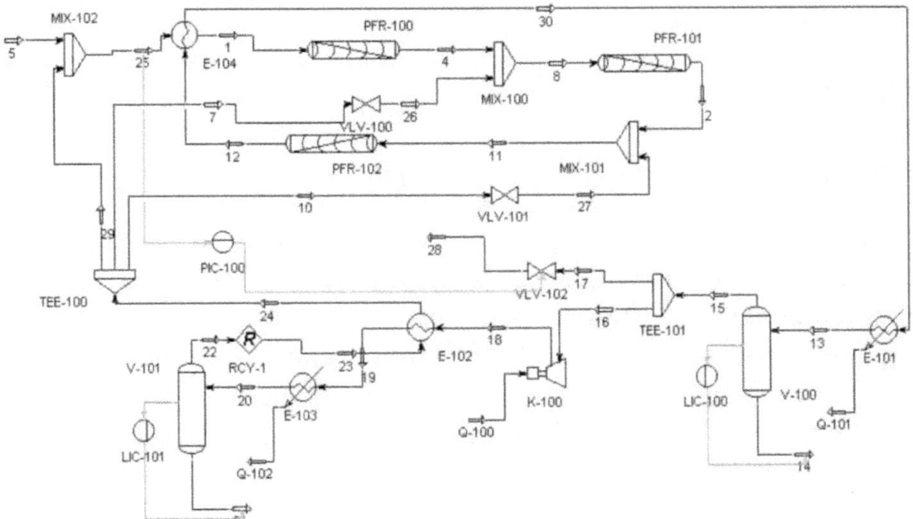

Fig. 9. The ammonia production process

Technical changes on process equipments were taken into account and some other issues such as economical costs, changes in pipes, environment impact, etc. were not considered. To illustrate the performance only the ammonia production process is used as case study.

First, the data is extracted from the Hysys simulator (Hysys 04), and then the "roles" of chemical substances are asked to the human designer. The first level of the process (abstraction level 0) is shown in Figure 10.

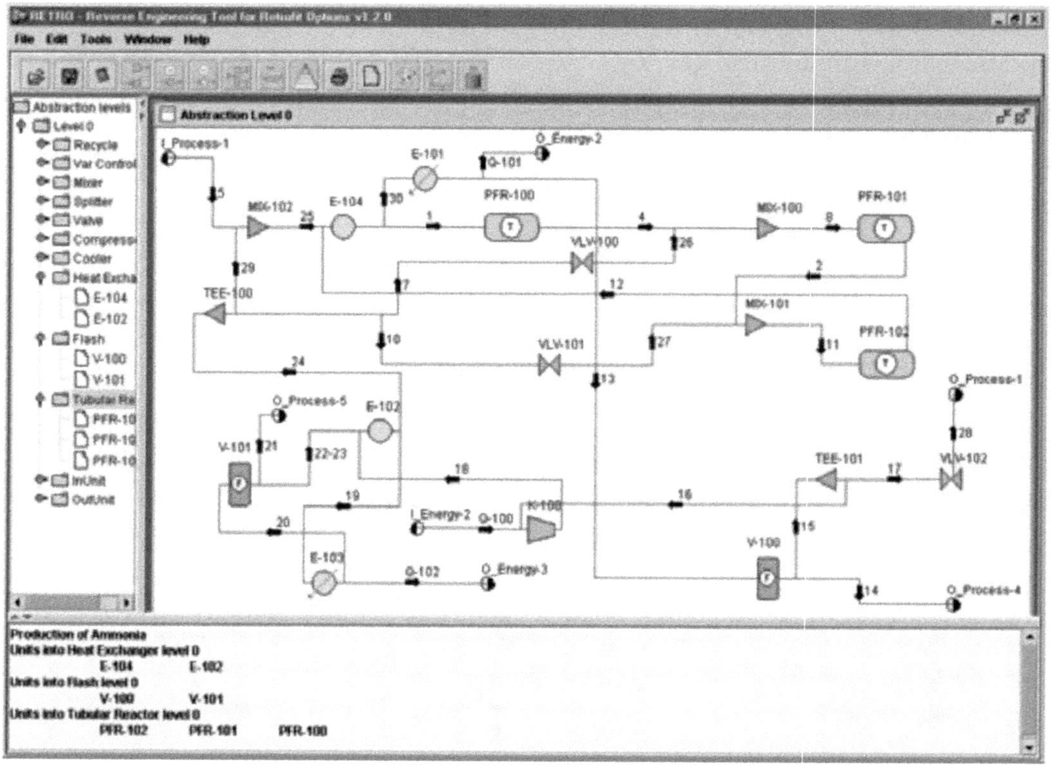

Fig. 10. First representation of the ammonia production process.

This GUI allows the user to interact in two ways, through menus and panels. The diagram panel (upper-right panel) allows the user to manipulate the process layout; the user can organise the components of the process according to its needs. The navigation panel (upper-left panel) is used to navigate into the levels of the process; abstraction levels and its corresponding components. The information panel (bottom panel) displays information about operations carried out in the prototypes. Table 1 summarises the type of equipments and its corresponding functions.

For the generation of meta-units, the flow change equipments (mixers and splitters) are grouped to other ones in the abstraction level 1, as it is shown in Figure 11.

Table 1. Equipment and functions in the ammonia production process.

General Function	Specific Function	Working Function	Label
Flow change	Flow increment	Mixer	MIX-100
			MIX-101
			MIX-102
	Flow decrement	Splitter	TEE-100
			TEE-101
Pressure change	Pressure increment	Compressor	K-100
	Pressure decrement	Valve	VLV-100
			VLV-101
			VLV-10
Temperature change	Temperature increment	Cooler	E-101
			E-103
	Temperature exchange	Heat Exchanger	E-102
			E-104
Separation	Distillation	Flash Separator	V-100
			V-101
Reaction	Reaction	Tubular Reactor	PFR-100
			PFR-101
			PFR-102

The different abstract levels were automatically generated until the final abstract models were created at level 6 (see Figure 12). The abstraction process continues until the whole process is represented by just one meta-unit. Only new units and meta-units are represented to illustrate how the functional groups are created. In this sense, connections between units and meta-units in the same level have not been represented. The scheme represents the process by means of groups of the general class of type of equipment; its general function can be deduced from them. Figure 12 represents all the unit and meta-units in each level, in similar way that they are presented to the designer.

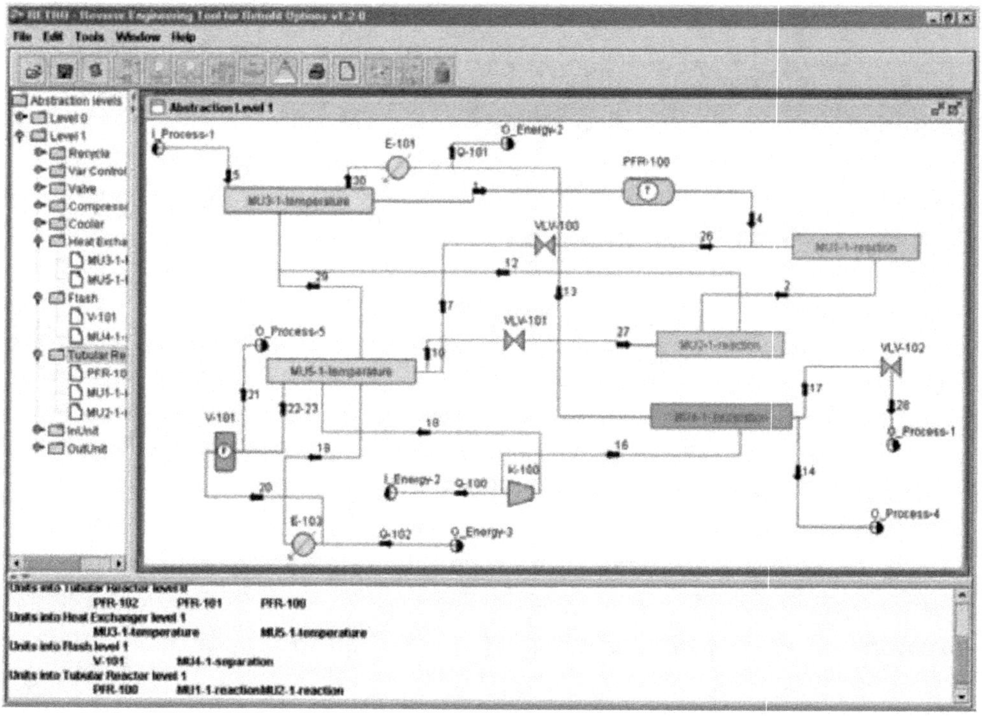

Fig. 11. Grouping of flow change units.

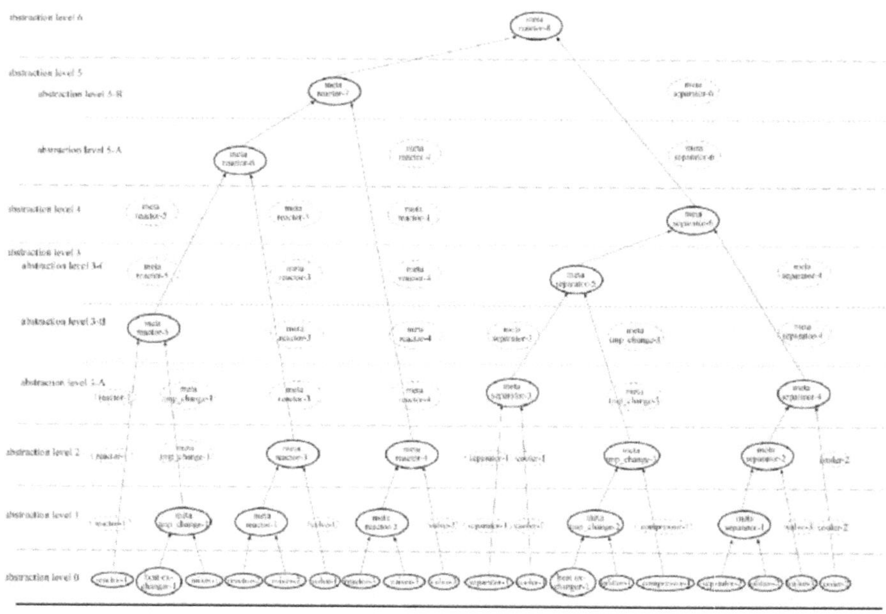

Fig. 12. Hierarchical representation of the ammonia production process in bottom-up direction.

4.2 Identification of candidates

According to the new design objective(s) the process must fulfill, modifications on reactors are used to illustrate the results of the framework:

Requirement

The redesign problem is the increase of the production of ammonia by 15% in the plant represented by the scheme of Figure 10.

The human designer first needs to identify the variables that may affect directly the production of ammonia. Then the designer identifies that the increase of production can be achieved by modifying any of the next conditions:

- pressure
- temperature
- concentration

This gives an idea on the types of equipment the diagnosis must focus on. Assuming that the concentration variable is selected, this is affected by reactors and separators. Initially, reactors affect the concentration of product because they produce the main product, and separators affect it in secondary manner by incrementing the purity of the product.

Therefore, the focus will be on reactors, where the ammonia is originated. Since all the roles of the chemical substances are known, the diagnostic module focuses on the concentration of ammonia (which has the main product role). Thus all values related to this substance are analysed.

We are interested in finding where the main product is produced. The search starts at the highest level in the hierarchy -abstraction level 6- following the flow direction, from left to right. As result of the search, all the units and meta-units affecting the concentration variable have been identified, as shows Table 2.

Table 2. Identified candidates

Abstraction level	Identified components
6	meta-reactor-8
5	meta-reactor-7
4	meta-reactor-6, meta-reactor-4
3	meta-reactor-5, meta-reactor-3, meta-reactor-2
2	reactor-1, meta-reactor-1, reactor-3
1	reactor-2

Since modifications to the process can be performed only at ground level, the cause and consequence units are searched in this level. To illustrate the cause and consequence identification, assume that we focus on the meta-reactor-3, which at ground level includes the units PFR-101, MIX-100, and VLV-100 (Figure 13). In the state analysis, the state conditions are propagated to the units connected to the meta-reactor-3 in the flow path. The analysis in backward/forward stream directions finishes when a closer primary function is reached.

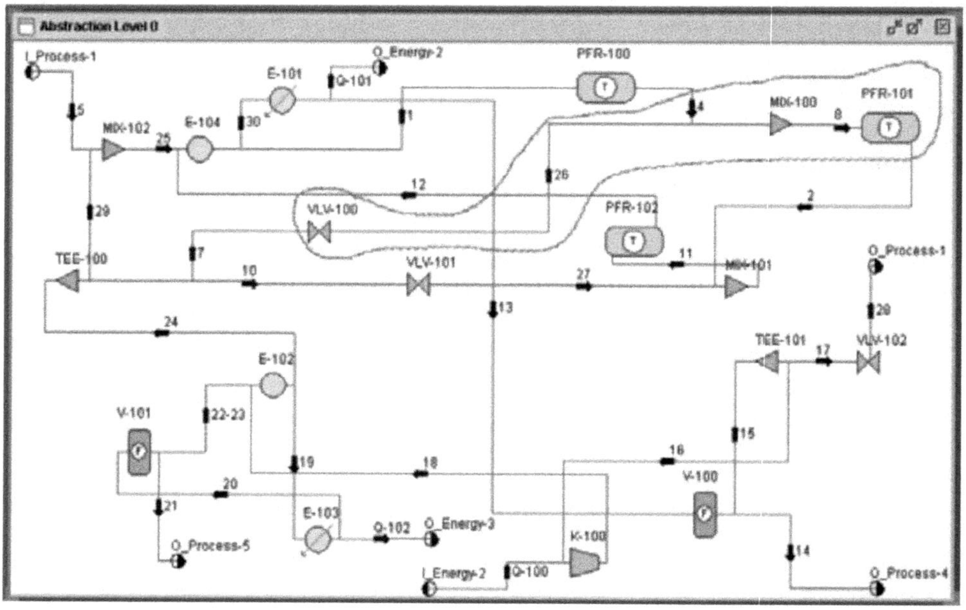

Fig. 13. Units composing the meta-reactor-3.

The state of the meta-reactor-3 is set to low capacity because the production of the main product is not enough. Since this meta-unit is not an initial unit in the path, its state may be originated by the effect of the performance of other units. Then back units are analysed. Considering the stream-4, the function to analyse is a source (PFR-100), which directly affects the concentration variable. Perhaps other back units in the same direction may affect the variable, but this is the closer primary function affecting the concentration variable. It may have low volume state, which originates the low capacity of meta-reactor-3. Therefore, the unit associated to this function is identified as cause unit and this branch of backward analysis in this direction finishes. Considering the stream-7, the function to analyse is a balance (TEE-100), which does not affect the variable. Then, the next function is analysed, which is a balance of temperature (E-102) that again does not affect directly the variable. The next function is storage (V-101), which affects the variable. This is another primary function affecting the concentration variable; it may have low volume state. Then its associated unit is a cause unit and this branch of backward analysis finishes.

Now, forward analysis is carried out following the output stream of the functional group. The low capacity state of meta-reactor-3 originates a low flow state and a low volume state, and consequently affects the following source functions producing a low capacity state in such function. Thus, considering the stream-2, the function to analyse is a balance (MIX101), which does not affect the concentration variable. The next function is source (PFR-102), which affects the variable; it may have low capacity state originated for the low capacity state of the meta-reactor-3. Then, the unit associated with this function is a consequence unit. Since is the closer primary function affected in the forward stream direction, the forward analysis finishes.

Therefore, the identified cause and consequence units for meta-reactor- 3 are shown in Table 3 , which also represents the cause and consequence units for all the meta-units.

Table 3. Cause and consequence units.

Candidate	Cause units	Consequence units
reactor-1 (PFR-100)	separator-2 (V-101)	reactor-2 (PFR-101)
reactor-2 (PFR-101)	reactor-1 (PFR-100)	reactor-3 (PFR-102)
reactor-3 (PFR-102)	reactor-2 (PFR-101)	separator-1 (V-101)
meta-reactor-1	reactor-1 (PFR-100) separator-2 (V-101)	reactor-3 (PFR-102)
meta-reactor-2	reactor-2 (PFR-101)	separator-1 (V-101)
meta-reactor-3	reactor-1 (PFR-100) separator-2 (V-101)	reactor-3 (PFR-102)
meta-reactor-4	reactor-2 (PFR-101)	separator-1 (V-101)
meta-reactor-5	separator-2 (V-101)	reactor-2 (PFR-101)
meta-reactor-6	separator-2 (V-101)	reactor-3 (PFR-102)
meta-reactor-7	separator-2 (V-101)	separator-1 (V-101)

4.3 Generation of alternatives

A CBR module is utilized to acquire alternative units/meta-units that can be included into the ammonia process. The process representation illustrated in Figure 12 signifies the composition of meta-units, whereas the representation in Figure 13 indicates the presence of units/meta-units at each level of abstraction. Based on a threshold of 14 items, the most analogous source cases derived from the computation of global similarities are encapsulated in Table 4. It displays the percentage of similarity, the designated function, and the intake and outlet functions of the source case.

Table 4. Result of the global similarity computation for meta-reactor-3.

Rank	Similarity	Function	Inlet Function	Outlet Function
1	56%	meta-reactor	reaction	separation
2	43%	meta-reactor	reaction	flow change
3	37%	tubular reactor	tmp change	separation
4	31%	plug flow reactor	pres change	tmp change
5	30%	meta-reactor	separation	pres change
6	29%	tubular reactor	reaction	separation
7	29%	meta-reactor	tmp change	tmp change
8	27%	meta-reactor	pres change	separation
9	25%	tubular reactor	tmp change	tmp change
10	23%	tubular reactor	tmp change	separation
11	20%	plug flow reactor	flow change	pres change
12	20%	meta-reactor	pres change	tmp change
13	16%	meta-reactor	reaction	tmp change
14	15%	meta-reactor	flow change	tmp change

The modelling of the 50 processes generated 1590 cases in the case library. Therefore, the software prototypes were continually enhanced according to the needs of these processes.

The aspects considered in the evaluation of the framework were:

1. Modelling of the process

- use of simplified models
- suitable grouping of equipment/sections
- intuitive goal-driven approach
- comprehensive and clear representations of equipment/sections
- easy and intuitive graphical interface
- transparent integration with the numerical simulator

2. Identification of candidates

- clear and easy search over simple but consistent concepts

- module easy to use
- intuitive interpretation of results

3. Suggestion of equipment/sections

- suggestions according to purpose-driven strategy
- appropriate guidelines for modification/substitution
- reuse of past design solutions
- easy access to abstract and detailed data of proposed solutions
- rapid response making agile the creation of alternative prototypes

5. Concise Summary

This chapter delineates a system for redesign support in technical processes grounded in hierarchical modeling. This hierarchical modeling is predicated on means-end and whole-part dimensions. The hierarchical structure improves the thinking process for identifying items to be altered and potential alternatives. The framework concentrated on conceptual redesign challenges utilizing abstract models. The processes are organized hierarchically according to their functions and objectives. The framework comprises four stages: design-description acquisition, candidate identification for redesign, alternative generation, and alternative adaptation and evaluation. The framework was implemented in the discipline of Chemical Engineering.

The redesign framework integrates model-based reasoning with case-based reasoning methodologies. This framework allows the designer to engage directly with the conceptual design of an operational process to automatically produce abstract multiple models, which may be adjusted to create alternative process designs. The practice might be seen as a reverse engineering method of "replay and modify". This model-based methodology effectively integrates hierarchical and functional modeling to depict and analyze intricate technological processes. The hierarchical case-based methodology offers a structured means of reutilizing components from prior procedures.

The framework enhances the application of Multimodelling and Multilevel Flow Modelling methodologies to incorporate mental abstract models concerning process behavior in redesign operations. These models offer a more natural perspective on reasoning for each task, hence enhancing the redesign processes.

The research possesses several limitations, among which the principal ones are:

- The framework was executed solely within a single domain. The concepts can be utilized in a different domain; however, a new implementation will be required.
- The framework underwent testing with simulated plants. We lacked access to actual plant data; however, the results were corroborated by a team of chemical engineers specializing in process design.
- The framework's implementation is unmanageable for inexperienced users due to the necessity of critical decisions made by human designers.

6. References

(Aamodt 94) A. Aamodt & E. Plaza. Case-based reasoning: Foundational issues, methodological variations and system approaches. AI Communications, vol. 7, pages 39-59, 1994.

(Arana 00) I. Arana H. Ahriz & P. Fothergill. Improving re-design support. In Proceedings of the Fifth World Conference on Integrated Design and Process Technology (IDPT2000), 2000.

(Arana 01) I. Arana H. Ahriz & P. Fothergill. Redesign Knowledge Analysis, Representation and Reuse. Industrial Knowledge Management A Micro-level Approach, 2001. Springer-Verlag.

(Akin 82) O. Akin. Representation and architecture. Silver Spring, O. Akin and E.F. Weinel eds., Maryland, Information Dynamics Inc., 1982.

(Alberts 92) L.K. Alberts P.M. Wognum & N.J. Mars. Structuring design knowledge on the basis of generic components. Artificial Intelligence in Design (AID'92), vol. In: Gero, J.S. (Ed.), pages 639-656, 1992. Kluwer, Dordrecht.

(Alberts 93a) L.K. Alberts. YMIR: An ontology for engineering design. PhD thesis, University of Twente, 1993. Netherlands.

(Alberts 93b) L.K. Alberts R.R. Bakker D. Beekman & P.M. Wognum. Model-based redesign of technical systems. Proceedings of the 4th international workshop on principles of diagnosis, 1993.

(Bakker 94) R.R. Bakker S.J. van Eldonk P.M. Wognum & N.I. Mars. The use of model-based diagnosis in redesign. In Proc. Reasoning about physical systems. 11th European Conference on Artificial Intelligence, pages 647-651, A. Cohn Ed., 1994.

(Ball 92) N.R. Ball & F. Bauert. The Integrated Design Framework: supporting the design process using a blackboard system. Artificial Intelligence in Design (AID'92), vol. Gero, J.S. (Ed.), pages 327-348, 1992. Kluwer academic publishers.

(Berker 96) I. Berker & D.C. Brown. Conflicts and Negotiation in Single Function Agent Based Design Systems. Concurrent Engineering: Research and Applications, Journal, Special Issue: Multi Agent Systems in Concurrent Engineering, vol. 4, no. 1, pages 17-33, 1996. Brown, D.C., Landes, S.E. and Petrie, C.J. (Eds.), Technomic Publishing Inc.

(Bertalanffy 50) L.von Bertalan.y. An Outline of General Systems Theory. British Journal for the Philosophy of Science, vol. 1, no. 2, 1950.

(Bhatta 94) S. Bhatta A. Goel & S. Prabhakar. Innovation in Analogical Design: A ModelBased Approach. Proceedings of Artificial Intelligence in Design, pages 57-74, 1994. Kluwer Academic Publishers, Dordrecht, The Netherlands.

(Bo 99) Y. Bo & F. Salustri. Function Modeling Based on Interactions of Mass, Energy and Information. Florida Artificial Intelligence Research Society, 1999.

(Borner 96) K. Borner E. Pipping E. Tammer & C. Coulon. Structural Similarity and Adaptation. Proceedings of the Third European Workshop on Case-Based Reasoning, pages 58-75, 1996. Springer-Verlag.

(Bras 92) B.A. Bras. Foundation for designing decision-based design processes. PhD thesis, University of Houston, 1992.

(Brazier 94) F.M. Brazier P.H. Van Langen Zs. Ruttkay & J. Treur. On formal speci.-cation of design tasks. Arti.cial Intelligence in Design (AID'94), vol. Gero, J.S. and Sudweeks, F. (Eds.), pages 535-552, 1994. Kluwer Academic Publishers, Dordrecht.

(Brazier 96) F.M. Brazier P.H. Van Langen J. Treur & N.J. Wijngaards. Redesign and reuse in compositional knowledge-based systems. Knowledge Based Systems, Special Issue on Models and Techniques for Reuse of Designs, vol. 9, no. 2, pages 105-119, 1996.

(Bridge 97) C. Bridge. From design to redesign: Revisiting design models to highlight similarities and differences. Seminar Presentation, 1997. Department of Architectural and Design Science, University of Sydney, Australia.

(Brown 89) D.C. Brown & B. Chandrasekaran. Design problem solving: knowledge structures and control strategies. Morgan Kaufmann, 1989.

(Brown 97) D.C. Brown & W.P. Birmingham. Understanding the Nature of Design. IEEE Expert, vol. 12, no. 2, pages 14-16, 1997.

(Brown 98) D.C. Brown. Intelligent Computer-Aided Design. AI in Design Group, Com-puter Science Department, WPI, September 1998. Worcester, MA, USA.

(Buckingham 91) S. Buckingham Shum. Cognitive Dimensions of Design Rationale. In People and Computers VI: Proceedings of HCI'91, pages 331-344, Cambridge Univer-sity Press. Cambridge, 1991.

(Chang 2003) Chang WC, Van YT. Researching design trends for the redesign of product form. Design Studies 24(2):173-180.

(Chandrasekaran 90) B. Chandrasekaran. Design problem solving: a task analysis. AI Magazine, vol. 11, no. 4, pages 59-71, 1990.

(Chandrasekaran 93) B. Chandrasekaran A. Goel & Y. Iwasaki. Functional Representation as Design Rationale. IEEE Computer, no. Special Issue on Concurrent Engineering, pages 48-56, 1993.

(Chandrasekaran 00) B. Chandrasekaran & J.R. Josephson. Function in Device Representation. Engineering with Computers, Special Issue on Computer Aided Engineering, vol. 16, pages 162-177, 2000.

(Checkland 81) P. Checkland. Systems thinking, systems practice. John Wiley and Sons, 1981.

(Chen 2009) Li, S. and Chen, L., 2009, Pattern-based Reasoning for Rapid Redesign: A Proactive Approach. Research in Engineering Design, Vol. 21, pp. 25-42.

(Chittaro 93) L. Chittaro G. Guida C. Tasso & E. Toppano. Functional and Teleological Knowledge in the Multimodeling Approach for Reasoning about Physical Systems: A case study in diagnosis. IEEE Transactions on Systems Man. and Cybernetics., vol. 23, no. 6, pages 1718-1751, 1993.

(Chittaro 98) L. Chittaro & A.N. Kumar. Reasoning about function and its applications to engineering. Artificial Intelligence in Engineering, vol. 12, pages 331-336, 1998.

(Culley 99) S.J. Culley. Final Report -Future Issues For Design Research Workshop. Technical Report, Faculty of Engineering and Design, University of Bath, 1999.

(Das 94) D. Das S. Gupta & D. Nau. Reducing setup cost by automated generation of redesign suggestions. Proc. ASME Computers in Engineering Conference, pages 159-170, 1994.

(Daube 89) F. Daube & B. Hayes-Roth. A Case-Based Mechanical Redesign System. In Proceedings of the 11th International Joint Conference on Artificial Intelligence (IJCAI-89) Vol. 2, pages 1402-1407, Morgan Kaufmann Publishers, 1989.

(de Kleer 79) J. de Kleer. Causal and Teleological Reasoning in Circuit Recongnition. PhD thesis, Massachusetts Institute of Technology, 1979.

(de Silva Garza 96) A. Gomez de Silva Garza & M.L. Maher. Design by interactive exploration using memory based techniques. Knowledge based systems, vol. 9, no. 1, pages 151-161, 1996.

(DeLoach 04) S.A. DeLoach. The MaSE Methodology. Methodologies and Software Engineering for Agent Systems. The Agent-Oriented Software Engineering Handbook Series : Multiagent Systems, Artificial Societies and Simulated Organizations, vol. 11, 2004. Bergenti, F., Gleizes, M.P., Zambonelli, F. (Eds.) Kluwer Academic Publishing.

(Dixon 89) J.R. Dixon M.J. Guenette R.K. Irani E.H. Nielsen M.F. Orelup & R.V. Welch. Computer-based models of design processes: the evaluation of design for redesign. NSF Engineering Design Research Conference, pages 491-506, 1989. University of Massachusetts, USA.

(Douglas 88) J.M. Douglas. Conceptual design of chemical processes. Mc Graw Hill, New York, 1988.

(Dove 2001) Dove R. Response ability: the language, structure, and culture of the agile enterprise. Wiley, New York.

(Dunskus 95) B.V. Dunskus D.L. Grecu D.C. Brown & I. Berker. Using Single Function Agents to Investigate Con.ict. Artificial Intelligence in Engineering Design and Manufacturing (AIEDAM), Special Issue: Conflict Management in Design, vol. 9, no. 4, pages 299-312, 1995.

(Eldonk 96) S.J. van Eldonk L.K. Alberts R.R. Bakker F. Diker & P.M. Wognum. Redesign of technical systems. Knowledge-based Systems, Special Issue on Models and Techniques for Reuse of Designs, vol. 9, pages 93-104, 1996.

(Falkenhainer 91) B. Falkenhainer & K.D. Forbus. Compositional Modeling: Finding the Right Model for the Job. Artificial Intelligence, vol. 51, pages 95-143, 1991.

(Falkenhainer 92) B. Falkenhainer & K.D. Forbus. Composing task-specific models. Automated Modelling, ASME, vol. 41, pages 1-9, 1992.

(Fensel 01b) D. Fensel. Ontologies: A silver bullet for knowledge management and electronic commerce. Heidelberg, Germany, 2001.

(Fischer 87) G. Fischer A.C. Lemke & C. Rathke. From design to redesign. In Proceedings of the 9th International Conference on Software Engineering, pages 369-376, IEEE Computer Society Press, 1987.

(Fischo. 78) B. Fischo. P. Slovic & S. Lichtenstein. Fault trees: Sensitivity of estimated failure probabilities to problem representation. Journal of Experimental Psychology: Human Perception and Performance, vol. 4, 1978.

(Forbus 84) K.D. Forbus. Qualitative Process Theory. Artificial Intelligence, vol. 24, pages 85-168, 1984.

(Forster 96) J. Forster I. Arana & P. Fothergill. Redesign knowledge representation with DEKLARE. In Proceedings of KEML'96: 6th Workshop on Knowledge Engineering: Methods and Languages, Paris, France, 1996.

(Forster 97b) J. Forster P. Fothergill & I. Arana. Enabling intelligent variant design using constraints. IEE Intelligent Design Systems Colloquium, vol. Digest No: 97/016, 1997. London.

(Fothergill 95) P. Fothergill J. Forster J. A. Lacunza F. Plaza & I. Arana. DEKLARE: A methodological approach to re-design. Proceedings of Conference on Integration in Manufacturing, pages 109-122, 1995. K. R. von Barisani, P. A. MacConaill and K. Tierney (Eds.), Vienna. IOS Press.

(Franke 92) D.W. Franke. A theory of teleology. PhD thesis, University of Texas at Austin, 1992.

(French 85) R. French & J. Mostow. Toward Better Models of the Design Process. AI Magazine, vol. 6, no. 1, pages 44-57, 1985.

(French 93) M.J. French R.V. Chaplin & P.M. Langdon. A creativity aid for designers. International Conference on Engineering Design, ICED'93, pages 53-59, 1993. The Hague, Netherlands.

(Fricke 2005) Fricke E, Schulz A P. Design for changeability (DfC): principles to enable changes in systems throughout their entire lifecycle. Systems Engineering 8(4):342359. Wiley Periodicals.

(Gero 90a) J.S. Gero. Design prototypes: a knowledge representation schema for design. AI Magazine, vol. 11, no. 4, pages 26-36, 1990.

(Gero 90b) J.S. Gero & A. Rosenman. A conceptual framework for knowledge-based de-sign research at Sydney University's design computing unit. Artificial Intelligence in Engineering, vol. 5, no. 2, pages 65-77, 1990.

(Gero 04) J.S. Gero & U. Kannengiesser. The situated Function-Behaviour-Structure framework. Design Studies, vol. 25, no. 4, pages 373-391, 2004.

(GIA 04) GIA. Greener Industry Association: Ammonia Process. http://www.uyseg.org/greener industry/index.htm, 2004.

(Goel 89) A. Goel & B. Chandrasekaran. Functional representation of designs and re-design problem solving. Proc. Eleventh International Joint Conference on Arti.cial Intelligence, pages 1388-1394, 1989. Los Altos, California: Morgan Kaufmann Publishers.

(Goel 91) A. Goel. A model-based approach to case adaptation. In In Proceedings of the 13th Annual Conference of the Cognitive Science Society (CogSci'91), pages 143-148, Chicago, Illinois, 1991.

(Goel 92) A. Goel & B. Chandrasekaran. Case-Based Design: A Task Analysis. Artificial Intelligence Approaches to Engineering Design, Innovative Design, vol. II, no. 6, pages 165-184, 1992. Academic Press, Tong y D. Sriram (editors).

(Goel 94b) A. Goel & S. Prabhakar. A control architecture for redesign and design veri.-cation. Proc. Second Australian and New Zealand Conference on Intelligent Information Systems Conference, pages 377-381, 1994. Brisbane, Qld., Australia.

(Goel 97a) A. Goel A. Gomez de Silva Garza N. Grue J.W. Murdock & M.M. Recker. Functional Explanations in Design. IJCAI-97 Workshop on Modeling and Reasoning, 1997.

(Gomez-Perez 04) A. Gomez-Perez R. Gonzalez-Cabero & M. Lama. Development of Semantic Web Services at the Knowledge Level. In The European Conference on Web Service (ECOWS'04), pages 72-86, 2004.

(Grossmann 00) I.E. Grossmann & A.W. Westerberg. Research Challenges in Process Systems Engineering. AICHE Journal, vol. 46, no. 9, pages 1700-1703, 2000.

(Gruber 93) T.R. Gruber. A translation approach to portable ontology specifications. Knowledge Acquisition, vol. 5, no. 2, pages 199-220, 1993.

(Gundersen 90) T. Gundersen. Retrofit Process Design -Research and Applications of Systematic Methods. Foundations of Computer-Aided Process Design, pages 213240, 1990. J.J. Siirola, I. E. Grossmann and G. Stephanopoulos (Eds.).

(Han 95) Ch. Han J.M. Douglas G. Stephanopoulos. Agent-Based Approach to a Design Support System For the Synthesis of Continuous Chemical Processes. Computers in Chemical Engineering, vol. 19 (Supplement), pages S63-S69, 1995.

(Heo 98) D.H. Heo A.C. Parker & C.P. Ravikumar. An Evolutionary Approach to System Redesign. Proc. Eleventh International Conference on VLSI Design, pages 359-362, 1998. Chennai, India.

(Hertwig 01) T.A. Hertwig A. Xu A.B. Nagy R.W. Pike J.R. Hopper & C.L. Yaws. A prototype system for economic, environmental and sustainable optimization of a chemical complex. Proc. 11th European Symposium on Computer Aided Process Engineering, pages 1017-1022, 2001. R. Gani and S.B. Jorgensen eds, Elsevier.

(Hoover 91) S.P. Hoover J.R. Rinderle S. Finger. Models and abstractions in design. In Proceedings of the International Conf. on Engineering Design, ICED'91, pages 4657, Zurich, 1991.

(Howe 86) A. Howe P. Cohen J. Dixon & M.D. Simmons. A domain-independent program for mechanical engineering design. In Proceedings of the 1st International Conference on Applications of AI in Engineering Problems, pages 289-299, Southampton University, U.K., 1986.

(Hysys 04) Hysys. Introduction to hysys.plant ver. 3.1. AEA Technology Engineering Software, Hyprotech, Calgary, Canada, 2004.

(JESS 08) JESS. Jess 7.1 manual. http://herzberg.ca.sandia.gov/jess/docs/index.shtml, 2008.

(Keuneke 91) A.M. Keuneke. Device Representation: The Significance of Functional Knowledge. IEEE Expert, vol. 6, no. 2, pages 22-25, 1991.

(Kim 93) G.Y. Kim & G. Bekey. Construting Design Plans for DFA Redesign. Proc. IEEE International Conference on Robotics and Automation, pages 312-318, 1993.

(Kirkwood 88) R.L. Kirkwood M.H. Locke & J.M. Douglas. A prototype expert system for synthesizing chemical process flowsheets. Comp. Chem. Eng., vol. 12, no. 4, pages 329-343, 1988.

(Kitamura 98) Y. Kitamura & R. Mizoguchi. Functional Ontology for Functional Understanding. Proceedings of The Twelfth International Workshop on Qualitative Reasoning, AAAI Press, pages 77-87, 1998. Cape Cod, USA.

(Kitamura 99) Y. Kitamura & R. Mizoguchi. Towards Redesign based on Ontologies of Functional Concepts and Redesign Strategies. In Second International Workshop On Strategic Knowledge And Concept Formation, page JSPS8, Iwate, Japan, 1999.

(Kolodner 93) J. Kolodner. Case-based reasoning. Morgan Kaufman Publishers, Inc., 1993.

(Kraslawski 00) R. Ben-Guang H. Fan-Yu A. Kraslawski & L. Nystrom. Review: study on the methodology for retrofitting chemical processes. Chemical Engineering Technology, vol. 23, no. 6, pages 479-484, 2000.

(Kuraoka 03) K. Kuraoka & R. Batres. An Ontological Approach to Represent HAZOP Information. Technical Report TR-2003-01, Process Systems En-gineering Laboratory, Tokyo Institute of Technology, 2003.

(Lander 97) S.E. Lander. Issues in Multi-agent Design Systems. IEEE Expert, vol. 12, no. 2, pages 18-26, 1997.

(Leveson 00) N.G. Leveson. Intent specifications: an approach to building human-centered specifications. IEEE Transactions on Software Engineering, vol. 26, no. 1, pages 1535, 2000.

(Lind 90) M. Lind. Representing Goals and Functions of Complex Systems. Technical Report 90-D-38, Institute of Automatic Control Systems, Technical University of Denmark, 1990.

(Lind 94) M. Lind. Modelling goals and functions of complex industrial plants. Applied Artificial Intelligence, vol. 8, pages 259-284, 1994.

(Lind 96) M. Lind. Status and challenges of intelligent plant control. Annual Reviews in Control, vol. 20, pages 23-41, 1996.

(Lind 99) M. Lind. Plant Modeling for Human Supervisory Control. Transactions of the Institute of Measurement and Control, vol. 21, no. 4/5, pages 171-180, 1999.

(Linnho. 88) B. Linnho. G.T. Polley & V. Sahdev. General Process Improvements Through Pinch Technology. Chem. Eng. Progr., vol. 84, pages 51-58, 1988.

(Maher 95) M.L. Maher B. Balachandran & D.M. Zhang. Case-based reasoning in design. Lawrence Erlbaum Associates, Sydney, 1995.

(Maher 97a) M.L. Maher & A. Gomez de Silva Garza. Case-Based Reasoning in Design. IEEE Expert, vol. 12, no. 2, pages 34-41, 1997.

(Maher 97b) M.L. Maher S. Simno. & J. Mitchell. Formalising building requirements using an activity/space model. Automation in construction, vol. 6, pages 77-95, 1997.

(Maher 01) M.L. Maher & A. Gomez de Silva Garza. GENCAD: A Hybrid Analogical/Evolutionary Model of Creative Design. Proc. of the 4th International Conference on Computational Models of Creative Design, 2001. J.S. Gero and M.L. Maher (eds).

(Maulik 92) P.C. Maulik M.J. Flynn D.J. Allstot & L.R. Carley. Rapid Redesign of Analog Standard Cells Using Constrained Optimization Techniques. Proc. IEEE Custom Integrated Circuits Conference, pages 8.1.1-81.3, 1992.

(Mitchell 83) T.M. Mitchell L.I. Steinberg S. Kedar-Cabelli V.E. Kelly J. Shul-Man & T. Weinrich. An intelligent aid for circuit redesign. In Proceedings of the 3th National Conference on Artificial Intelligence (AAAI-83), Washington, USA, 1983.

(Mostow 89) J. Mostow. Design by derivational analogy: Issues in the automated replay of design plans. Artificial Intelligence, vol. 40, no. 1-3, pages 119-184, 1989.

(Nelson 90) D.A. Nelson & J.M. Douglas. A systematic procedure for retro.tting chemical plants to operate utilising different reaction paths. Ind. Eng. Chem. Res., vol. 29, pages 819-829, 1990.

(Niles 01) I. Niles & A. Pease. Towards a Standard Upper Ontology. In 2nd International Conference on Formal Ontology in Information Systems (FOIS-2001), pages 2-9, 2001.

(Ohsuga 97) S. Ohsuga. Strategic Knowledge Makes Knowledge Based Systems Truly Intelligent. In Proceedings of the First International Workshop on Strategic Knowledge and Concept Formation, pages 1-24, Lutchi Research Centre, 1997.

(Ozone 08) Ozone. Ozone Developers' Guide. http://www.ozone-db.org/frames/documentation/overview.html, 2008.

(Papoulias 83) S.A. Papoulias & I.E. Grossman. A Structural Optimization Approach in Process Synthesis II: Heat Recovery Networks. Computers and Chemical Engineering, vol. 7, pages 707-721, 1983.

(Pasanen 01) A. Pasanen. Phenomenon-Driven Process Design Methodology. PhD the-sis, Department of Process and Environmental Engineering of the University of Oulu, Finland, 2001.

(Pistikopoulos 87) E.N. Pistikopoulos & I.E. Grossman. Optimal Retrofit Design for Improving Process Flexibility in Linear Systems. Technical Report EDRC-06-24-87, Engineering Design Research Center. Carnegie Mellon University, 1987.

(Pos 97) A. Pos. Problem Solving for Redesign. 10th European Knowledge Workshop on Knowledge Acquisition. Modeling and Management, pages 205-220, 1997. Sant Feliu de Guíxols, Catalonia.

(Price 97) C.J. Price I.S. Pegler M.B. Ratcli.e & A. McManus. From troubleshooting to process design: closing the manufacturing loop. In 2nd International Conference on Casebased Reasoning Research and Development, pages 114-121, London, UK, 1997. LNCS Vol. 1266, Springer-Verlag.

(Price 98) C.J. Price. Function-directed electrical design analysis. Artificial Intelligence in Engineering, vol. 12, pages 445-456, 1998. Elsevier.

(Price 03) C.J. Price N.A. Snooke & S.D. Lewis. Adaptable Modeling of Electrical Systems. In 17th International Workshop on Qualitative Reasoning, pages 147-153, 2003. Brasilia, Brazil.

(Qian 92) L. Qian & J.S. Gero. A design support system using analogy. Proceedings of the Second International Conference on AI in Design, pages 795-813, 1992. Kluwer Academic Publishers.

(Rapoport 94) H. Rapoport R. Lavie & E. Kehat. Retrofit design of new units into an existing plant: Case study: Adding new units to an aromatics plant. Computers and Chemical Engineering, vol. 18, no. 8, pages 743-753, 1994.

(Rasmussen 85) J. Rasmussen. The Role of hierarchical knowledge representation in decision making and system management. IEEE Transactions on Systems, Man and Cybernetics, vol. SMC-15, no. 2, 1985.

(Rasmussen 86) J. Rasmussen. Information processing and human-machine interaction: An approach to cognitive engineering. North Holland, 1986.

(Salomons 95) O.W. Salomons. Computer Support in the Design of Mechanical Products. PhD thesis, University of Twente, 1995.

(Sasajima 95) M. Sasajima Y. Kitamura M. Ikeda & R. Mizoguchi. FBRL: A Function and Behavior Representation Language. Proceedings of IJCAI, pages 1830-1836, 1995.

(Schoen 91) E. Schoen. Intelligent Assistance for the Design of Knowledge based Systems. PhD thesis, Stanford University, 1991.

(Sembugamoorthy 86) V. Sembugamoorthy & B. Chandrasekaran. Functional Representation of Devices and Compilation of Diagnostic Problem-Solving Systems. Experience, Memory and Reasoning, J.L. Kolodner and C.K. Riesbeck, eds, Lawrence Erlbaum, Hillsdale, N.J., pages 47-73, 1986.

(Sylvester 00) R.W. Sylvester W.D. Smith & J. Carberry. Information and modelling for greener process design. AIChE Symp. Series, vol. 96, no. 323, pages 26-30, 2000.

(Simon 96) H.A. Simon. The sciences of the artificial. MIT Press, Cambridge, Massachusetts, 1996.

(Skuce 93) D. Skuce. A multi functional knowledge management system. Knowledge Acquisition, vol. 5, pages 305-346, 1993.

(Smith 87) R. Smith & B. Linnho. Process integration using pinch technology. Proceedings of ATEE Symp. Energy Management in Industry, 1987. Paris.

(Smyth 96) B. Smyth & M.T. Keane. Using adaptation knowledge to retrieve and adapt design cases. Knowledge-based Systems, Special Issue on Models and Techniques for Reuse of Designs, vol. 9, no. 2, pages 127-136, 1996.

(Smyth 01) B. Smyth M.T. Keane & P. Cunningham. Hierarchical Case-Based Reasoning Integrating Case-Based and Decompositional Problem-Solving Techniques for Plant-Control Software Design. IEEE Trans. on Knowledge and data engineering, vol. 13, no. 5, pages 793-812, 2001.

(Sowa 95) J. Sowa. Distinctions, Combinations and Constraints. In Proceedings of IJCAI 95 Workshop on Basic Ontological Issues in Knowledge Sharing, Montreal, Canada, 1995.

(Steier 91) D. Steier. Automating Algorithm Design within a General Architecture for Intelligence. Automating Software Design, pages 577-602, 1991. Lowry, M.R. and Mccartney, R.D. (Eds.).

(Steinberg 85) L. Steinberg & T. Mitchell. The Redesign System: A Knowledge-base approach to VLSI CAD. IEEE Design and Test of Computers, vol. 2, no. 45-54, 1985.

(Stephanopoulos 90a) G. Henning G. Stephanopoulos & H. Leone. MODEL.LA A modelling framework language for process engineering II. Multifaced modelling of processing systems. Computers and Chemical Engineering, vol. 14, no. 8, pages 847-869, 1990.

(Stephanopoulos 90b) G. Stephanopoulos G. Henning & H. Leone. MODEL.LA A mod-elling framework language for process engineering I. The formal framework. Computers and Chemical Engineering, vol. 14, no. 8, pages 813-846, 1990.

(Stroulia 92a) E. Stroulia M. Shankar A. Goel & L. Penberthy. A Model-Based Approach to Blame Assignment in Design. Proceedings of the Second International Conference on AI in Design, pages 519-537, 1992. J.S. Gero (editor).

(Stroulia 92b) E. Stroulia & A. Goel. Generic Teleological Mechanisms and their use in Case Adaptation. Proceedings of the Fourteenth Anual Conference of the Cognitive Science Society, pages 319-324, 1992. Indiana, USA.

(Struss 91) P. Struss. A theory of model simplification and abstraction for diagnosis. In Proceedings of the Fifth International Workshop on Qualitative Reasoning, pages 25-57, 1991.

(Subba-Rao 99) S. Subba-Rao A. Nahm Z. Shi X. Deng & A. Syamil. Artificial intelligence and expert systems applications in new product development-a survey. Journal of Intelligent Manufacturing, vol. 10, no. 3-4, pages 231-244, 199. Kluwer Academic Publishers.

(Sumi 97) Y. Sumi. Supporting the Acquisition and Modelling of Requirements in Software Design. In Proceedings of the First International Workshop on Strategic Knowledge and Concept Formation, pages 205-216, Lutchi Research Centre, 1997.

(Sun 08) Sun Microsystems. Java Language Specification. http://java.sun.com/docs/index.html, 2008.

(Sycara 92) K. Sycara. CADET: A case-based synthesis tool for engineering design. International Journal for Expert Systems, no. 2, pages 157-188, 1992.

(Takeda 90a) H. Takeda P. Verkaamp T. Tomiyama & H. Yoshikawa. Modelling Design Processes. AI Magazine, vol. 11, no. 4, pages 37-48, 1990.

(Takeda 94a) H. Takeda. Abduction for design. In Proceddings of Formal Design Methods for CAD, pages 221-243, Tallinn, Estonia, 1994.

(Tay 2003) Tay FEH, Gu J. A methodology for evolutionary product design. Engineering with Computers 19:160-173.

(Thornton 93) A.C. Thornton & A. Johnson. Constraint specification and satisfaction in embodiment design. In Procedings of the International Conference on Engineering Design, ICED 93, pages 1319-1326, The Hague, 1993.

(Tjoe 86) T.N. Tjoe & B. Linnho. Using pinch technology for processes retrofit. Chem. Eng., vol. 93, pages 47-60, 1986.

(Tomiyama 87) T. Tomiyama & H. Yoshikawa. Extended General Design Theory. In Proceedings of the IFIP WG 5.2 Working Conference on Design Theory for CAD, pages 95-125, 1987.

(Treur 89) J. Treur. A logical analysis of design tasks for expert systems. International Journal of Expert Systems, vol. 2, pages 233-253, 1989.

(Turton 98) R. Turton R.C. Bailie W.B. Whiting & J.A. Shaeiwitz. Analysis, synthesis and design of chemical processes. Prentice-Hall, New Jersey, 1998.

(Uerdingen 01) E. Uerdingen U. Fischer & K. Hunderbuhler. A screening method for identifying economic improvement potentials in retrofit design. In ESCAPE-11 (European Symposium on Computer Aided Process Engineering), pages 573-578, 2001.

(Ullman 91) D.G. Ullman. Design histories: archiving the evolution of products. In Proceedings of the DARPA Workshop on Manufacturing, Salt Lake City, USA, 1991.

(Umeda 90) Y. Umeda T. Tomiyama & H. Yoshikawa. Function, Behaviour and Structure. Applications of Artificial Intelligence in Engineering V, pages 177-193, 1990. J.S. Gero (Ed).

(Umeda 92) Y. Umeda T. Tomiyama & H. Yoshikawa. A design methodology for a self-maintenance machine based on funtional redundancy. Design Theory and Methodology - DTM'92, pages 317-324, 1992.

(Umeda 94) Y. Umeda T. Tomiyama H. Yoshikawa & Y. Shomimura. Using Functional maintenance to improve fault tolereance. IEEE Expert, vol. 9, no. 3, pages 25-31, 1994.

(Umeda 97) Y. Umeda & T. Tomiyama. Functional Reasoning in Design. IEEE Expert, vol. 12, no. 2, pages 42-48, 1997.

(Vaselenak 87) J.A. Vaselenak I.E. Grossman & A.W. Westerberg. Optimal Retro.t De-sign of Multiproduct Batch Plant. Industrial and Engineering Chemistry Research, vol. 26, no. 4, pages 718-726, 1987.

(Vescovi 93) M. Vescovi & Y. Iwasaki. Device design as functional and structural re.nement. In Working Notes of the IJCAI'93 Workshop on AI in Design, pages 55-60, 1993.

(Vicente 92) K.J. Vicente & J. Rasmussen. Ecological interface design: Theoretical foundations. IEEE Trans. on Systems, Man and Cybernetics, vol. 22, no. 4, 1992.

(Wetzel 00) B. Wetzel. Selection Engine. http://selectionengine.sourceforge.net/, 2000.

(Wielinga 97) B.J. Wielinga & A.Th. Schreiber. Con.guration-design problem solving. IEEE Expert, vol. 12, no. 2, pages 49-56, 1997.

(Wood 01) M. Wood & S.A. DeLoach. An Overview of the Multiagent Systems Engi-neering Methodology. Agent-Oriented Software Engineering, LNAI, vol. 1957, 2001. P. Ciancarini, M. Wooldridge, (Eds.).

(Xin 01) W. Xin & X. Guangleng. Supporting design reuse based on integrated design rationale. IEEE International Conference on Systems, Man and Cybernetics, vol. 3, pages 1909-1912, 2001.

(Yoshikawa 91) H. Yoshikawa F. Arbab & T. Tomiyama. Intelligent CAD III: Selected and Reviewed Papers and Reports. In Third International Workshop on Computer Aided Design, 1991.

Chapter-5

Prospects for Modification of Food Proteins

1. Introduction: An Overview

Proteins ares vital in foods, serving not only their nutritional purpose but also as modulators of structure and sensory perception of a food product. The functional behavior of a protein is naturally influenced by physico-chemical circumstances such as pH, ionic strength, temperature, and pressure, rendering it both unpredictable and advantageous in food production. Proteins are typically expensive to produce, and the growing global population and rising standards of living intensify worries regarding the availability of protein for food consumption. To achieve a more sustainable utilization of protein sources, several approaches have been pursued over the past decades that significantly enhanced protein availability: (i) improved production or protein refinement techniques, (ii) utilization of alternative protein sources, and (iii) optimized application of protein functionality. In wheat production, correlations between genetic expression and functional product behavior have enabled breeders to enhance cultivars for specific geographic locations (e.g. Payne et al., 1984). Alternatively, one can express particular proteins in non-native sources, such as human milk proteins in plants like rice (e.g. Lönnerdal, 2002). Targeted modifications in the genomes of food-producing organisms can result in mutations in the main sequences of pertinent proteins, potentially introducing novel functionalities. If adequate amounts of the novel protein are produced and integrated with the baseline protein levels in the food, the functional qualities of the food system (texture formation) may be enhanced. The altered protein may be extracted for utilization as a food additive. Recently, several proteins from unconventional sources have been recognized as human dietary ingredients that are beginning to be utilized, such as algae, leaves, insects, and other seeds. The effective use of these novel proteinaceous materials has been constrained, necessitating advancements in their extractability, digestibility, nutritional value, and overall functional and sensory characteristics. The regulation of protein functional behavior occurs later in the process, specifically at the constituent level. This can occur through physical-chemical processes, enzymatic reactions, or chemical engineering methods. This review will concentrate on the application of chemical engineering to investigate or enhance protein functioning in food products. The rationale for utilizing chemical modification of proteins will be examined regarding its significance in comprehending the fundamental principles of proteins as structural agents in food. This encompasses enhancing shelf life and sensory attributes, along with the innovation of novel capabilities of dietary proteins, including the utilization of plant proteins as meat texturizers. We will further examine how these findings have thus far facilitated a more sustainable usage of protein, encompassing factors such as consumer acceptance and current or evolving regulations around its use as ingredients.

2. Functional role of proteins as food components

This paragraph will succinctly outline the functional features of proteins as food components. The molecular mechanisms of their functions in food products will be emphasized. The subsequent paragraph will examine these molecular pathways as targets for chemical engineering. Extensive reviews have been published on each of these topics, which will be referenced.

2.1 Surface properties: Emulsions and foams

Surfaces or interfaces in food products are abundantly present in terms of emulsions (oilwater interface) or foams (air-water interface). Examples of foams and emulsions in food products include ice cream, dressings or margarine. In foams and emulsions respectively air cells or oil droplets are dispersed or in an aqueous phase; the latter case also water in oil emulsions exist. As a result of their amphiphilic character, proteins, being composed of polar and non-polar amino acids, can contribute to the formation and stability of such dispersions by adsorbing to the interface and developing stabilizing films by coating the oil or air droplets and hence act as emulsifiers. The ability of proteins to induce film formation depends on a number of parameters of molecular nature which have been studied into detail in the past and is collectively governed by a net energy gain from absorbing at an interface. Milk proteins have been identified as good foaming agents as a result of their aggregation state, molecular stability and flexibility, electrostatics, and (surface) exposed hydrophobicity (Hunter et al., 1991; Luey et al., 1991; Shirahama et al., 1990; Suttiprasit et al., 1992; Waniska & Kinsella, 1985). Chemical modifications have been employed in the past to improve surface activity of less performing proteins. These studies and consequences of the used modifications will be discussed in more detail in the following paragraph. A wide range of methods has been employed in literature to study the chemical and molecular properties of proteins adsorbed at interfaces which lead to a detailed understanding of the principle forces of importance to surface activity. These methods include ellipsometry and infrared reflection absorption spectroscopy (IRRAS) which provide information on denaturation and concentration of adsorbed proteins (de Jongh & Wierenga, 2006; Grigoriev et al., 2007; Martin et al., 2003; McClellan et al., 2003). Stabilization of an air-water or oilwater interface is governed by a multiple step process. First, absorption at the interface requires proteins to diffuse to the interface and their retention at the interface is governed by the kinetic barrier of absorbance which, in turn, is influenced by factors such as exposed hydrophobicity (Wierenga et al., 2003), and net charge (Wierenga et al., 2005). Effective absorption onto the interface is followed by rearrangement of the protein molecules to form a thermodynamically stable but dynamic monolayer of molecules coating the droplets (de

Jongh & Wierenga, 2006; Graham & Phillips, 1979a, 1979b). Some publications report that the rearrangement process coincides with structural extension of the protein molecules (reviewed in MacRitchie, 1978) while other results suggest that secondary structure does not undergo variation subsequent to adsorption onto the interface (de Jongh & Wierenga, 2006; Graham & Phillips, 1979c). Despite the finding that local concentrations of 150 – 300mg/ml can be reached at the interface (Meinders et al., 2001), proteins are still able to laterally diffuse as was shown by thiolated ovalbumin (de Jongh & Wierenga, 2006), illustrating that chemical modification has been used to underpin the molecular mechanisms of the surface activity of proteins.

2.1.1 Emulsions

Emulsions consist of two immiscible liquids, oil and water, in which the droplets are termed dispersed phase and the liquid surrounding the droplets is called continuous phase. Depending on the concentrations of each liquid and the environmental conditions, oil-inwater emulsions or water-in-oil emulsions can be formed. These consist of oil droplets in a continuous water phase and water droplets in a continuous oil phase, respectively. Examples of food related emulsions are milk, vinaigrette, and mayonnaise. Emulsions are often unstable by nature and will phase separate or coalesce over extended time, or storage. To stabilize emulsions, so-called emulsifiers can be used which form a small layer on the surface of the dispersed phase, thereby physically separating the continuous phase from the dispersed phase. Such treatment will allow the incorporation of emulsified solutions in food products which can be stored over an extended time frame. Amphiphilic proteins, containing both hydrophilic and hydrophobic regions, are effective emulsifiers. These proteins adsorb onto the interface between the oil and water phase, and stabilize the oil and water phase by selective interaction with both surfaces, thereby preventing the individual droplets from coalescing (reviewed in Rodríguez Patino et al., 2008). Other types of emulsifiers used in food products include lipids, phospholipids, surfactants or polysaccharides (Dickinson, 1992; McClements, 2005). Apart from a texture perspective, the absorption of proteins at the oil-water interface is of interest to the delivery of nutrients (reviewed in Malaki Nik et al., 2010). Upon digestion, changes occur at the oil-water interface of emulsified food products as a function of emulsifier stability, affecting the digestibility and subsequently the availability of encapsulated nutrients. The amphiphilic character of proteins can be modulated for example by the covalent attachment of lipid chains (see paragraph 3.2). Attachment of lipid chains of various lengths to a protein renders it more hydrophobic resulting in an improved ability to stabilize emulsions and/or foams (Aewsiri et al., 2011a, 2011b). Apart from exposed hydrophobicity, other factors have also been identified to determine the affinity of a protein for an oil-water interface. Aggregation or molecular weight of proteins for example adversely affects the emulsifying activity of proteins (Baldursdottir et al., 2010; Corzo-Martínez, 2011). That the aggregation effect is more complex than originally postulated was shown by single molecule total internal reflectance fluorescence microscopy of bovine serum albumin showing that the orientation at which the aggregates absorb to the interface largely determine the rate of absorbance to the interface (Walder & Schwartz, 2010). The identification of factors contributing to the surface activity of proteins allows the improvement of emulsifying activity by means of targeted chemical modification.

2.1.2 Foams

Foams consist of gas bubbles dispersed into a liquid. The stability of the air bubbles in a foam is determined by the foaming agent which forms a layer of adsorbed molecules separating the air bubbles from the continuous liquid phase, similar to the emulsifying activity described in paragraph 2.1.1 (reviewed in Halling, 1981; Wilde, 2000). Adsorption of a protein to the air-water interface induces partial dehydration of the molecule promoting protein-protein interactions. This effect is further amplified by the finding that local protein concentrations at the air-water interface can reach up to 150 to 300mg/ml (Meinders et al., 2001). The rate of absorption to the air-water interface has been reported to depend largely on the hydrophobic nature of the protein under investigation (Kudryashova et al., 2003; Wierenga et al., 2003). Increasing the exposed hydrophobicity of proteins by means of conjugation with lipid chains was shown to increase the adsorption rate to the air-water interface (Wierenga et al., 2003). Net charge is a second parameter of interest determining adsorption kinetics, in which higher net charge slows down the adsorption process due to the electric repulsive forces involved (de Jongh et al., 2004; Kudryashova et al., 2005; Le Floch-Fouéré et al., 2011). Highly aggregated heat-treated ovalbumin was further shown to induce a ten-fold decrease the diffusion rate of proteins to the interface compared to the native protein (Kudryashova et al., 2005). However, the tendency of aggregated ovalbumin to remain adhered to the interface upon first interaction is significantly larger than for nonaggregated protein, which was found to rapidly desorb from the surface after absorption (Kudryashova et al., 2005). Collectively, it has been shown that surface activity of proteins is not determined by a single molecular characteristic but rather depends on a combination of factors. Hence, several types of chemical modification can be employed to improve the airwater interface activity of proteins.

2.2 Aggregation and gelation

Protein aggregation is a major topic in the field of food science, the regulation of which is believed to markedly affect the texture of food products (Zhou et al., 2008; reviewed in Doi, 1993). Aggregated protein can act as a nucleation prerequisite to induce gelation, albeit at high protein concentration (Alting et al., 2003; Barbut & Foegeding, 1993; Ju & Kilara, 1998). Processing conditions and storage can induce protein aggregation, even at ambient temperatures (Promeyrat et al., 2010; Santé-Lhoutellier et al., 2008). The resulting protein aggregates can vary widely in size and morphology as a result of the environmental conditions under which they were formed, among other factors. For example, upon inspection using electron microscopy, amylin (Patil et al., 2011), hen egg white lysozyme (Arnaudov & de Vries, 2005), and β-lactoglobulin (Arnaudov et al., 2003; Veerman et al. 2002) were found to form negatively stained and long fibrillar aggregates at pH values far away from the isoelectric point of the respective proteins. Near the isoelectric pH (Arnaudov & de Vries, 2005), or at high salt concentration (Arnaudov & de Vries, 2006; Veerman et al. 2002), spherical or amorphous aggregates are formed. That aggregate morphologies cannot always be categorized as either fibrillar or amorphous in a clear-cut manner was shown by various groups observing substantial heterogeneity in aggregate morphology, also called polymorphism, within the same preparation (Bauer et al., 1995; Jiménez et al., 2002, 1999). As a consequence, gels formed under these conditions of β-lactoglobulin are particulate and the particle size depends on heating temperature and heating rate (Bromley et al., 2006).

These observations lead to the understanding that net charge plays a major role in determining aggregate and gel morphology (Krebs et al., 2009; Langton & Hermansson, 1992). Cryo-EM investigation of aggregates formed from ovalbumin which had been succinylated to various degrees provided further evidence that net charge dominantly determines aggregate morphology (Weijers et al., 2008). The propensity of proteins to aggregate, or rate of aggregation, has been shown to vary as a function of protein conformational stability (Chiti et al., 2000; Hurle et al., 1994; Kelly, 1998; Quintas et al., 1997; Ramirez-Alvarado et al., 2000; Siepen & Westhead, 2002), rate of unfolding (Broersen et al., 2007a), net charge (Calamai et al., 2003; DuBay et al., 2004), and secondary structure propensity (Fernandez-Escamilla et al., 2004). Exposed hydrophobicity (Calamai et al., 2003) and the possibility to form disulfide bonds naturally affect the aggregation and gelation propensity of proteins as these two forces are primarily driving the assembly process. This has been exemplified in a study which showed that β-lactoglobulin A modified with *N* ethylmaleimide largely resisted aggregation induced by heating (Kitabatake et al., 2001). Extensive knowledge of the molecular factors driving the aggregation process of proteins has lead to the development of a number of algorithms able to predict protein aggregation with high fidelity (Chiti et al., 2003; DuBay et al., 2004; Fernandez-Escamilla et al., 2004; Maurer-Stroh et al., 2010).

The energetics and kinetics of protein aggregation have been subject of many publications to date and has been reviewed on numerous occasions (e.g. Luheshi & Dobson, 2009; Straub & Thirumalai, 2011). It has been recognized that, regardless of primary sequence or physicochemical properties, all proteins have an inherent tendency to form aggregates in vitro under certain conditions (reviewed in Chiti & Dobson, 2006; Dobson, 1999). Protein aggregation is a multiple step complex process which can be viewed as a cascade of steps of assembly which may vary in molecular detail as a function of the protein studied or the environmental conditions. Nevertheless, the aggregate growth mechanisms of many different proteins share essential characteristics which have been elucidated both by experimental and computational methods (Teplow et al., 2006). The onset of aggregation often requires the (partial) unfolding or conformational rearrangement of proteins (Calamai et al., 2003; Dobson, 1999; Kelly, 1996; Rochet & Lansbury, 2000). Using human lysozyme (Canet et al., 1999) and HypF-N (Marcon et al., 2005) it has been shown that a population of less than 1% of partially folded protein can be sufficient to trigger the onset of the aggregation process. The resulting exposure of hydrophobic regions which normally reside in the core of a folded protein drives the self-assembly process to form small oligomers. These oligomers, or nuclei, are metastable and their transient and short-lived nature dictates that they can dissociate into monomeric protein, which has been shown for various proteins. Many efforts in the field of protein aggregation suggest that the nucleus has to reach a critical size which then allows further assembly by monomer addition to ultimately form mature fibrils or aggregated networks (Jarrett & Lansbury, 1993; Lomakin et al., 1996, 1997; Sorci et al., 2011). An alternative scenario is the formation of intermediate protofibrils along the pathway which subsequently assemble into fibrils (Harper et al., 1997; Walsh et al., 1997). Mature fibrils have classically been viewed as the stable end-stage of the aggregation process which are not susceptible for dissociating conditions. Recently, using calorimetric methods (Morel et al., 2010), molecular dynamics simulations (Zidar & Merzel, 2011), and mechanical deformation studies (Paparcone & Buehler, 2011; Paparcone et al., 2010; Xu et al., 2010) it has been observed that fibrils can be dissociated albeit at high temperature.

Detailed knowledge of the molecular parameters determining aggregation propensity, rate and morphology resulted in the ability to tune protein aggregation through chemical engineering. The attachment of sugar chains to proteins has been shown to inhibit selfassociation (Marquardt & Helenius, 1992; reviewed in Helenius et al., 1997; Land & Braakman, 2001; Song et al., 2001). This effect was largely attributed to covalently linked sugar moieties affecting kinetic partitioning between folding and aggregation from an (partially) unfolded state. For example, glycosylation was found to increase the folding rate of the protein rapidly shielding exposed hydrophobic regions which could potentially act as a driving force for aggregation at ambient temperature (Broersen et al., 2007b; ShentalBechor & Levy, 2008; Wang et al., 1996). Interestingly, high temperatures induced more rapid aggregation of glycosylated proteins compared to their non-glycosylated counterparts (Broersen et al., 2007b; Chobert et al., 2006). However, another study involving glycated bovine serum albumin concluded that glycation of the protein inhibited its aggregation upon incubation at moderate temperatures (Rondeau et al., 2010, 2007). Glycosylation of proteins has also been shown to affect gel properties: attachment of a ketohexose to ovalbumin by means of the Maillard reaction resulted in the formation of gels with enhanced breaking strength (Sun et al., 2004). The effects of various types of chemical protein engineering on the physico-chemical functionality of proteins will be discussed in more detail in paragraph 3 of this review.

2.3 Protein structural integrity

Proteins in food products can lose their native structure as a result of processing conditions including storage, heat treatment, acidification, dehydration, mechanical processing or shear, and microbial hydrolysis. For example, long-term storage of milk powder has been found to induce lactosylation of the proteins present in the preparation which, in turn, results in affected powder solubility and emulsifying and foaming properties (reviewed in Thomas et al., 2004). Unfolded or hydrolyzed protein molecules can exert very different functionality to food products compared to folded proteins, a classical example being the boiling of an egg which converts the liquid-like transparent egg white into an opaque semisolid structure with very different textural properties. This paragraph will shortly discuss the principles of protein folding and structure and the forces that are implied. It was first recognized by Anfinsen (1973) that the primary sequence of a protein dictates the specific folded, or native, conformation a protein will assume to allow functional activity. Following urea-induced denaturation of ribonuclease A , the protein was allowed to refold by removal of urea. The protein was found to regain its native structure and functionality after this treatment suggesting that proteins can adopt their native conformation spontaneously (Anfinsen, 1973). This finding was awarded with the Nobel Prize in Chemistry in 1972 and opened up an avenue of experimental and theoretical work in the field of protein folding and unfolding. The structural insights into the folding and unfolding processes of many proteins have since then been explored using a vast range of biophysical instrumentation, both at the ensemble (reviewed in Buchner et al., 2011; Sanchez-Ruiz, 2011) as well as at single molecule (reviewed in Borgia et al., 2008; Ferreon & Deniz, 2011) level. Protein conformational stability can be defined as the ability of the natively folded structure of a protein to resist unfolding. Two types of stability can be distinguished: the difference in energy content between the folded state and unfolded state of a protein is termed thermodynamic stability. Boltzmann's distribution law defines the distribution of folded and unfolded protein molecules at a certain time. The rate at which the unfolded protein collapses to a folded state is reflected by the folding rate and represents the kinetic stability of a protein. Energy landscape theories and the folding funnel hypothesis have both been used as models to understand the energetic barriers of a protein separating the folded from the unfolded state (reviewed in Onuchic et al., 1997; Plotkin & Onuchic, 2002; Wolynes, 2005). Both models start from a similar principle in which there is an energy difference between the folded state of a protein and its unfolded state and that the folded state is defined as the favored entropic state of a protein. Local energy minima in the folding process can result in the accumulation of transient intermediate structures, which are neither folded, nor unfolded, to a larger or lesser extent, depending on the environmental conditions of folding or the primary sequence of the protein (reviewed in Baldwin, 2008; Englander et al., 2007). These intermediate structures, which often lack biological activity, are sometimes sufficiently stable to allow substantial accumulation (reviewed in Englander et al., 2007). Stable folding intermediates are related to a high propensity of aggregation as the hydrophobic core is not sufficiently shielded while opposing assembly forces are absent (reviewed in Ferreira et al., 2006). Kinetic protein stability is defined by a variety of molecular parameters including the proximity of native contacts in the primary sequence of a protein (Cieplak et al., 2004; Plaxco et al., 1998, 2000), internal friction or the energetic of intrachain interactions, energy barriers to backbone rotations and long-range residue interactions (Pabit et al., 2004; Qiu & Hagen, 2004), rate of diffusional motion of an unfolded peptide chain through the solvent (Pabit et al., 2004), and the presence or absence of intermediate state(s) (Baumketner, 2003; Onuchic et al., 1997). Forces that contribute to thermodynamic stability are the strength of intramolecular hydrogen bonds and solvent-protein interactions, both enthalpic in nature. The entropic contribution is mainly defined by the hydrophobic effect of folding through an increase in disorder of water molecules upon folding. Many of the forces retaining a protein structure intact can be disrupted, removed or introduced by chemical modification. For example, reaction of proteins with sugars by means of the Maillard reaction can lead to distinct changes in protein stability. The stability of proteins in food products can be affected by using proteins which have been modified by means of glycosylation or charge modification. Succinylation with the aim to increase net charge of soy protein hydrolysate lead to improved digestibility of the protein as investigated by a multienzyme method involving trypsin, chymotrypsin, and peptidase (Achouri & Zhang, 2001). This finding suggests that the protein had undergone structural rearrangement as a result of the succinylation process. Succinylation also lead to destabilization of Faba bean legumin (Schwenke et al., 1998). This group used differential scanning calorimetry (DSC) to study protein stability of legumin and found a decreased specific enthalpy for unfolding upon succinylation. Succinylation also resulted in an increased surface hydrophobicity of the protein suggesting at least partial unfolding of the molecule. Kosters and colleagues (2003) compared the effects of many different types of chemical modification on protein stability, including lipophilization using capric acid, glycosylation and succinylation of ovalbumin. Ovalbumin stability was probed by DSC, tryptophan fluorescence and circular dichroism (CD). Lipophilization resulted in a decreased denaturation temperature of ovalbumin reflected in an enthalpy decrease and a lower stability upon guanidine titration. Glycosylation was found to stabilize structural integrity of ovalbumin. A similar finding using CD and DSC to study temperature-induced unfolding has been reported upon glucosylation (Broersen et al., 2004; van Teeffelen et al., 2005) and fructosylation (Broersen et al., 2004) of β lactoglobulin or glucosylation of codfish parvalbumin (de Jongh et al., 2011). Interestingly, assays performed at ambient temperature but involving denaturant-induced unfolding reported a decrease in protein stability upon conjugation of glucose to β-lactoglobulin. This phenomenon, which appeared unique for a glycosylation reaction, has been studied in further detail by van Teeffelen and colleagues (2005). The observations could be explained in terms of a decreased change in heat capacity upon unfolding as a result of glucosylation indicating that the hydration pattern of proteins upon glycosylation is significantly affected.

2.4 Shelf life

Proteins can affect shelf life and stability of food products by enhancing antioxidant activity, affecting gas exchange, antimicrobial activity or by stabilization of emulsion or foam-based food products (del Rosario Moreira et al., 2011; Emmambux et al., 2004; Mendis et al., 2005). Protein films can be used as as packaging biomaterials as a result of their ability to form networks with rheologically advantageous characteristics (Arvanitoyannis, 1999; Audic & Chaufer, 2005, Longares et al., 2005). However, mixtures of for example proteins and polysaccharides have been found to exert superior functional properties compared to proteins or polysaccharides in isolation (reviewed by Pogaku et al., 2007). For example, the application of an edible coating of storage proteins obtained from sorghum, called kafirins, has been shown to extend the shelf life of freshly harvested pears (Buchner et al., 2011). The shelf life of meat was shown to be extended upon application of a collagen and gelatin coating and led to reduced decoloration, antioxidant activity, and reduction of microbial spoilage (Havard & Harmony, 1869). Such coatings extend quality and shelf-life by acting as a slow-release gas barrier (Baldwin, 1994; Buchner et al., 2011; Park, 1999). Nanobiocomposites of maize prolamin protein zein have also been employed as a gas barrier by coating tomatoes (Park et al., 1994), and apples (Bai et al., 2003). Even though collagen and gelatin coatings were reported to both effectively retain water in meat products (Antoniewski et al., 2007; Farouk et al., 1990), extensive moisture loss of kafirin coated pears compared to the uncoated product left them unacceptable toward consumers (Buchner et al., 2011). The authors (Buchner et al., 2011) therefore suggested to prepare wax or triglycerides/kafirin mixtures instead of pure kafirin coatings to prevent moisture loss as kafirin films themselves do not function effectively as water barriers (Emmambux et al., 2004; Gillgren & Stading, 2008). Because lipids form a very suitable moisture barrier as a result of their hydrophobic character, lipophilized proteins possibly form more effective coatings for fruit. However, to date, no work has been published to demonstrate the effect of lipid-incorporation into proteins to prepare stable coatings for fruit.

Some proteins and peptides are known to have antimicrobial activity (Nizet, 2006; reviewed by Wimley, 2010). Some of these are also applied as food preservatives such as nisin, which is a potent antibacterial 34 amino acid peptide containing a number of uncommon amino acids. Nisin has been employed as an approved food preservative in cheese (Martins et al., 2010), fish, meat, and beverages (reviewed in Lubelski et al., 2008). Another known antimicrobial peptide is ε-poly-L-lysine which exhibits antimicrobial activity against bacteria and fungi and is used as a food preservative (reviewed in Hamano, 2011). This asset has also been explored within a food based environment, for example by applying mixed formulations of chitosan, a linear polysaccharide, and casein polymers to a number of food products including carrot, cheese, and salami (del Rosario Moreira et al., 2011). Pure caseinate films applied to squash slices showed limited antimicrobial activity (Ponce et al., 2008). While chitosan alone exerts significant anti-microbial activity, the inclusion of casein polymers into the formulation improved bactericidal properties even more (del Rosario Moreira et al., 2011) as a result of the ionic interaction between the two biopolymers (Pereda et al., 2008, 2009). Cao-Hoang and colleagues (2010) produced a nisin-containing sodium caseinate film to investigate the antimicrobial activity of both surface- and in-depth Listeria innocua inoculated soft cheese. The presence of the film reduced surface contamination with L. Innocua significantly, while antimicrobial activity within the cheese matrix depended on the distance from the film-coated surface. Antimicrobial films prepared from a mixture of oregano oil and whey protein isolate showed inhibition of growth of lactic acid bacteria, reduction of pseudomonads, total flora and growth rates when applied to fresh beef (Zinoviadou et al., 2009). Even though in many cases complex formulae have been employed, containing both a protein component as well as a carbohydrate or oil component, no publications are known that show the effects of covalently linked lipid or sugar to protein films as a potential edible film.

2.5 Sensory: Color, flavor, odor, texture

Sensory aspects of food products include sensation of flavor, odor, color, and texture. These factors play a large role in consumer acceptation of food products and the effects of various types of protein chemical engineering and their applications will be discussed in this paragraph. Many types of modification target the amino groups of lysine residues, including succinylation, lipidation and glycosylation through the Maillard reaction. Textural properties, including emulsifying, foam and gelling capacities, have been discussed in detail in paragraphs 2.1 and 2.2.

2.5.1 Flavor

Even though most proteins are tasteless, ingestion of a small number of proteins is perceived as sweet. These include thaumatin (Ohta et al., 2008; van der Wel & Loeve, 1972), monellin (Morris & Cagan, 1972), brazein (Ming & Hellekant, 1994), and lysozyme (Masuda et al., 2001). The sweetness of lysozyme results from the abundance of lysine residues which was shown by alanine substitution in lysozyme (Masuda et al., 2005a). It is therefore perceivable that modification of lysine residues by conjugation of a chemical group has consequences for the sweetness of the protein. Extensive acetylation and phosphopyridoxylation of lysine residues of lysozyme decreased the perceived sweetness of the protein further demonstrating that lysine residues play a major role in sensory aspects of this protein (Masuda et al., 2005, Kaneko & Kitabatake, 2001). No other reports on flavor modulating aspects as a result of chemical protein engineering have been reported.

2.5.2 Odor

One publication which studied the effect of acylation by acetic and succinic anhydride of flaxseed protein isolates reported no off-odors upon modification, although no results were presented (Wanasundara & Shahidi, 1997).

2.5.3 Color

Three types of modifications have been reported to affect the color of the protein preparation. First, succinylation was reported to convert the color of soy isolate from tan to chalk-white upon visual inspection (Franzen & Kinsella, 1976a). Upon measurement of surface reflectance using a colorimeter, increasing degrees of succinylation of flax seed protein lead to brighter protein preparations (Wanasundara & Shahidi, 1997). A similar observation has been reported for succinylation of fish muscle (Groninger, 1973), alfa-alfa leaf protein (Franzen & Kinsella, 1976b), and soy bean protein preparations in a U.S. patent for coffee whitener (Melnychym & Stapley, 1973). Acylation by treatment of protein with acetic anhydride also lead to brighter flaxseed protein isolate, although the effect of succinylation of the same protein was stronger (Wanasundara & Shahidi, 1997). Franzen and Kinsella (1976a) showed no effect on color upon acetylation of soy isolate. A third type of modification with a strong effect on color is glycosylation through the Maillard reaction. This reaction is a complex cascade of reactions initiated by the interaction of a reducing sugar with an amino group. Colored products are formed only at later stages of the reaction pathway and include aldols and melanoidins which are high molecular weight compounds (reviewed in Zamora & Hidalgo, 2006). Paragraph 3.1 further extends on the formation of browning products and Amadori compounds related to the Maillard reaction.

2.5.4 Texture

Texture is a complex consumer perception of mouthfeel, tastants and afterfeel. A number of most relevant texture attributes, like 'spreadability' (essential for attributes like 'creamy'), 'crumbliness' or 'separating/wateriness' have been shown to be directly related to the energy household in protein-based products when energy is exerted onto the system. This applied energy may arise from oral processing, exposure to heat, gravity or applied pressure during for example industrial processing. The energy balance in protein-based food structures has been proposed by van Vliet and Walstra (1995): $W_{applied} = W_{stored} + W_{dissipated} + W_{fracture.}$ This model implies that all energy applied to the gel can be used either for fracture, can dissipate or can be stored in the network (and regained after release of exerted forces). At a microstructural level this translates into fracture nucleation points and fracture propagation modes, whereas energy dissipation is often assumed to be controlled by serum flow properties. This latter factor is directly determined by the porosity of the gel as set-up by the microstructure and the pore deformation propensity when the system is put under strain. Van den Berg and colleagues (2008) showed that an attribute like 'spreadability' is directly related to directing as much energy as possible to fracture. When energy can be efficiently stored in the protein matrix, this directs the attribute 'crumbliness'. The effective interaction between protein-building blocks that make up the product matrix can be engineered. Strengthening this interaction, using for example transglutaminase (e.g. Dondero et al., 2006), will direct the energy flow from fracture to storage and gels will become less spreadable and more crumbly. Sala and colleagues (2008) demonstrated that altering the interaction between a filler (such as a fat particle) and the protein matrix significantly influences the texture of these protein gels. The equilibrium between the active and inactive properties of the filler can be meticulously regulated through chemical alteration.

3. Categories of chemical protein engineering - Investigating potential functionalities

Proteins can undergo chemical modification in vitro through covalent cross-linking with a target molecule. Numerous processes occur post-translationally within the tightly regulated cellular milieu, including the phosphorylation of cAMP-dependent protein kinases, which contributes to the augmentation of glycogen breakdown (Soderling, 1975). This paragraph will concentrate exclusively on the chemical engineering designed to connect certain chemicals to proteins that serve as functional elements in meals. These compounds alter the protein's function and are mostly theorized to impart features to the protein that are minimally present in the unmodified form, such as enhanced foaming properties, aggregation suppression, or improved surface activity. The rationale for the chemical alteration of proteins is multifaceted but can be classified into four primary categories.

i. Waste control: For example the re-use of fish gelatin from waste requires less natural resources for their production. Other examples include the production of a peptide with anti-oxidant activity from algae (Sheih et al., 2009). Protein rich by-products are also recovered upon electrocoagulation of wastewater resulting from egg processing (Xu et al., 2002). Chemical modification can be used to increase the functional properties of these proteins.

ii. Health considerations: An example of this is the replacement of meat or soy proteins by (other) vegetable proteins (O'Kane et al., 2004; Pedroche et al., 2004; Vioque et al., 1999; reviewed in Moure et al., 2006). Similar to proteins obtained as by-product from waste material, unmodified plant proteins often have limited functional applicability. Hence, chemical modification of plant proteins can be used to improve molecular functionality.

iii. Cost effectivity by extending molecular functionality of a protein. Chemical modification can be employed to enhance the functional properties of a protein, such that less material is required to obtain a product with similar structural characteristics.

iv. Structure-function relationships: Chemical modification is often used to investigate the contribution of specific molecular parameters, such as surface hydrophobicity, to functionality of the protein at ingredient level. Ample examples of protein functional extension are discussed in the rest of this paragraph.

The modifications detailed in the literature are numerous and encompass phosphorylation (the addition of a phosphate group to serine, tyrosine, or threonine), methylation (the addition of a methyl group to arginine or the protein's N-terminus), glycosylation (the addition of carbohydrates to lysine or the N-terminus), acetylation (the addition of an acetyl group to an amino group such as lysine or the N-terminus), among others. A multitude of publications and book chapters have examined the diverse modifications and their effects on the molecular behavior of proteins (Feeney et al., 1982; Means & Feeney, 1971; Tawfik, 2002). This paragraph intends to briefly examine the contemporary advancements of distinct reactive groups in proteins that can be addressed by chemical engineering. The chemical foundation of these modifications will be examined, and literature applications will be read solely to demonstrate the extensive range of uses of the chemical modification process.

3.1 Glycosylation and deglycosylation

The methods of food storage and preparation, including heating through pasteurization or sterilization, frequently create conditions that trigger spontaneous and unplanned reactions between lowering carbohydrates in the food matrix and proteins. The Maillard reaction, a complicated series of reactions, is responsible for the creation of browning products and significantly influences the flavor and color of food items. The Maillard reaction commences with a condensation reaction between the ε-amino group of lysine and the reducing group of a sugar, resulting in the formation of Amadori or Heyn's rearrangement products through N-substituted glycosylamine. In the later phases of this reaction, the Amadori and Heyn's rearrangement products undergo degradation through several pathways (Mossine et al., 1994; Röper et al., 1983). The last phases of the Maillard reaction encompass significant protein cross-linking and the synthesis of melanoidins (Pellegrino et al., 1999). The diverse array of intermediate chemical structures generated allows for the utilization of many intermediates as indicators of the Maillard reaction, facilitating the assessment of the reaction's extent in food products for quality control purposes. The production of ε-N-(furoylmethyl)-L-lysine (furosine), an intermediate in the Maillard reaction, was demonstrated to result from lactosylation during the storage of milk powder (Le et al., 2011). Extended storage of high-protein nutrition bars exhibited nonenzymatic Maillard browning due to the interaction between whey protein isolate and high-fructose corn syrup or sorbitol syrup (McMahon et al., 2009). The bread baking procedure influences color development as indicated by the quantities of furosine and hydroxymethylfurfural (Ramírez-Jiménez et al., 2000). The alteration of baking temperature and dough formulation influences the degree of furosine development and degradation, the release of acid-labile lysine, and the synthesis of carboxymethyllysine in cookies (Charissou et al., 2007). The Maillard reaction has been deliberately utilized for glycosylation to investigate the implications of covalent sugar attachment to proteins regarding factors such as protein stability and aggregation (Feeney et al., 1975). The glycosylation of proteins via the Maillard reaction has been noted to both promote and safeguard against aggregation. The incubation of hazelnut proteins with glucose led to the production of high molecular weight protein aggregates, as identified by SDS-PAGE (Cucu et al., 2011). Simultaneously, other studies have demonstrated that Maillard reaction with glucose impeded the aggregation of cod fish parvalbumin (de Jongh et al., 2011). The Maillard reaction, being a multi-step process, raises the possibility that the different intermediates exhibit varying degrees of resistance to aggregation, with some potentially offering protection while others may promote aggregation. Enhanced thermal stability following glycosylation was seen in numerous proteins, including the apple allergen Mal d3 during interaction with glucose (Sancho et al., 2005), and β-lactoglobulin with reaction with glucose and fructose (Broersen et al., 2005). Conversely, deglycosylation was demonstrated to cause denaturation and aggregation of ovalbumin (de Groot et al., 2007). In addition to the Maillard reaction, protein glycosylation can occur through other alternative pathways. N-glycosylation occurs by the alteration of the side chains of asparagine or arginine (Kornfeld & Kornfeld, 1985). O-glycosylation occurs by the alteration of serine, threonine, or tyrosine (Hart, 1992). Numerous glycan structures have been recognized as integral to these two forms of modification and are frequently essential to the biological activity of the protein (Rudd et al., 2001). The next two forms of glycosylation primarily occur due to post-translational protein processing in vivo and are not employed to induce protein glycosylation in the food sector. Therefore, N- and O-glycosylation will not be addressed in this paragraph.

3.2 Lipophilization

Covalent linkage of lipids to proteins results in increased hydrophobic exposure of a protein with interesting applications related to altered surface properties. Increased exposed hydrophobicity of proteins has for example been related to an improved capacity to form and stabilize emulsions and foams which is the result of improved potential to interact with hydrophobic surfaces, both the air-water and oil-water interface, and including (model)membranes (Nakai, 1983; Wierenga et al., 2003; reviewed in Wilde, 2000; Wilde et al., 2004). Various saturated and unsaturated fatty acids have been employed to induce lipophilization of proteins including caproic acid (Liu et al., 2000), capric acid (Aewsiri et al., 2010; Kosters et al., 2003; Liu et al., 2000), lauric acid (Aewsiri et al., 2010), myristic acid (Aewsiri et al., 2010; Ibrahim et al., 1993; Liu et al., 2000), palmitic acid (Haque et al., 1982; Haque & Kito, 1983a, 1983b; Ibrahim et al., 1991), stearic acid (Djagny et al., 2001; Ibrahim et al., 1993), and oxidized forms of linoleic acid (Aewsiri et al., 2011a, 2011b), and the efficiency of the lipophilization reaction was found to be inversely proportional to the length of the lipid chains used (Liu et al., 2000). Reaction of 28% of the available free amino groups of ovalbumin with activated capric acid was shown to result in retained secondary structure while inducing oligomerization and destabilization of the protein structure as a result of lowering the enthalpy for unfolding (Kosters et al., 2003). The presence of acyl chains was thought to cause significant dehydration of the protein. In another study, hen egg white lysozyme was lipophilized with short and middle chain saturated fatty acids including caproic (C6:0), capric (C10:0), and myristic (C14:0) acid (Liu et al., 2000). Lipophilization of lysozyme was reported to decrease the thermal stability of lysozyme as a result of partial loss of α-helical content of the protein, and this molecular destabilization appeared to be proportionally related to the chain length and the number of bound fatty acids. The lysine residues involved in the modification were thought to be located in the helical region and to subsequently induce partial unfolding of the α-helical region surrounding these residues (Liu et al., 2000). Lysozyme has also been chemically modified using palmitic acid (Ibrahim et al., 1991, 1993) with the primary aim to study the effect of lipophilization on the antimicrobial effect of the protein. Even though increasing extents of covalent linkage with palmitoyl residues lead to insoluble protein, as spectrophotometrically determined by solution turbidity at 500 nm , foaming stability and emulsifying activity were progressively improved by linkage of palmitic acid to the protein molecule. More groups showed that the foaming or emulsifying activities of a wide range of proteins could be improved upon lipophilization, including soybean glycinin (Haque et al., 1982), α_{s1}-casein (Haque & Kito, 1983b), and cuttlefish skin gelatin (Awesiri et al., 2011a, 2011b). A further effect resulting from the incorporation of myristic and stearic acids into lysozyme was related to antimicrobial activity and stearic and palmitic acid conjugation resulted in more effective antimicrobial agents against E. coli, than the attachment of myristic acid or the unmodified protein (Ibrahim, 1993). Myristoylation was found to induce lysozyme aggregation resulting in concurrent loss of antimicrobial function. The effects of palmitoylation on the structural and functional properties of α_{s1}-casein have also been explored (Haque & Kito, 1983a, 1983b). The conjugation of the ε-amino groups of α_{s1}-casein with palmitic acid lead to micelle formation as a result of increased hydrophobicity while negative net charge was increased (Haque & Kito 1983a). Further work by this group showed that palmitic acid linkage to α_{s1}-casein did not lead to large scale structural rearrangement of the molecule, both at a secondary and a tertiary structure level using circular dichroism. Interestingly, Aewsiri and colleagues have also investigated the antioxidative activity of cuttlefish skin gelatin modified with a combination of oxidized linoleic acid and oxidized tannic acid, a potent antioxidant (Aewsiri et al., 2011a). Oxidation of lipids and proteins in foams primarily takes place at the air-water interface and the addition of a hydrophilic antioxidant alone reduces surface activity (Aewsiri et al., 2011a). The co-conjugation of tannic acid and linoleic acid with gelatin enhances the migration of the protein to the air-water or oil-water interface, hence increasing foaming and emulsifying activities, while preserving antioxidant properties (Aewsiri et al., 2011a).

3.3 Chemical-reactive groups

Sulfhydryl groups play an important role in regulating the self-assembly of proteins as well as their stability driven by disulfide interchange reactions (Sawyer, 1968). Hence, the presence of these groups has substantial impact on the aggregation and gelation behavior of a wide variety of proteins which has been confirmed by many researchers (Arntfield et al., 1991; Broersen et al., 2006; Graña-Montes et al., 2011; Hayakawa & Nakai, 1985; Hoffmann & van Mil, 1997; Margoshes, 1990; Mine, 1992; Sawyer, 1968; Shimada & Cheftel, 1989). A variety of modifications can be performed targeting sulfhydryl groups, which are part of the cysteine residues. Sulfhydryl groups are highly reactive against various reactants and are thus suitable targets for modification. Sulfhydryls can be blocked to prevent cross-linking by S-methyl methanethiosulfonate (MMTS), *N*-ethylmaleimide (NEM) (Kitabatake et al., 2001), or iodoacetamide (Anson, 1940; Huggins & Jensen, 1949; Smythe, 1936), or additional sulfhydryl-groups can be attached to primary amines (SATA) of proteins. Further, Nhydroxysuccinimide esters can react irreversibly with primary amines releasing N hydroxysuccinimide.

Disulfide bonds are thought to play a crucial role determining the stability of proteins (Betz, 1993; Zavodszky et al., 2001; reviewed by Creighton, 1988) as well as to impact on the aggregation process and gel formation of various proteins including ovalbumin (Broersen et al., 2006; Kato et al., 1983), vicilin (Arntfield et al., 1991), and *β*-lactoglobulin (Sawyer, 1968). Aggregates and gel networks are often the result of combined action of hydrophobic and electrostatic interactions and covalent interactions, in the form of disulfide bonds, are sometimes present (Kato et al., 1983; Koseki et al., 1989; Sun & Hayakawa, 2002). Thiolation of ovalbumin mediated through the reaction of S-acetylmercaptosuccinic anhydride (SAMSA) with primary amines results in the formation of acetylthio groups and the acetyl group can be cleaved off to yield reactive sulfhydryl groups by the addition of hydroxylamine (Klotz & Heiney, 1962). A range of modification degrees can be obtained by varying the S-AMSA:lysine ratio (Broersen et al., 2006). Next to the linkage of a sulfhydryl group, additional carboxyl groups are conjugated through this reaction introducing additional charge variation which can lead to an additional parameter which can induce variation in aggregation, gelation or stability of a protein. To circumvent this variation, proteins with activated sulfhydryl groups are best compared with similarly modified proteins with blocked (i.e. not reacted with hydroxylamine) acetylthio groups, rather than directly with the unmodified protein (Broersen et al., 2006). Thiolation of ovalbumin in this way lead to limited changes at a secondary and tertiary structure level at high degrees of modification suggesting that the original molecular fold was largely retained upon modification. High degrees of thiolation resulted in a decrease of thermal stability of ovalbumin while fibril morphology was affected. Interestingly, the rate of aggregate formation was not modified by the presence of additional sulfhydryl groups available for disulfide formation. It was concluded that disulfide formation does not represent the prime driving force for aggregation of ovalbumin which was further illustrated by the finding that at room temperature, where significant sulfhydryl groups are already exposed upon modification, no disulfide bond aggregation was observed. Final aggregate morphology, gel formation and stability are affected as a result of rapid covalent network formation which does not allow rearrangement into more stable networks, as illustrated by lower gel Young's moduli obtained upon thiolation (Broersen et al., 2006).

3.4 Modification of charge through methylation and succinylation

Net charge and local charge densities of proteins have been implicated in the regulation of protein stability, aggregation, and aggregate morphology affecting the visual appearance of food products. These hypotheses have been substantiated by a range of observations which involved charge introduction, removal or reversal through succinylation and methylation reactions (Broersen et al., 2007a; Weijers et al., 2008). The reactions of succinylation and acetylation both lead to blockage of the reactive amino groups of proteins with an acyl residue and are hence collectively termed acylation reactions. The rate of acylation reaction depends on the rate of nucleophilic attack. Succinylation leads to increased net negative charge by the covalent linkage of succinate anions to the cationic amino groups of a protein thereby converting a cationic group into an anionic residue having implications for the distribution of net charge of a protein. Upon acetylation, ammonium cations are replaced by neutral acetyl groups resulting in electrostatically neutral groups. Large extents of succinylation have been reported to affect the integrity of secondary and tertiary structure of soy protein hydrolysate as shown by intrinsic tryptophan fluorescence and circular dichroism (Achouri & Zhang, 2001). Similar conformational rearrangements have been reported upon succinylation of whey protein isolate (Gruener & Ismond, 1997), bovine serum albumin (Jonas & Weber, 1970), canola protein (Lakkis & Villota, 1992), Faba bean legumin (Schwenke et al., 1998), rapeseed 12S globulin (Gueguen et al., 1990), and winged bean protein (Narayana & Rao, 1991). As a result of co-incubation of soy protein hydrolysate with succinic anhydride, which is a common compound used to succinylate proteins, heterogeneous reaction mixtures were obtained. Next to the aimed amine groups, this method of succinylation also commonly results in *O*-succinylation, involving threonine or serine hydroxyl groups or tyrosine succinylation (Achouri & Zhang, 2001; Chang & Sun, 1978; Schwenke et al., 1998). This latter reaction was found to be reversible upon treatment with hydroxylamine (Habeeb & Atassi, 1969), but, when present, to induce substantial expansion of Faba bean legumin as observed by viscometric studies (Schwenke et al., 1998). It was postulated that the high accumulation of negative charge upon extensive succinylation leads to dissociation and expansion of the individual subunits legumin is composed of (Schwenke et al., 1998). Other functional properties are equally affected by succinylation. For example, protein solubility has been reported to increase upon succinylation as has been demonstrated for rapeseed preparations (Dua et al., 1996), flax protein isolate (Wanasundara & Shahidi, 1997), oat protein isolate (Mirmoghatadaie et al., 2009), and soy protein isolate (Franzen & Kinsella, 1976a). Improved solubility has been related to the ability of proteins to perform more efficiently as stabilizers in emulsions and foams (Nakai & Li-Chan, 1988; Waniska & Kinsella, 1979), which, in turn, is greatly affected by their ability to absorb at the air-water interface (Wierenga et al., 2005). It has indeed been shown that treatment of a variety of proteins with succinic anhydride leads to increased foam capacity (Dua et al., 1996; Franzen & Kinsella, 1976a; Mirmoghatadaie et al., 2009), although others suggest that succinylation leads to decreased foam expansion capacity (Wanasundara & Shahidi, 1997). These seemingly contradictive findings may be explained by the demonstration by Wierenga and colleagues (2005) that the likelihood of a protein molecule to adsorb at an interface is the result of a balance between hydrophobic and steric effects: highly charged molecules may be adsorbing to the interface as a result of hydrophobic interaction, but the density at which protein molecules continue adsorbing to the interface is mainly determined by the repulsive nature of the charged proteins. Other effects observed upon charge modification of proteins are related to emulsification properties: methylation increased while succinylation decreased the emulsifying activity of rapeseed preparations (Dua et al., 1996). Other studies show an increased emulsifying activity and stability upon succinylation of soy protein (Franzen & Kinsella, 1976a), and oat protein isolate (Mirmoghatadaie et al., 2009). In terms of gelation, an increase in net charge lead to more transparent gels upon gelation of ovalbumin which is related to the morphology of the aggregated network making up the gel structure (Weijers et al., 2008). Overall, many and detailed efforts have been made employing net charge modification of proteins in the field of food science. These studies have lead to in-depth knowledge of the role of electrostatics to common protein functionalities such as emulsification, foaming, and aggregation propensity.

3.5 PEGylation

The covalent attachment of a polyethylene glycol (PEG) polymer chain to a protein, known as 'PEGylation,' is predominantly utilized in pharmaceutics, since the conjugation of non-toxic PEG confers significant benefits for drug administration. (reviewed in Damodaran & Fee, 2010; DeSantis & Jones, 1999; Francis et al., 1998). The protein targets for PEG modification are regarded as non-specific and include the ε-amino groups of lysine and other nucleophilic groups such as glutamic acid, aspartic acid, threonine, serine or tyrosine on the surface of the protein resulting in highly heterogeneous protein-PEG conjugates upon modification (Losso & Nakai, 2002). Commercially available PEG is available as mixtures of different oligomer sizes in various molecular weight ranges enabling the variation of exposed hydrophobicity of proteins. Conjugation of hydrophilic PEG to a hydrophobic protein generally results in an increase in hydrodynamic size and water solubility (Damodaran & Fee, 2010). From a pharmaceutical viewpoint, PEGylation has been reported to enhance circulation life of bovine liver catalase in the blood of mice while the presence of PEG does not induce an immune response upon injection (Abuchowski et al., 1977). Modification of peroxidase from turnip was shown to enhance catalytic activity of the enzyme with increased stability in organic solvents as well as increased temperature resistance (Quintanilla-Guerrero et al., 2008). Similar results were found upon PEGylation of trichosanthin, which showed prolonged plasma half-life and reduced immunogenicity (He et al., 1999). PEGylation of lysozyme similarly lead to stabilization of the protein against pH and temperature variation as well as resistance against proteolysis (Silva Freitas & AbrahãoNeto, 2010). Most of these studies have been carried out in the context of pharmaceutical application. The study of PEGylation as a potential route to bring about modification of physicochemical parameters of proteins applied in food products has been less well explored. The only known effort in the field of food science combined oxidative sulfitolysis with conjugation of 5000 dalton activated PEG to investigate the impact on the emulsifying properties of β-lactoglobulin A (Losso & Nakai, 2002). PEG molecules were found to cover the entire surface of unmodified and sulfitolyzed β-lactoglobulin and PEGylation alone did not improve emulsifying activity or emulsion stability. The combination of sulfitolysis and PEGylation however lead to more stable emulsions and improved emulsion activity index as a result of better stabilization of individual droplets against coalescence by the absorption of PEG onto the surface of the sulfitolyzed β-lactoglobulin (Losso & Nakai, 2002). The small number of available studies precludes direct conclusions regarding the applicability of PEGylation to enhance the functionality of proteins in food products.

3.6 Deamidation

Deamidation entails the breakdown of the amino acids glutamine and asparagine into glutamic and aspartic acid, accomplished through acid, alkaline, or enzymatic treatment. (Liao et al., 2009; Shih, 1990; reviewed in Wright & Urry, 1991). Consequently, deamidated protein is often obtained as a by-product of food processing. For example, the extrusion of wheat flour induces deamidation of wheat proteins (Izzo et al., 1993). Deamidation has been shown to affect protein functionality. Functional properties, such as solubility, emulsifying and foaming properties, of gluten have been reported to improve upon low levels of deamidation brought about by mild acid hydrolysis (Hamada & Marshall, 1989; Matsudomi et al., 1982, 1985). Deamidation also was reported to increase exposed hydrophobicity of gluten induced by a conformational change and, subsequently, to increase surface activity (Matsudomi et al., 1982). A number of studies investigated the molecular mechanism for protein structural destabilization upon deamidation. For example, the deamidation treatment has direct implications for charge density and, in turn, affects electrostatic interactions the protein may undergo by interacting with water or upon self-assembly (Finley, 1975; reviewed in Riha et al., 1996). This role of electrostatics to deamidationinduced disruption of protein structure was supported by further observations on wheat gluten: both acetic acid and HCl induced deamidation had substantial consequences for the secondary structure of wheat gluten. It was thus postulated that strong deamidation induced protein unfolding as a result of electrostatic repulsion (Liao et al., 2010). Acetic acid induced deamidation of wheat gluten was further found to inhibit SDS-stable aggregate formation whilst largely retaining its ability to form disulfide bonds (Liao et al., 2010). The rate of the deamidation reaction has been found to depend on primary sequence and pH under which the reaction takes place, but was independent of ionic strength for model peptides (Patel & Borchardt, 1990a, 1990b; Robinson & Rudd, 1974; Tyler-Cross & Schirch, 1991), soy protein and egg white lysozyme (Zhang et al., 1993).

3.7 Cross-linking

A range of aldehydes, including as glutaraldehyde, formaldehyde, and α-hydroxyadipaldehyde, have been employed to facilitate the chemical cross-linking of proteins. Among these chemicals, glutaraldehyde had the highest efficiency in cross-linking bovine serum albumin, resulting in the formation of extensive insoluble networks. (Hopwood, 1969). Also for other proteins gluteraldehyde has been reported as the most efficient cross-linking agent yielding thermally and chemically stable cross-links (Bowes & Cater, 1968; Nimni et al., 1987). Primary amino groups have been reported to act as prime target to initiate the aldehydeinduced cross-linking reaction (Quiocho & Richards, 1966), next to aromatic amino acids (Hopwood et al., 1970). Depending on environmental conditions, gluteraldehyde can bring about cross-linking through a wide variety of reaction mechanisms (reviewed in Migneault et al., 2004). This is caused by the large number of different molecular structures gluteraldehyde can assume in solution (Hardy et al., 1969; Korn et al., 1972; Richard & Knowles, 1968; reviewed in Migneault et al., 2004) although the mechanistic details for this is unknown. Due to its documented toxicity, gluteraldehyde-induced cross-linking has not been utilized in food technology, except as a means to facilitate the study of intermediates in the aggregation route or to immobilize proteins on a surface for further examination.

3.8 Assessing the extent of alteration

Researchers predominantly assess the efficacy of the chemical engineering process on proteins to evaluate the impact of a particular alteration on the functional features under investigation. Consequently, specialized chemical and biophysical assays have been established and are now extensively utilized. These quantitative tests primarily rely on the generation of a chromogenic or fluorogenic product through particular interactions with reactive protein groups. Certain assays will yield data on the mean extent of alteration throughout the total population of protein molecules in a solution.Examples of such assays are Ellman's reagent (Ellman, 1959) or the sulfhydryl-disulfide exchange (SEI) index (Owusu Apenten et al., 2003). Others are also useful to obtain information on the distribution of the degree of modification obtained, such as mass spectrometry. The tests created can be classified based on the specific conjugated chemical group of the protein they target, including amine groups, thiol groups, or carboxyl groups. Other studies employ technologies that specifically investigate the attached molecule, such as gas-liquid chromatography (GLC) to ascertain the extent of lipid incorporation. (Haque et al., 1982).

3.8.1 Amine groups

Amine groups in proteins derive from either free amino groups or N-terminal residues of the proteins. (Skraup & Kaas, 1906; Chibnall, 1942). From 1906 it was recognized that lysines in proteins were largely responsible for the free amino groups present in proteins (Skraup & Kaas, 1906). This finding triggered the development of various assays to study accessible ε-amino groups of lysines in proteins (Gurin & Clarke, 1934; Sanger, 1945). Three of the most commonly used assays in the field of protein engineering to determine the number of available amino groups of a chemically modified protein involve the use of chemicals 2,4,6-trinitrobenzenesulphonic acid (TNBS) (Fields, 1971), ninhydrin (Yemm & Cocking, 1955; Schilling et al.; 1963; Samejima et al., 1971), and ortho-phtaldialdehyde (OPA) (Roth, 1971). The latter assay is based on the reaction of the OPA compound with free amino groups in proteins in the presence of β-mercaptoethanol under alkaline conditions. This reaction results in the formation of highly fluorescent alkyl-iso-indole derivatives which emit at a wavelength of 455 nm upon excitation at 340 nm . High concentrations (i.e. 10%) of sodium dodecyl sulfate (SDS) are often added to the protein solution to aid the exposure of all amino groups which are sometimes buried in the folded protein. The extinction coefficient for the formed adducts of both α - and ε-amino groups are similar with an absorptivity of $6000 M^{-1}\ cm^{-1}$. The OPA assay was further developed to evaluate the degree of proteolysis of dairy proteins by determining the number of α-

amino groups released upon hydrolysis (Church et al., 1983). The assay can also be used to quantify all types of reactions involving the modification of lysine and, hence, includes lipidation of ovalbumin (Kosters et al., 2003; Wierenga et al., 2003), lysozyme (Liu et al., 2000), and gelatin (Aewsiri et al., 2010, 2011a, 2011b). Other modifications which can be quantified using the OPA assay are succinylation (Kosters et al., 2003; Wierenga et al., 2005), methylation (Kosters et al., 2003), glycosylation (Broersen et al., 2004; Kosters et al., 2003), and thiolation of proteins (Broersen et al., 2006).

The second frequently used method to quantify available amino groups upon modification involving the chemical TNBS was developed by Okuyama and Satake (1960) and Satake and colleagues (1960). This method was first employed to study free amino acid groups in trypsin and chymotrypsin inhibitors (Haynes et al., 1967) and for routine screening of protein concentrates for animal feeds (Hall et al., 1973). The chemical TNBS reacts with high preference to free amino groups resulting in the formation of trinitrophenyl derivatives. The reaction product can be quantified spectrophotometrically at 335 nm . A disadvantage of this method is that TNBS also reacts with free sulfhydryl groups, albeit at a slower rate than with amino groups and to form a labile product (Kotaki et al., 1964). The TNBS assay has been used for various types of protein modification including fatty acid incorporation (Andersson et al., 1971; Ibrahim et al., 1991, 1993), glycosylation of proteins as a result of the Maillard reaction (Sun et al., 2004), succinylation (El-Adawy, 2000; Schwenke et al., 1998; Zhao et al., 2004a, 2004b), and acetylation of faba bean legumin (Krause et al., 1996).

The compound 2,2-dihydroxyindane-1,3-dione (ninhydrin) reacts with ε-amino groups and ammonium ions into a blue-purple Schiff base product called Ruhemann's purple, that can be colorimetrically detected at a wavelength of 440 nm (Yemm & Cocking, 1955; Schilling et al., 1963; Samejima et al., 1971). The ninhydrin assay has been used to quantify degrees of succinylation of soy protein (Franzen & Kinsella, 1976a), and acylation of flax protein isolates (Wanasundara & Shahidi, 1997).

The comparison of TNBS, OPA, and ninhydrin for the determination of α-amino groups in pea protein isolates and their hydrolysates concluded that TNBS and OPA yielded equivalent findings, however ninhydrin identified just half of the amino groups found by TNBS and OPA. (Panasiuk et al., 1998).

3.8.2 Sulfhydryl groups

Two commonly used assays are available to evaluate the successful conjugation or blockage of sulfhydryl groups in proteins. The assay which was developed first by Ellman (1959) involves the reaction of 5,5'-dithiobis(2-nitrobenzoic acid) (DTNB) or Ellman's reagent with free sulfhydryl groups yielding colored 3-carboxylato-4-nitrothiophenolate

(CNT). Spectrophotometric absorbance intensity at 412 nm provides for a direct measure of the concentration of CNT in solution and cysteine is commonly used as calibration standard. Ellman's assay has been used before to determine the extent of *N*-ethylmaleimide modification of β-lactoglobulin A (Kitabatake et al., 2001; Wada & Kitabatake, 2001), thiolation of ovalbumin (Broersen et al., 2006), or acylation of soy protein sulfhydryl groups (Franzen & Kinsella, 1976a). However, the sizeable DTNB at 400 dalton may not be able to detect sulfhydryl groups which are buried inside the folded structure of intact proteins or aggregated proteins (reviewed by Visschers & de Jongh, 2005). A useful alternative for the Ellman's assay has been developed by Owusu Apenten and colleagues (2003) for β - lactoglobulin and bovine serum albumin. Chemical reactivity of thiol groups is an absolute requirement to enable covalent cross-linking through disulfide bond formation (Hillier et al., 1980). This assay, called the sulfhydryl-disulfide exchange (SEI) index, therefore provides a direct measure of the chemical reactivity of thiol groups as it determines the conversion of substrate in time and kinetically relates the conversion to that of fully exposed thiol groups (Owusu Apenten et al., 2003). Chemical activity of introduced sulfhydryl groups by modification of ovalbumin using the SEI index has been verified for example by Wierenga and colleagues (2006).

3.8.3 Carboxylic acid groups

The carboxylic acid content of proteins is primarily investigated using the compound 2-ethyl-5-phenylisoxazolium-3'-sulfonate or Woodward's reagent K (Woodward & Olofson, 1961; Woodward et al., 1961; Sinha & Brewer, 1985). The activity of Woodward's reagent K is the result of a multistep process. First, Woodward's reagent K is converted into ketoketenimine at neutral pH . The intermediate compound is then either further disintegrated to form ketoamide or interacts with carboxylic acid groups of a protein. The latter interaction results in the formation of an enol ester (Pétra, 1971) which absorbs at 340 nm with a molar extinction coefficient of $7000 M^{-1} cm^{-1}$ (Sinha & Brewer, 1985). At a later stage, Kosters and de Jongh (2003) revised the extinction coefficient of the product to 3150 M^{-1} cm^{-1} at 269 nm to improve specificity of the reaction and eliminate the substantial contribution of side reactions with other nucleophiles in proteins (Llamas et al., 1986), and histidine and cysteine (Bustos et al., 1996; Johnson & Dekker, 1996) to the absorbance at 269 nm . This revised version of the assay employing Woodward's reagent K to estimate the number of carboxylic acid groups of chemically modified proteins has been used by a number of researchers. Wierenga and colleagues (2005) used Woodward's reagent K to estimate the degree of modification of succinylated ovalbumin to study the relation between protein net charge and adsorption to air-water interfaces. Similar net charge variation induced by succinylation was used to investigate colloidal versus conformational stability of ovalbumin to aggregation (Broersen et al., 2007a). To investigate the stability of ovalbumin, the protein was modified by succinylation, methylation, glycosylation, and lipophilization and the degrees of modification were validated using Woodward's reagent K and the OPA assay (paragraph 3.8.1).

3.8.4 Conjugated groups

An alternative route to obtain information on the degree of protein modification is to selectively probe the conjugated group. This can be achieved for example by incorporating an isotopically labeled reagent or inclusion of a chromophore or fluorophore which can then be quantified by read-out of fluorescence intensity using a fluorimeter or simple absorbance measurements using a standard spectrophotometer. Raman spectroscopy was shown to provide direct insight into degrees of succinylation and acetylation of a range of proteins originating from soy, egg white or whey by distinct contributions of the conjugated groups at 1737 cm^{-1} and 1420 cm^{-1} (Zhao et al., 2004a, 2004b). The peak intensities at these wavelengths could be directly converted to obtain information on the degree of modification. Degrees of palmitoylation of soybean glycinin (Haque et al., 1982) and α_{s1} casein (Haque & Kito, 1983a) have been determined using gas liquid chromatography.

Chromatography has been further explored in the shape of cation exchange chromatography to validate the degree of methylation of ovalbumin which was found to provide comparable read-outs as the revised version of Woodward's reagent K method to determine carboxylic acid groups (Kosters & de Jongh, 2003). All these techniques provide insight in the ensemble average degree of modification. Chemical engineering inherently implies the rise of heterogeneous species of proteins. Mass spectrometry, often employed as Matrix-assisted laser desorption/ionization-time of flight (MALDI-tof) mass spectrometry, has proven a powerful method to specifically obtain insight into the distribution of the modification reactions. This method has been employed to derive information on modification distributions of *N*-ethylmaleimide modified β-lactoglobulin A (Wada & Kitabatake, 2001), and glycosylation of β-lactoglobulin (Broersen et al., 2004; van Teeffelen et al., 2005). These studies demonstrated that degrees of modification obtained upon chemical engineering of proteins are rather broad and show a Gauss distribution profile rather than a single well-defined modification degree (Broersen et al., 2004; van Teeffelen et al., 2005). Some attempts have been made to isolate modified protein fractions with more defined degrees of modification, for example by using ion exchange chromatography of succinylated ovalbumin (Wierenga et al., 2005).

4. Potential applications of dietary protein modification

It is evident that considerable effort has been devoted in recent decades to enhancing the understanding and regulation of protein behavior and the development of protein-based microstructures through chemical engineering methodologies. What is the extent of the influence these findings have had on the advancement of new food applications? Numerous established product categories have benefited from altered protein functionality, resulting in enhanced product attributes. In the early 1970s, Unilever developed new margarine lines that exhibited enhanced performance in spreadability, extended storage stability, and baking, attributed to the acetylation of milk proteins which improved fat emulsification (Evans & Irons, 1970). Modification of egg yolk proteins through N-succinylation enhanced the quality of mayonnaises and salad dressings (Evans & Irons 1970). An other instance of enhanced protein functionality is the utilization of succinylation to augment the solubility and dispersibility characteristics of soy proteins during the extraction and refining process (Melnychyn & Stapley, 1973). It is intriguing to assess through a patent literature review the frequency with which the diverse technology options outlined in section 3 for improved protein behavior management have resulted in distinctive market offerings.

4.1 Contemporary utilization of protein modifications in food and feed products

Figure 1 illustrates a landscape representation of patents (worldwide), filed in the last decade, in the area of food and feed where protein modification has played a crucial role in deriving a new type of product functionality. The height of the contour indicates the activity in that particular area. The distance between patents reflects their commonality. In total only 445 relevant patents (grouped in 157 families) can be found. In comparison, a search on any biobased-product (including products with protein-based technical polymers as in coatings, paints, paper, etc.) revealed more than 8600 patents. Clearly the role of protein engineering to derive new food product specifications are very limited, especially in view of the potential commercial impact. The observation that the patent-families found are spread rather constant over the plot illustrates that these patents are not directly linked to each other in terms of engineering approach or application area. When evaluating the patent filings of the last ten years five areas can be distinguished with a relatively high patenting activity and these will be discussed below in more detail. These are numbered 1 to 5 in the figure.

1. Refinery of seed storage proteins (12 patents). There is an increasing interest to use readily abundant and relatively cheap seed storage proteins as nutritional

component in food and feed. The major difficulty of this protein-source is to obtain functional proteins after the refinery steps. Loss of functionality occurs especially when the protein is used as powdered ingredient, because of difficulties in resolubilization and unpredictable caking of the powder during storage/transport. A number of patents have been filed that use mild engineering tools, like Maillardation, or enzymetreatments to preserve functional proteins during refinery steps (e.g. patent EP1370157B1: "Highly soluble, high molecular weight soy protein"). Interestingly, all patents filed in this area pay attention to the in principle reversibility of the modification applied.

2. Nano-particles (23 patents). To better direct properties of protein-based nano-particles enzymatic introduction of lipidic groups like small fatty acids or PEGylation has been employed to encapsulate bioactives, typically as microemulsions (e.g. patent US20070154907A1: "Microemulsions as precursors to solid nanoparticles"). These applications are considered food-grade or have passed medical-ethical approval in their testing/and or usage. There are no indications that these application have found their way into the food/feed product market.

3. Nutritional availability in feeds (13 patents). A considerable number of patents can be found in the application area to increase the nutritional availability of amino acids in feed (14 patents). Most patent-positions are dealing with destabilizing proteinstructures to promote their digestibility and their (proposed) nutritional uptake. Typical patented approaches are de-amidation and Maillardation (e.g. patent WO2004020977A3: "De-amidation to promote nutritional value of rumen in feed"). No patents can be found that link nutritional value and protein modification in foods.

4. Protein-based emulsion-stabilizers (two times 7 patents). A few patents are found where specific protein modification is used to improve emulsification of (oil in water) products. Especially enzymatic glycosylation and lipidation approaches have been used (e.g. patent US7126042B1: "Recombinant oleosins from cacao and their use as flavoring or emulsifying agents"). In view of the small scale examples provided in these patents, it is not likely that these inventions have been implemented in a commercial product.

5. Edible coatings (12 patents). A whole family of patents is present on enzymatic protein cross-linking (mainly by transglutaminase) in relation to the production of edible coatings. Modifications typically act on a microstructural level, and not so much on the protein molecular level, to strengthen the spatial network formed (e.g. patent EP963704B1: "Food containing proteinaceous material treated with a

transglutaminase and an oxidoreductase"). As these interventions in a product do not necessarily need to be labeled on the product, it is difficult to evaluate whether they have resulted in product development.

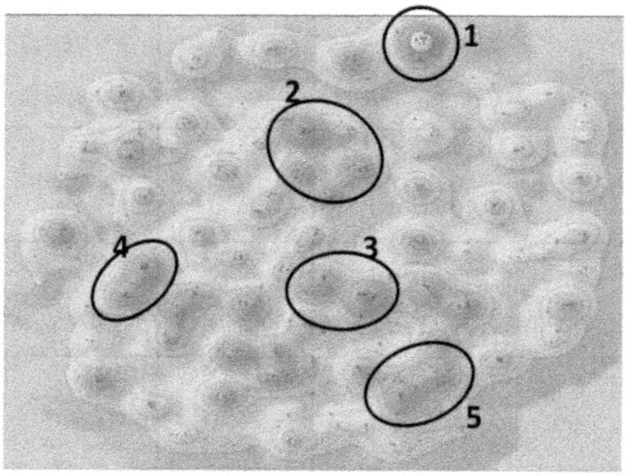

Fig. 1. Landscape representation of patent (families) filed between 2000 and 2011 where engineering/modification of protein functionality has been used to derive new material/product characteristics. The small black dots indicate the position of a patentfamily. The numbered circles represent the areas with the highest activity in patents. The numbers are explained in the body text.

4.2 Food legislation

While paragraph 3 illustrated the active and advancing comprehension of protein activity in complex systems during recent decades, it is notable how limited the contributions to novel applications from these insights appear to be. To see this more thoroughly, one must consider that technology advancements coincided with an escalation in legislative measures. This is represented in the United States by the Food and Drug Administration (FDA). The European Food Safety Authority (EFSA) serves as its European equivalent. The FDA is tasked with safeguarding and advancing public health by regulating and overseeing food safety. It accomplishes this by establishing regulations that delineate the parameters for the incorporation of novel ingredients, procedures, or formulations pertaining to food. Several legislations have significantly influenced the food industry. The Nutrition Labeling and Education Act was enacted in 1990. This mandated the labeling of food goods regarding their composition, facilitating the traceability of ingredients to their origin. This legislation also modified the requirement that all nutrient content claims (e.g., 'high fiber', 'low fat', etc.) must comply with the standards established by the FDA. Ultimately, this necessitated that each modified protein obtain a new label and receive prior recognition and approval from the FDA.

Exceptions included those altered elements deemed to arise from a 'natural process' or those that were 'reversibly transformed' (thus transitory). The FDA Modernization Act of 1997 aimed to expedite the clearance process for new pharmaceutical medications while also influencing food technology by recognizing the advancements in technological, commerce, and public health concerns. A 'new' food item must be assessed in the intricate context of its synthesis and digestion within the food system. The Amendments Act of 2007 further modified this legislation, resulting in more precisely defined standards for what constitutes safety in food products. In 2010, the Food Safety Modernization Act was enacted. This statute mandates food producers to facilitate the tracking and tracing of all utilized materials. The utilization of altered proteins, whether through genetic modification or chemical alteration, need distinct certification for market clearance. In summary, the FDA and EFSA recognize that future food production necessitates advances both at the ingredient level and through processing methods, and they are generally receptive to protein engineering approaches. Simultaneously, it mandates that functionally enhanced ingredients are scrutinized across the entire supply chain, from refinery to nutritional efficacy and human health, considering the product's complexity.

4.3 Opportunities in the domain of food product sustainability

The escalating global population and improved welfare are significantly boosting the demand for protein as a nutritious component in diet. The recognition of proteins as fundamental components in non-food applications, aimed at fostering a more sustainable economy, has intensified the demand for advances in the production, refinement, and utilization of proteins from a broader range of sources than those now utilized in the economy. The formulation of a vision for food quality, particularly the significance of nutritional impact, must be contextualized within the fundamental requirements established by both consumer expectations and the involved sectors. This can be illustrated by the Pyramid of Food Innovation, as depicted in figure 2. At the forefront are the foods that must be developed in an exceedingly sustainable manner; achieving this goal necessitates adherence to reduced levels of restrictions, limitations, and associated scientific obstacles. Food safety forms the most fundamental aspect here. Terminology like for example 'nanotechnology' cannot count on consumer acceptance and also the inclusion of genetically modified ingredients gives rise within Europe to hesitance in applicability by food producing companies. Especially for traceability and chemical characterization the demands become exceedingly higher. This is the level though where new protein engineering routes could be contributing most. A major bottleneck today in reformulation-strategies is the occurrence of (sensory) differentiations relative to the original product. Exceptions are reduced sugar or salt products, but from a marketing-technical perspective moderations of structuring components are preferably performed within the frame as novel food. A continuously on-going drive to elucidate structure-texture-product acceptance relations is essential, just like innovations in the area of optimized processing tools to deliver products that are acceptable with retained food safety. Typically the efforts on applying protein modifications are focusing in deriving food structures more efficiently or to provide new functional building blocks to create new food structures. In an economically global society there is a strong pressure on both ingredient-prices and commercial acceptable processing. Industrial entrepreneurship is essential and the availability of second-line ingredients for products hampers the implementation of new food production strategies. Especially within the EU there is an increasing attention to social inequality issues.

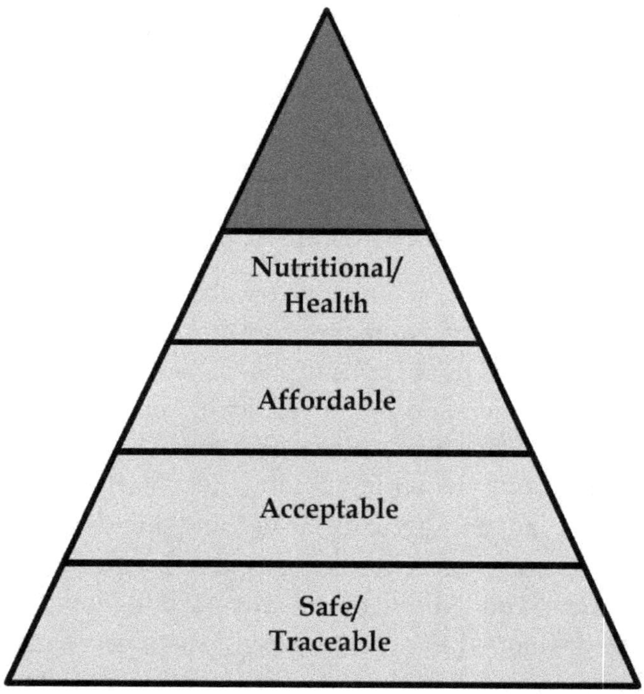

Fig. 2. The Pyramid of Food Innovation

From a governmental regulatory point of view there is increasing support for open innovations and shared responsibilities to produce products of good quality that are affordable to all social classes. It is unclear what the role of optimized protein behavior by engineering approaches can be. Consumers are susceptible to additional health-aspects when safe and acceptable good quality products are available for reasonable prices. In this information technology era consumers are capable to evaluate the added value of products far better, setting higher demands for health marks of food products. The use of added value to products, like bioactives, has not led to major winners in the food sector yet. The aspiration and necessity to innovate in food product creation may be fueled by the extensive information regarding the relationship between product functioning and microstructural morphologies, which are, in turn, determined by the molecular properties of particular proteins. Conversely, there exist strong methodologies in protein engineering that enable the directed modulation of protein behavior and their tendencies to (self)assemble into spatial networks. Nonetheless, it is essential to acknowledge that these two facets have yet to converge to enhance the sustainability of protein synthesis and utilization in f Considerations regarding the chemical modification of proteins Food.

4.4 Considerations of chemical modification of proteins

Although genetic modification of proteins often yields a homogeneous output, chemical engineering methods for protein modification are recognized for producing heterogeneity. An exemplary case is the utilization of the Maillard reaction using β-lactoglobulin (Broersen et al., 2004) or fish parvalbumin (de Jongh et al., 2011). Typically, it can be observed that 6 out of the 12 available lysine residues are glycosylated, with mass analysis revealing a substantial population of protein molecules exhibiting 5 or 7 sugar moieties. Additionally, some molecules may possess 4 or 8 groups, and traces of molecules with 3 and 9 groups are also present. This typically arises from differences in reactivity among the various amino groups within a particular protein structure; that is, certain amino acids are more exposed and hence more susceptible to fast change. Buried amino acids do not constitute a straightforward target for alteration, as most compounds cannot penetrate the folded or aggregated protein structure. Another factor that likely influences the degree of variability is the positioning of target amino acids within the main sequence. The alteration of an amino acid adjacent to a negatively charged amino acid by succinic anhydride will be obstructed due to steric hindrance. Efforts have been undertaken to fractionate protein molecules with varying levels of modification. Five ovalbumin preparations with varying degrees of modification have been purified by ion exchange chromatography (Wierenga et al., 2005). The diverse characteristics of chemically modified proteins significantly affect their approval as novel additives under existing food legislation. Chemical heterogeneity necessitates that each transformed species be individually assessed for toxicity to be deemed safe for use as a novel food additive.

4.5 Health risks

This paragraph will address the potential health implications associated with chemically modified proteins in food products, including allergic reactions, the presence of antinutritional factors related to proteins, and the formation of potentially toxic compounds as a byproduct of the modification process. Concerns exist regarding the health dangers associated with products having modified protein constituents due to protein modification. Bernstein and colleagues (2003) outlined three potential mechanisms through which novel food additives may cause adverse health effects. These encompass toxicity, nutritional impairment, and food allergies.

4.5.1 Toxicity

One of the known toxic compounds to be formed upon chemical engineering of proteins is acrylamide, which results from the Maillard reaction between reducing sugars and asparagine or methionine (Mottram et al., 2002; Stadler et al., 2002). The high solubility of this compound induces rapid absorption and metabolism in the body (reviewed in Dearfield et al., 1988). Following absorption, acrylamide can bind to DNA aiding a genotoxic and carcinogenic response which has been demonstrated in animals (Rudén, 2004). The occurrence of carcinogenic acrylamide in foods has been related to frying and cooking of food products (Rosen & Hellenäs, 2002; Tareke et al., 2002). Two communications published in Nature in 2002 demonstrated the requirement of the reaction between asparagine or methionine and a reducing sugar to intermediate formation of the dicarbonyl reactant followed by Strecker degradation (Mottram et al., 2002; Stadler et al., 2002). Particularly plant proteins are rich in asparagine suggesting that glycosylation of proteins from plant-origin using the Maillard reaction should be carefully considered in terms of the known toxic effects of acrylamide. Other toxic side effects of chemical modification of proteins have not been reported.

4.5.2 Impaired nutrition

Chemical engineering can theoretically have far-reaching consequences for the nutritional value of proteins. In a limited number of cases protein engineering has been intentionally employed to modify the nutritional quality, for example to increase the nutritional value of plant proteins (Liao et al., 2010). However, the main objective of most studies employing protein engineering is to investigate the consequences of the modification procedure for a range of functional properties. Only a small number of studies investigates nutritional aspects of chemical protein engineering which is mostly regarded as a convenient or inconvenient side effect of the modification procedure. Nutritional aspects covered in literature are exclusively based on in vitro studies and cover protein digestibility, availability of essential amino acids or the presence of anti-nutritional factors.

4.5.2.1 Anti-nutritional factors

Anti-nutritional factors are related to reduced protein digestibility and amino acid availability (reviewed by Gilani et al., 2005; Salunkhe et al., 1982) and are commonly present in large concentrations in plant products (Kay, 1979; Liener, 1980). One of the studies in this field reports on the nutritional quality of mung bean isolate following the exposure to varying concentrations of acetic or succinic anhydride to induce acylation (El-Adawy, 2000). The concentrations of anti-nutritional factors tannin, phytic acid, and trypsin inhibitor showed a significant loss with increasing degrees of modification suggesting that this type of modification can positively impact the effect of anti-nutritional factors. The concentration of trypsin inhibitor even decreased with 70% of the original level of trypsin inhibitor in unmodified protein. The introduction of bulky and/or negatively charged side groups was postulated to affect the extent of protein-tannin (El-Adawy, 2000), protein-mineral-phytic acid (Dua et al., 1996; El-Adawy, 2000), or protein-phenol (Loomis (1974) interactions. Loomis (1974) further showed that the flour and protein production processes provide for optimal conditions for the conversion of polyphenols into quinone oxidation products which, in turn, may bind covalently with sulfhydryl groups of cysteine and ε-amino groups of lysine and N-terminal amino groups. Further support was provided by Dua and colleagues (1996) who showed that acylation and methylation of rapeseed meal and its water-soluble fraction resulted in loss of anti-nutritional factors polyphenol, glucosinolates and phytic acid. However, the methylation procedure employed by Dua and colleagues (1996) resulted in very limited degrees of modification compared to the succinylation and acetylation process suggesting that other factors than the chemical conjugation itself may play a role in the loss of anti-nutritional factors upon chemical modification. As El-Adawy (2000) comments, the extensive dialysis of the protein following the acylation procedure may well be primarily responsible for the loss of water-soluble anti-nutritional factors upon modification. No studies are known to date that report on the loss of anti-nutritional factors upon dialysis.

4.5.2.2 In vitro digestibility

In vitro digestibility of modified proteins is often assayed through exposure of the proteins to a single or a mixture of enzymes including trypsin and pancreatin (Salgó et al., 1984), a combination of trypsin, chymotrypsin and peptidase (Hsu et al., 1977), or pepsin-pancreatin mixtures (Haque et al., 1982) to simulate (post)gastrointestinal digestion of food proteins. The small increase in digestibility of acylated mung bean protein isolate reported by ElAdawy (2000) was primarily correlated to the concurrent loss of tannin; tannins have been shown to play an important role in the reduction of protein digestibility

(Barroga et al., 1985). Alternative factors proposed to induce increased digestibility of modified proteins include the improved access of sites susceptible to enzymatic cleavage as a result of the dissociation of quaternary complexes of proteins or partial unfolding induced by the modification procedure (Achouri & Zhang, 2001). Protein unfolding as a result of modification has been shown for soy protein hydrolysate using techniques to study the secondary and tertiary structure content of the protein upon succinylation (Achouri & Zhang, 2001). However, the acylation of cotton seed flour did not improve in vitro protein digestibility (Rahma & Narasinga Rao, 1983) suggesting that results in this area are somewhat controversial and perhaps other factors play a role (see paragraph 4.5.2.1). In vitro protein digestibility using a multienzyme system containing trypsin, chymotrypsin and peptidase was not impaired for pea proteins upon acetylation (Johnson & Brekke, 1983) while Ma (1984) reported increased digestibility for acylated pea protein, similar to low degrees of succinylation of soy protein hydrolysate (Achouri & Zhang, 2001). Loss of in vitro digestibility has also been reported upon succinylation of a variety of proteins, particularly affecting the release of lysine (Matoba & Doi, 1979; Siu & Thompson, 1982; Wanasundara & Shahidi, 1997) or lipophilization of soy bean glycinin with palmitic acid (Haque et al., 1982). Data on in vivo digestibility of ingested proteins upon modification have not been reported.

4.5.2.3 Availability of essential amino acids

A commonly used target for protein modification are the lysine or cysteine residues, lysine being classified as an essential amino acid, i.e. this amino acid cannot be synthesized de novo by humans and should therefore be ingested. Extensive modification of these amino acids can therefore result in a lower availability. Few studies report on the impact of lysine or cysteine modification on the availability of these amino acids, usually assayed through amino acid analysis. Overall only small decreases in lysine were reported upon succinylation of soy protein hydrolysate (Achouri & Zhang, 2001). Similar findings were reported for acylated soy proteins (Franzen & Kinsella, 1976a), acylated sunflower proteins (Kabirullah & Wills, 1982), and succinylation of oat proteins (Ma & Wood, 1987). It is not clear whether these reported effects of chemical engineering inducing the loss of lysine availability ultimately result in a noticeable and substantial loss of nutritional quality for the human population.

4.5.3 Food hypersensitivity and intolerance

Approximately 3-4% of children and 1-2% of adults in the industrialized world have an allergic reaction to one or more consumed dietary proteins (Baral & Hourihane, 2005; Jansen et al., 1994). Not all ingested proteins act as allergens; nonetheless, proteins associated with allergic responses frequently exhibit characteristics such as exceptional stability against heat, acid, or proteolytic digestion, a tendency to bind to lipids, and varying degrees of glycosylation (Lehrer et al., 2002; Metcalfe et al., 1996). Posttranslational changes, such as N-glycosylation and proline hydroxylation, have been proposed to influence IgE reactivity to Phl p 1 found in timothy grass pollen (Petersen et al., 1998). The exact mechanism by which glycosylation influences the allergic response remains unclear. Efforts have been undertaken to create and assess algorithms that forecast protein allergenic responses based on sequence homology (Aalberse & Stapel, 2001; Jameson & Wolf, 1988), structural similarity, and evolutionary connections (Jenkins et al., 2007), albeit their predictive efficacy remains constrained. The Food and Agriculture Organization (FAO) of the United Nations, in collaboration with the World Health Organization (WHO), formulated a decision tree to evaluate the allergenic potential of genetically modified food ingredients, utilizing sequence homology and subsequent serum screenings (FAO/WHO 2001, reviewed by Bernstein et al., 2003). Considering these findings and the absence of evidence for a common sequence pattern of linear IgE epitopes, it can be inferred that the mechanisms generating immunological sensitization to dietary proteins remain unclear. Assessing the possibility of chemically modified food proteins to provoke allergic reactions necessitates comprehensive understanding of the fundamental mechanisms of the allergic response. No known articles report on the allergenic effects of changed eaten dietary proteins in humans. In the pharmaceutical domain, PEGylation is employed to optimize the distribution of proteins or peptides. A study from the late 1970s indicated that covalent attachment of methoxypolyethylene glycols, either 1900 daltons or 5000 daltons, to bovine liver catalase did not elicit a modulated immune response after repeated injections of the modified enzymes into mice (Abuchowski et al., 1977).

5. Concise Summary

There exists a distinct aim to more effectively use fundamental knowledge regarding protein behavior in complex systems to sustainable food production in the future. There is a necessity to enhance flexibility in protein sources and to optimize the utilization of proteins' structural and nutritional capabilities. The requisite technology is predominantly available, as evidenced by the variety of books discussed in paragraph 3. The mechanistic understanding of the principles governing the relationship between modified protein functionality is remarkable. The advancement of novel applications utilizing chemical protein engineering to enhance protein functionality or facilitate protein source interchangeability remains inadequate. There are several causes for identification. The application of food protein modification has been confined to the realms of food chemistry and food physics. There is a lack of substantial research examining the fine-tuning of microstructure formation in product development at a sensory or physiological level in humans. In contrast to genetic alteration, chemically created proteins have intrinsic heterogeneity. The FDA must account for all degrees of heterogeneity in its approval processes rather than focusing solely on a singular chemical with modified functionality. This is a substantial and arduous task, since it necessitates the fractionation of derived ingredients and enough stability of the resultant products. Furthermore, numerous efforts to diminish variability in the protein material may result in non-food quality substances that cannot be examined for sensory or human physiological factors. The food safety authorities have revised that protein must be considered within the intricate context of the food product. This encompasses not only chemical analysis but also product stability, consumer approval (sensory evaluation), and human physiological data. Particularly in these latter two fields, there is a lack of accessible public literature to assist these authorities in forming impartial perspectives on tolerance. The scientific community plays a crucial role in uniting many disciplines to provide literature that informs authorities' legislative decisions.

6. References

Aalberse, R.C. & Stapel, S.O. (2001). Structure of food allergens in relation to allergenicity. Pediatr. Allergy Immunol. 12: 10-14.

Abuchowski, A.; McCoy, J.R.; Palczuk, N.C.; van Es, T. & Davis, F.F. (1977). Effect of covalent attachment of polyethylene glycol on immunogenicity and circulating life of bovine liver catalase. J. Biol. Chem. 252: 3582-3586.

Achouri, A. & Zhang, W. (2001). Effect of succinylation on the physicochemical properties of soy protein hydrolysate. Food Res. Int. 34: 507-514.

Aewsiri, T.; Benjakul, S.; Visessanguan W.; Wierenga, P.A. & Gruppen, H. (2010). Improvement of foaming properties of cuttlefish skin gelatin by modification with *N*-hydroxysuccinimide esters of fatty acid. Food Hydrocoll. In press.

Aewsiri, T.; Benjakul, S.; Visessanguan, W.; Wierenga, P.A. & Gruppen, H. (2011a). Emulsifying property and antioxidative activity of cuttlefish skin gelatin modified with oxidized linoleic acid and oxidized tannic acid. Food Bioprocess. Technol. In press.

Aewsiri, T.; Benjakul, S.; Visessanguan, W.; Wierenga, P.A. & Gruppen, H. (2011b). Surface activity and molecular characteristics of cuttlefish skin gelatin modified by oxidized linoleic acid. Int. J. Biol. Macromol. In press.

Alting, A.C.; Hamer, R.J.; de Kruif, C.G. & Visschers, R.W. (2003). Cold-set globular protein gels: interactions, structure and rheology as a function of protein concentration. J. Agric. Food Chem. 51: 3150-3156.

Andersson, L.-O.; Brandt, J. & Johansson, S. (1971). The use of trinitrobenezenesulfonic acid in studies on the binding of fatty acid anions to bovine serum albumin. Arch. Biochem. Biophys. 146: 428-440.

Anfinsen, C.B. (1973). Principles that govern the folding of protein chains. Science 181: 223230.

Anson, M.L. (1940). The reactions of iodine and iodoacetamide with native egg albumin. J. Gen. Physiol. 23: 321-331.

Antoniewski, M.N.; Barringer, S.A.; Knipe, L. & Zerby, H. (2007). The effect of a gelatin coating on the shelf-life of fresh meat. J. Food Sci. 72: E382-E387.

Arnaudov, L.N.; de Vries, R.; Ippel, H. & van Mierlo, C.P. (2003). Multiple steps during the formation of β-lactoglobulin fibrils. Biomacromolecules 4: 1614-1622.

Arnaudov, L.N. & de Vries, R. (2005). Thermally induced fibrillar aggregation of hen egg white lysozyme. Biophys. J. 88: 515-526.

Arnaudov, L.N. & de Vries, R. (2006). Strong impact of ionic strength on the kinetics of fibrilar aggregation of bovine β-lactoglobulin. Biomacromolecules 7: 3490-3498.

Arntfield, S.D.; Murray, E.D. & Ismond, M.A.H. (1991). Role of disulfide bonds in determining the rheological and microstructural properties of heat-induced protein networks from ovalbumin and vicilin. J. Agric. Food Chem. 39: 1378-1385.

Arvanitoyannis, I.S. (1999). Totally and partially biodegradable polymer blends based on natural and synthetic macromolecules: preparation, physical properties, and potential as food packaging materials. Polym. Rev. 39: 205-271.

Audic, J.L. & Chaufer, B. (2005). Influence of plasticizers and crosslinking on the properties of biodegradable films made from sodium caseinate. Eur. Polym. J. 41: 1934-1942.

Bai, J.; Alleyne, V.; Hagenmaier, R.D.; Mattheis, J.P. & Baldwin, E.A. (2003). Formulation of zein coatings for apples (Malus domestica Borkh). Postharv. Biol. Technol. 28: 259-268.

Baldursdottir, S.G.; Fullerton, M.S.; Nielsen, S.H. & Jorgensen, L. (2010). Adsorption of proteins at the oil/water interface - observation of protein adsorption by interfacial shear stress measurements. Colloids Surf. B Biointerfaces 79: 41-46.

Baldwin, R.L. (2008). The search for folding intermediates and the mechanism of protein folding. Annu. Rev. Biophys. 37: 1-21.

Baldwin, E.A. (1994). Edible coatings for fresh fruits and vegetables: past, present, and future. In: Edible Coatings and Films to Improve Food Quality. Ed. By Krochta, J.M., Bandwin, E.A., Nisperos-Carriedo, M.O., Technomic Publishing, Lancaster PA, pp. 25-64.

Baral, V.R. & Hourihane, J.O. (2005). Food allergy in children. Postgrad. Med. J. 81: 693-701.

Barbut, S. & Foegeding, E.A. (1993). Ca^{2+}-induced gelation of preheated whey protein solutions. J. Food Sci. 58: 867-871.

Barroga, C.F.; Laurena, A.C. & Mendoza, E.M.T. (1985). Polyphenol in mung bean (Vigna radiata L. Wilczek) determination and removal. J. Agric. Food Chem. 33: 1006-1009.

Bauer, H.H.; Aebi, U.; Häner, M.; Hermann, R.; Müller, M.; Arvinte, T. & Merkle, H.P. (1995). Architecture and polymorphism of fibrillar supramolecular assemblies produced by in vitro aggregation of human calcitonin. J. Struct. Biol. 115: 1-15.

Baumketner, A.; Jewett, A. & Shea, J.E. (2003). Effects of confinement in chaperonin assisted protein folding: rate enhancement by decreasing the roughness of the folding energy landscape. J. Mol. Biol. 332: 701-713.

Bernstein, J.A.; Bernstein, I.L.; Bucchini, L.; Goldman, L.R.; Hamilton, R.G.; Lehrer, S.; Rubin, C. & Sampson, H.A. (2003). Clinical and laboratory investigation of allergy to genetically modified foods. Environ. Health Persp. 111: 1114-1121.

Betz, S.F. (1993). Disulfide bonds and the stability of globular proteins. Protein Sci. 2: 15511558.

Borgia, A.; Williams, P.M. & Clarke, J. (2008). Single-molecule studies of protein folding. Annu. Rev. Biochem. 77: 101-125.

Bowes, J.H. & Cater, C.W. (1968). The interaction of aldehydes with collagen. Biochim. Biophys. Acta 168: 341-352.

Broersen, K.; Weijers, M.; de Groot, J.; Hamer, R.J. & de Jongh, H.H. (2007a). Effect of protein charge on the generation of aggregation-prone conformers. Biomacromolecules 8: 1648-1656.

Broersen, K.; Elshof, M.; de Groot, J.; Voragen, A.G.J.; Hamer, R.J. & de Jongh, H.H.J. (2007b). Aggregation of β-lactoglobulin regulated by glucosylation. J. Agric. Food Chem. 55: 2431-2437.

Broersen, K.; Voragen, A.G.; Hamer, R.J. & de Jongh, H.H. (2004). Glycoforms of betalactoglobulin with improved thermostability and preserved structural packing. Biotechnol. Bioeng. 86: 78-87.

Broersen, K.; van Teeffelen, A.M.; Vries, A.; Voragen, A.G.; Hamer, R.J. & de Jongh, H.H. (2006). Do sulfhydryl groups affect aggregation and gelation properties of ovalbumin? J. Agric. Food Chem. 54: 5166-5174.

Bromley, E.H.C.; Krebs, M.R.H. & Donald, A.M. (2006). Mechanisms of structure formation in particulate gels of β-lactoglobulin formed near the isoelectric point. Eur. Phys. J. E. 21: 145-152.

Buchner, S.; Kinnear, M.; Crouch, I.J.; Taylor, J. & Minnaar, A. (2011). Effect of kafirin protein coating on sensory quality and shelf-life of 'Packham's Triumph' pears during ripening. J. Sci. Food Agric. In press.

Buchner, G.S.; Murphy, R.D.; Buchete, N.V. & Kubelka, J. (2011). Dynamics of protein folding: probing the kinetic network of folding-unfolding transitions with experiment and theory. Biochim. Biophys. Acta 1814: 1001-1020.

Bustos, P.; Gajardo, M.I.; Gómez, C.; Hughes, G.; Cardemil, E. & Jabalquinto, A.M. (1996). Woodward's reagent K reacts with histidine and Cysteine residues in Escherichia coli and Saccharomyces cerevisiae phosphoenolpyruvate carboxykinases. J. Protein Chem. 15: 467-472.

Calamai, M.; Taddei, N.; Stefani, M., Ramponi, G. & Chiti, F. (2003). Relative influence of hydrophobicity and net charge in the aggregation of two homologous proteins. Biochemistry 42: 15078-15083.

Canet, D.; Sunde, M.; Last, A.M.; Miranker, A.; Spencer, A.; Robinson, C.V. & Dobson, C.M. (1999). Mechanistic studies of the folding of human lysozyme and the origin of amyloidogenic behavior of its disease-related variants. Biochemistry 38: 6419-6427.

Cao-Hoang, L.; Chaine, A.; Grégoire, L. & Waché, Y. (2010). Potential of nisin-incorporated sodium caseinate films to control Listeria in artificially contaminated cheese. Food Microbiol. 27: 940-944.

Chang, T.-S. & Sun, S.F. (1978). Structural studies on the succinylated bovine serum albumin. Int. J. Peptide Protein Res. 11: 65-72.

Charissou, A.; Ait-Ameur, L. & Birlouez-Aragon, I. (2007). Kinetics of formation of three indicators of the Maillard reaction in model cookies: influence of baking temperature and type of sugar. J. Agric. Food Chem. 55: 4532-4539.

Chibnall, A.C. (1942). Bakerian lecture: amino-acid analysis and the structure of proteins. Proc. Roy. Soc. B 131: 136-160.

Chiti, F. & Dobson, C.M. (2006). Protein misfolding, functional amyloid, and human disease. Annu. Rev. Biochem. 75: 333-366.

Chiti, F.; Stefani, M.; Taddei, N.; Ramponi, G. & Dobson, C.M. (2003). Rationalization of the effects of mutations on peptide and protein aggregation rates. Nature 424: 805-808.

Chiti, F.; Taddei, N.; Bucciantini, M.; White, P.; Ramponi, G. & Dobson, C.M. (2000). Mutational analysis of the propensity for amyloid formation by a globular protein. EMBO J. 19: 1441-1449.

Church, F.C.; Swaisgood, H.E.; Porter, D.H. & Catignani, G.L. (1983). Spectrophotometric assay using *o*-phthaldialdehyde for determination of proteolysis in milk and isolated milk proteins. J. Dairy Sci. 66: 1219-1227.

Cieplak, M. (2004). Cooperativity and contact order in protein folding. Phys. Rev. E Stat. Nonlin. Soft Matter Phys. 69: 031907.

Corzo-Martínez, M.; Soria, A.C.; Villamiel, M.; Olano, A.; Harte, F.M. & Moreno, F.J. (2011). Effect of glycation on sodium caseinate-stabilized emulsions obtained by ultrasound. J. Dairy Sci. 94: 51-58.

Creighton, T.E. (1988). Disulphide bonds and protein stability. Bioessays 8: 57-63.

Cucu, T.; Platteau, C.; Taverniers, I.; Devreese, B.; de Loose, M. & de Meulenaer, B. (2011). ELISA detection of hazelnut proteins: effect of protein glycation in the presence or absence of wheat proteins. Food Addit. Contam. Part A Chem. Anal. Control Expo. Risk Assess. 28: 1-10.

Damodaran, V.B. & Fee, C. (2010). Protein PEGylation: an overview of chemistry and process considerations. Eur. Pharm. Rev 15: 18-26.

Dearfield, K.L.; Abernathy, C.O.; Ottley, M.S.; Brantner, J.H. & Hayes, P.F. (1988). Acrylamide: its metabolism, developmental and reproductive effects, genotoxicity, and carcinogenicity. Mutat. Res. 195: 45-77.

De Groot, J.; Kosters, H.A. & de Jongh, H.H. (2007). Deglycosylation of ovalbumin prohibits formation of a heat-stable conformer. Biotechnol. Bioeng. 97: 735-741.

De Jongh, H.H.J.; Kosters, H.A.; Kudryashova, E.; Meinders, M.B.J.; Trofimova, D. & Wierenga, P.A. (2004). Protein adsorption at air-water interfaces; a combination of details. Biopolymers 74: 131-135.

De Jongh, H.H.J. & Wierenga, P.A. (2006). Assessing the extent of protein intermolecular interactions at air-water interfaces using spectroscopic techniques. Biopolymers 82 : 384-389.

De Jongh, H.H.; Taylor, S.L. & Koppelman, S.J. (2011). Controlling the aggregation propensity and thereby digestibility of allergens by Maillardation as illustrated for cod fish parvalbumin. J. Biosci. Bioeng. 111: 204-211.

Del Rosario Moreira, M.; Pereda, M.; Marcovich, N.E. & Roura, S.I. (2011). Antimicrobial effectiveness of bioactive packaging materials from edible chitosan and casein polymers: assessment on carrot, cheese, and salami. J. Food Sci. 76: M54-M63.

DeSantis, G. & Jones, J.B. (1999). Chemical modification of enzymes for enhanced functionality. Curr. Opin. Biotechnol. 10: 324-330.

Dickinson, E. (1992). An introduction to food colloids. Oxford, UK: Oxford University Press.

Djagny, K.B.; Wang, Z. & Xu, S. (2001). Conformational changes and some functional characteristics of gelatin esterified with fatty acid. J. Agric. Food Chem. 49: 2987-2991.

Dobson, C.M. (1999). Protein misfolding, evolution and disease. Trends Biochem. Sci. 24: 329332.

Doi, E. (1993). Gels and gelling of globular proteins. Trends Food Sci. Technol. 4: 1-5.

Dondero, M.; Figueroa, V.; Morales, X. & Curotto, E. (2006). Transglutaminase effects on gelation capacity of thermally induced beef protein gels. Food Chemistry 99, 546-554.

Dua, S.; Mahajan, A. & Mahajan, A. (1996). Improvement of functional properties of rapeseed (Brassica campestris Var. Toria) preparations by chemical modification. J. Agric. Food Chem. 44: 706-710.

DuBay, K.F.; Pawar, A.P.; Chiti, F.; Zurdo, J.; Dobson, C.M. & Vendrusculo, M. (2004). Prediction of the absolute aggregation rates of amyloidogenic polypeptide chains. J. Mol. Biol. 341: 1317-1326.

El-Adawy, T.A. (2000). Functional properties and nutritional quality of acetylated and succinylated mung bean protein isolate. Food Chem. 70: 83-91.

Ellman, G.L. (1959). Tissue sulfhydryl groups. Arch. Biochem. Biophys. 82: 70-77.

Emmambux, M.N.; Stading, M. & Taylor, J.R.N. (2004). Sorghum kafirin film property modification with hydrolysable and condensed tannins. J. Cereal Sci. 40: 127-135.

Englander, S.W.; Mayne, L. & Krishna, M.M. (2007). Protein folding and misfolding: mechanisms and principles. Q. Rev. Biophys. 40: 287-326.

Evans, M.T.A. & Irons, L. (1970). N-succinylated egg yolk proteins. German Patent 1951247.

FAO/WHO (2001). Evaluation of allergenicity of genetically modified foods. Report of a Joint FAO/WHO Expert Consultation of Allergenicity of Foods Derived from Biotechnology. Available: ftp://ftp.fao.org/es/esn/food/allergygm.pdf [accessed 08 August 2011].

Farouk, M.M.; Price, J.F. & Salih, A.M. (1990). Effect of an edible film overwrap on exudation and lipid oxidation in beef round steak. J. Food Sci. 55: 1510-1512, 1563.

Feeney, R.E.; Blankenhorn, G. & Dixon, H.B. (1975). Carbonyl-amine reactions in protein chemistry. Adv. Protein Chem. 29: 135-203.

Feeney, R.E.; Yamasaki, R.B. & Geoghegan, K.F. (1982). Chapter 1: Chemical modification of proteins: an overview. In 'Modification of Proteins', Advances in Chemistry; American Chemical Society, Washington DC. pp 3-55.

Fernandez-Escamilla, A.M.; Rousseau, F.; Schymkowitz, J. & Serrano, L. (2004). Prediction of sequence-dependent and mutational effects on the aggregation of peptides and proteins. Nat. Biotechnol. 22: 1302-1306.

Ferreira, S.T.; De Felice, F.G. & Chapeaurouge, A. (2006). Metastable, partially folded states in the productive folding and in the misfolding and amyloid aggregation of proteins. Cell Biochem. Biophys. 44: 539-548.

Ferreon, A.C. & Deniz, A.A. (2011). Protein folding at single-molecule resolution. Biochim. Biophys. Acta 1814: 1021-1029.

Fields, R. (1971). The measurement of amino groups in proteins and peptides. Biochem. J. 124: 581-590.

Fraenkel-Conrat, H. (1957). Methods for investigating the essential groups for enzyme activity. Methods Enzymol. 4: 247-269.

Finley, J.W. (1975). Deamidated gluten: a potential fortifier for fruit juices. J. Food Sci. 40: 1283-1285.

Francis, G.E.; Fisher, D.; Delgado, C.; Malik, F.; Gardiner, A. & Neale, D. (1998). PEGylation of cytokines and other therapeutic proteins and peptides: the importance of biological optimisation of coupling techniques. Int. J. Hematol. 68: 1-18.

Franzen, K.L. & Kinsella, J.E. (1976a). Functional properties of succinylated and acetylated soy protein. J. Agric. Food Chem. 24: 788-795.

Franzen, K.L. & Kinsella, J.E. (1976b). Functional properties of succinylated and acetylated leaf proteins. J. Agric. Food Chem. 24: 914-919.

Frister, H.; Meisel, H. & Schlimme, E. (1988). OPA method modified by use of N,N-dimethyl-2-mercaptoethylammonium chloride as thiol component. Fresenius' Z. Anal. Chem. 330: 631-633.

Gerbanowski, A.; Rabiller, C.; Larré, C. & Guéguen, J. (1999a). Grafting of aliphatic and aromatic probes to bovine serum albumin: influence on its structural and physicochemical characteristics. J. Protein Chem. 18: 325-335.

Gerbanowski, A.; Malabat, C.; Rabiller, C. & Guéguen, J. (1999b). Grafting of aliphatic and aromatic probes on rapeseed 2 S and 12 S proteins: influence on their structural and physicochemical properties. J. Agric. Food Chem. 47: 5218-5226.

Gilani, G.S.; Cockell, K.A. & Sepehr, E. (2005). Effects of antinutritional factors on protein digestibility and amino acid availability in foods. J. AOAC Int 88: 967-987.

Gillgren, T. & Stading, M. (2008). Mechanical and barrier properties of avenin, kafirin and zein films. Food Biophys. 3: 287-294.

Graham, D.E. & Phillips, M.C. (1979a). Proteins at liquid interfaces. I. Kinetics of adsorption and surface denaturation. J. Colloid Interface Sci. 70: 403-414.

Graham, D.E. & Phillips, M.C. (1979b). Proteins at liquid interfaces. II. Adsorption isotherms. J. Colloid Interface Sci. 70: 415-426.

Graham, D.E. & Phillips, M.C. (1979c). Proteins at liquid interfaces. III. Molecular structures at adsorbed films. J. Colloid Interface Sci. 70: 427-439.

Graña-Montes, R.; de Groot, N.S.; Castillo, V.; Sancho, J.; Velazquez-Campoy, A. & Ventura, S. (2011). Contribution of disulfide bonds to stability, folding and amyloid fibril formation: the PI3-SH3 domain case. Antioxid. Redox Signal. In press.

Grigoriev, D.O.; Derkatch, S.; Krägel, J. & Miller, R. (2007). Relationship between structure and rheological properties of mixed BSA/Tween 80 adsorption layers at the air/water interface. Food Hydrocoll. 21: 823-830.

Groninger, H.S. Jr. (1973). Preparation and properties of succinylated fish myofibrillar protein. J. Agric. Food Chem. 21: 978-981.

Gruener, L. & Ismond, M.A.H. (1997). Effects of acetylation and succinylation on the physicochemical properties of the canola 12S globulin. Food Chem. 60: 357-363.

Gueguen, J.; Bollecker, S.; Schwenke, K.D. & Raab, B. (1990). Effect of succinylation on some physicochemical and functional properties of the 12 S storage protein from rapeseed (Brassica napus L.). J. Agric. Food Chem. 38: 61-69.

Gurin, S. & Clarke, H.T. (1934). Allocation of the free amino groups in proteins and peptides. J. Biol. Chem. 107: 395-419.

Habeeb, A.F.S.A. & Atassi, M.Z. (1969). Enzymic and immunochemical properties of lysozyme II. Conformation, immunochemistry and enzymic activity of a derivative modified at tryptophan. Immunohistochemistry 6: 555-566.

Hall, R.J.; Trinder, N. & Givens, D.I. (1973). Observations on the use of 2,4,6 trinitrobenzenesulphonic acid for the determination of available lysine in animal protein concentrates. Analyst 98: 673-686.

Halling, P.J. (1981). Protein-stabilized foams and emulsions. Crit. Rev. Food Sci. Nutr. 15: 155203.

Hamada, J.S. & Marshall, W.E. (1989). Preparation and functional properties of enzymatically deamidated soy proteins. J. Food Sci. 54: 598-601.

Hamano, Y. (2011). Occurrence, biosynthesis, biodegradation, and industrial and medical applications of a naturally occurring ε-poly-L-lysine. Biosci. Biotechnol. Biochem. 75: 1226-1233.

Haque, Z.; Matoba, T. & Kito, M. (1982). Incorporation of fatty acid into food protein: palmitoyl soybean glycinin. J. Agric. Food Chem. 30: 481-486.

Haque, Z. & Kito, M. (1983a). Lipophilization of α_{s1}-casein. 1. Covalent attachment of palmitoyl residue. J. Agric. Food Chem. 31: 1225-1230.

Haque, Z. & Kito, M. (1983b). Lipophilization of α_{s1}-casein. 2. Conformational and functional effects. J. Agric. Food Chem. 31: 1231-1237.

Hardy, P.M.; Nicholls, A.C. & Rydon, H.N. (1969). The nature of gluteraldehyde in aqueous solution. J. Chem. Soc. Chem. Commun. 10: 565-566.

Harper, J.D.; Wong, S.S.; Lieber, C.M. & Lansbury, P.T. Jr. (1997). Atomic force microscopy imaging of seeded fibril formation and fibril branching by the Alzheimer's disease amyloid- β protein. Chem. Biol. 4: 951-959.

Hart, G.W. (1992). Glycosylation. Curr. Opin. Cell Biol. 4: 1017-1723.

Havard, C. & Harmony, M.X. (1869). Improved process of preserving meat, fowls, fish. June 8, U.S. Patent 90: 944.

Hayakawa, S. & Nakai, S. (1985). Contribution of the hydrophobicity, net charge and sulfhydryl groups to thermal properties of ovalbumin. Can. Inst. Food Sci. Technol. J. 18: 290-295.

Haynes, R.; Osuga, D.T. & Feeney, R.E. (1967). Modification of amino groups in inhibitors of proteolytic enzymes. Biochemistry 6: 541-547.

He, X.-H.; Shaw, P.-C. & Tam, S.-C. (1999). Reducing the immunogenicity and improving the inv vivo activity of trichosanthin by site-directed PEGylation. Life Sci. 65: 355368.

Helenius, A.; Trombetta, E.S.; Hebert, D.N. & Simons, J.F. (1997) Calnexin, calreticulin and the folding of glycoproteins. Trends Cell Biology 7: 193-200.

Hillier, R.M.; Lyster, R.L.J. & Cheeseman, G.C. (1980). Gelation of reconstituted whey powders by heat. J. Sci. Food Agric. 31: 1152-1157.

Hoagland, P.D. (1968). Acylated β-caseins. Effect of alkyl group size on calcium ion sensitivity and on aggregation. Biochemistry 7: 2542-2546.

Hoffmann, M.A.M. & van Mil, P.J.J.M. (1997). Heat-induced aggregation of β-lactoglobulin: role of the free thiol group and disulfide bonds. J. Agric. Food Chem. 45: 2942-2948.

Hopwood, D. (1969). A comparison of the crosslinking abilities of gluteraldehyde, formaldehyde and α-hydroxyadipaldehyde with bovine serum albumin and casein. Histochemie 17: 151-161.

Hopwood, D., Callen, C.R. & McCabe, M. (1970). The reactions between gluteraldehyde and various proteins. An investigation of their kinetics. Histochem. J. 2: 137-150.

Hsu, H.W.; Vavak, D.L.; Satterlee, L.S. & Miller, G.A. (1977). A multienzyme technique for estimating protein digestibility. J. Food Sci. 42: 1269-1273.

Huggins, C. & Jensen, E.V. (1949). Thermal coagulation of serum proteins; the effects of iodoacetate, iodoacetamide, and thiol compounds on coagulation. J. Biol. Chem. 179: 645-654.

Hunter, J.R.; Carbonell, R.G. & Kilpatrick, P.K. (1991). Coadsorption and exchange of lysozyme/ β-casein mixtures at the air/water interface. J. Colloid Interface Sci. 143: 37-53.

Hurle, M.R.; Helms, L.R.; Li, L.; Chan, W. & Wetzel, R. (1994). A role for destabilizing amino acid replacements in light-chain amyloidosis. Proc. Natl. Acad. Sci. U.S.A. 91: 54465450.

Ibrahim, H.R.; Kato, A. & Kobayashi, K. (1991). Antimicrobial effects of lysozyme against gram-negative bacteria due to covalent binding of palmitic acid. J. Agric. Food Chem. 39: 2077-2082.

Ibrahim, H.R.; Kobayashi, K. & Kato, A. (1993). Length of hydrocarbon chain and antimicrobial action to gram-negative bacteria of fatty acylated lysozyme. J. Agric. Food Chem. 41: 1164-1168.

Izzo, H.V.; Lincoln, M.D. & Ho, C.-T. (1993). The effect of temperature, feed moisture, and pH on protein deamidation in an extruded wheat flour. J. Agric. Food Chem. 41: 199202.

Jameson, B.A. & Wolf, H. (1988). The antigenic index: a novel algorithm for predicting antigenic determinants. Comp. Appl. Biosci. 4: 181-186.

Jansen, J.J.N.; Kardinaal, A.F.M.; Huijbers, G.; Vlieg-Boerstra, B.J.; Martens, B.P.M. & Ockhuizen, T. (1994). Prevalence of food allergy and intolerance in the adult Dutch population. J. Allergy Clin. Immun. 93: 446-456.

Jarrett, J.T. & Lansbury, P.T. Jr. (1993). Seeding "one-dimensional crystallization" of amyloid: a pathogenic mechanism in Alzheimer's disease and scrapie? Cell 73: 1055-1058.

Jenkins, J.A.; Breiteneder, H. & Mills, E.N.C. (2007). Evolutionary distance from human homologs reflects allergenicity of animal food proteins. J. Allergy Clin. Immunol. 120: 1399-1405.

Jiménez, J.; Nettleton, E.J.; Bouchard, M.; Robinson, C.V.; Dobson, C.M. & Saibil, H. (2002). The protofilament structure of insulin amyloid fibrils. Proc. Natl. Acad. Sci. U.S.A. 99: 9196-9201.

Jiménez, J.; Guijarro, J.I.; Orlova, E.; Zurdo, J.; Dobson, C.M.; Sunde, M. & Saibil, H.R. (1999). Cryo-electron microscopy structure of an SH3 amyloid fibril and model of the molecular packing. EMBO J. 18: 815-821.

Johnson, E.A. & Brekke, J. (1983). Functional properties of acylated pea protein isolates. J. Food Sci. 48: 722-725.

Johnson, A.R. & Dekker, E.E. (1996). Woodward's reagent K inactivation of Escherichia coli Lthreonine dehydrogenase: increased absorbance at 340 – 350 nm is due to modification of cysteine and histidine residues, not aspartate or glutamate carboxyl groups. Protein Sci. 5: 382-390.

Jonas, A. & Weber, G. (1970). Partial modification of bovine serum albumin with dicarboxylic anhydrides. Physical properties of the modified species. Biochemistry 9: 4729-4735.

Ju, Z.Y. & Kilara, A. (1998). Aggregation induced by calcium chloride and subsequent thermal gelation of whey protein isolate. J. Dairy Sci. 81: 925-931.

Kabirullah, M. & Wills, R.B.H. (1982). Functional properties of acetylated and succinylated sunflower protein isolate. J. Food Technol. 17: 235-249.

Kaneko, R. & Kitabatake, N. (2001). Structure-sweetness relationship in thaumatin: importance of lysine residues. Chem. Senses 26: 167-177.

Kato, A.; Ibrahim, H.; Takagi, T. & Kobayashi, K. (1990). Excellent gelation of egg white preheated in the dry state is due to the decreasing degree of aggregation. J. Agric. Food Chem. 38: 1868-1872.

Kato, A.; Nagase, Y.; Matsudomi, N. & Kobayashi, K. (1983). Determination of molecularweight of soluble ovalbumin aggregates during heat denaturation using low-angle laser-light scattering technique. Agric. Biol. Chem. 47: 1829-1834.

Kay, D.E. (1979). Crop and product digest. No. 3. Food legumes. London, Tropical Products Institute, 435 p .

Kelly, J.W. (1998). The alternative conformations of amyloidogenic proteins and their multistep assembly pathways. Curr. Opin. Struct. Biol. 8: 101-106.

Kitabatake, N.; Wada, R. & Fujita, Y. (2001). Reversible conformational change in β lactoglobulin A modified with *N*-ethylmaleimide and resistance to molecular aggregation on heating. J. Agric. Food Chem. 49: 4011-4018.

Klotz, I.M. & Heiney, R.E. (1962). Introduction of sulfhydryl groups into proteins using acetylmercaptosuccinic anhydride. Arch. Biochem. Biophys. 96: 605-612.

Korn, A.H.; Feairheller, S.H. & Filachione, E.M. (1972). Gluteraldehyde: nature of the reagent. J. Mol. Biol. 65: 525-529.

Kornfeld, R. & Kornfeld, S. (1985). Assembly of asparagines-linked oligosaccharides. Annu. Rev. Biochem. 54: 631-664.

Koseki, T.; Kitabatake, N. & Doi, E. (1989). Irreversible thermal denaturation and formation of linear aggregates of ovalbumin. Food Hydrocoll. 3: 123-134.

Kosters, H.A. & de Jongh, H.H. (2003). Spectrophotometric tool for the determination of the total carboxylate content in proteins; molar extinction coefficient of the enol ester from Woodward's reagent K reacted with protein carboxylates. Anal. Chem. 75: 2512-2516.

Kosters, H.A.; Broersen, K.; de Groot, J.; Simons, J.W.; Wierenga, P. & de Jongh, H.H. (2003). Chemical processing as a tool to generate ovalbumin variants with changed stability. Biotechnol. Bioeng. 84: 61-70.

Kotaki, A.; Harada, M. & Yagi, K. (1964). Reaction between sulfhydryl compounds and 2,4,6-trinitrobenzene-1-sulfonic acid. J. Biochem. 55: 553-561.

Krause, J.-P.; Mothes, R. & Schwenke, K.D. (1996). Some physicochemical and interfacial properties of native and acetylated legumin from faba beans (Vicia faba L.). J. Agric. Food Chem. 44: 429-437.

Krebs, M.R.H.; Devlin, G.L. & Donald, A.M. (2009). Amyloid fibril-like structure underlies the aggregate structure across the pH range for β-lactoglobulin. Biophys. J. 96: 50135019.

Kudryashova, E.V.; Meinders, M.B.J.; Visser, A.J.W.G.; van Hoek, A. & de Jongh, H.H.J (2003). Structural properties and rotational dynamics of egg white ovalbumin adsorbed at the air/water interaface. Eur. J. Biophys. 32: 553-562.

Kudryashova, E.V.; Visser, A.J.W.G., & de Jongh, H.H.J. (2005). Reversible self-association of ovalbumin at air-water interfaces and the consequences for the exerted surface pressure. Protein Science 14: 483-493.

Lakkis, J. & Villota, R. (1992). Effect of acylation on substructural properties of proteins: a study using fluorescence and circular dichroism. J. Agric. Food Chem. 40: 553-560.

Land, A. & Braakman, I. (2001). Folding of the human immunodeficiency virus type 1 envelope glycoprotein in the endoplasmatic reticulum. Biochimie 83: 783-790.

Langton, M. & Hermansson, A.-M. (1992). Fine-stranded and particulate gels of β lactoglobulin and whey protein at varying pH . Food Hydrocoll. 5: 523-539.

Le, T.T.; Bhandari, B. & Deeth, H.C. (2011). Chemical and physical changes in milk protein concentrate (MPC80) powder during storage. J. Agric. Food Chem. 59: 5465-5473.

Le Floch-Fouéré, C.; Pezennec,S.; Pézolet, M.; Rioux-Dubé, J.F.; Renault, A. & Beaufils, S. (2011). Unexpected differences in the behavior of ovotransferrin at the air-water interface at pH 6.5 and 8.0. J. Colloid Interface Sci. 356: 614-623.

Lehrer, S.B.; Ayuso, R. & Reese, G. (2002). Current understanding of food allergens. Ann. N. Y. Acad. Sci. 964: 69-85.

Liao, L.; Zhao, M.; Ren, J.; Zhao, H.; Cui, C. & Hu, X. (2010). Effect of acetic acid deamidation-induced modification on functional and nutritional properties and conformation of wheat gluten. J. Sci. Food Agric. 90: 409-417.

Liener, I.E. (1980). Toxic constituents of plant foodstuffs. New York and London, Academic Press, 502 p.

Liu, S.-T.; Sugimoto, T.; Azakami, H. & Kato, A. (2000). Lipophilization of lysozyme by short and middle chain fatty acids. J. Agric. Food Chem. 48: 265-269.

Llamas, K.; Ownes, M.; Blakeley, L. & Zerner, B. (1986). N-ethyl-5-phenylisoxazolium-3'sulfonate (Woodward's reagent K) as a reagent for nucleophilic side chains of proteins. J. Am. Chem. Soc. 108: 5543-5548.

Lomakin, A.; Chung, D.S.; Benedek, G.B.; Kirscher, D.A. & Teplow, D.B. (1996). On the nucleation and growth of amyloid β-protein fibrils: detection of nuclei and quantitation of rate constants. Proc. Natl. Acad. Sci. U.S.A. 93: 1125-1129.

Lomakin, A.; Teplow, D.B.; Kirschner, D.A. & Benedek, G.B. (1997). Kinetic theory of fibrillogenesis of amyloid β-protein. Proc. Natl. Acad. Sci. U.S.A. 94: 7942-7947.

Longares, A.; Monahan, F.J.; O'Riordan, E.D. & O'Sullivan, M. (2005). Physical properties of edible films made from mixtures of sodium caseinate and WPI. Int. Dairy J. 15: 1255-1260.

Lönnerdal, B. (2002). Expression of human milk proteins in plants. J. Am. Coll. Nutr. 21: 218221

Loomis, W.D. (1974). Overcoming problems of phenolics and quinones in the isolation of plant enzymes organelles. In: Methods in Enzymology; Fleischer, S., Packer, L., Eds.: Academic Press: New York, p. 528.

Losso, J.N. & Nakai, S. (2002). Stabilization of oil-in-water emulsions by β-lactoglobulinpolyethylene glycol conjugates. J. Agric. Food Chem. 50: 1207-1212.

Lubelski, J.; Rink, R.; Khusainov, R.; Moll, G.N. & Kuipers, O.P. (2008). Biosynthesis, immunity, regulation, mode of action and engineering of the model lantibiotic nisin. Cell. Mol. Life Sci. 65: 455-476.

Luey, J.-K.; McGuire, J. & Sproull, R.D. (1991). The effect of pH and NaCl concentration on adsorption of β-lactoglobulin at hydrophilic and hydrophobic silicon surfaces. J. Colloid Interface Sci. 143: 489-500.

Luheshi, L.M. & Dobson, C.M. (2009). Bridging the gap: from protein misfolding to protein misfolding diseases. FEBS Lett. 583: 2581-2586.

Ma, C.-Y. (1984). Functional properties of acylated oat protein. J. Food Sci. 49: 1128-1131.

Ma, C.-Y. & Wood, D.F. (1987). Functional properties of oat proteins modified by acylation, trypsin hydrolysis or linoleate treatment. J. Am. Oil Chem. Soc. 64: 1726-1731.

MacRitchie, F. (1978). Proteins at interfaces. Adv. Protein Chem. 32: 283-326.

Malaki Nik, A.; Wright, A.J. & Corredig, M. (2010). Interfacial design of protein-stabilized emulsions for optimal delivery of nutrients. Food Funct. 1: 141-148.

Marcon, G.; Plakoutsi, G.; Canale, C.; Relini, A.; Taddei, N.; Dobson, C.M.; Ramponi, G. & Chiti, F. (2005). Amyloid formation from HypF-N under conditions in which the protein is initially in its native state. J. Mol. Biol. 347: 323-335.

Margoshes, B.A. (1990). Correlation of protein sulfhydryls with the strength of heat-formed egg white gels. J. Food Sci. 55: 1753.

Martin, A.H.; Meinders, M.B.J.; Bos, M.A.; Cohen Stuart, M.A. & van Vliet, T. (2003). Conformational aspects of proteins at the air/water interface studied by infrared reflection-absorption spectroscopy. Langmuir 19: 2922-2928.

Martins, J.T.; Cerqueira, M.A.; Souza, B.W.; Carmo Avides, M. & Vicente, A.A. (2010). Shelf life extension of ricotta cheese using coatings of galactomannans from nonconventional sources incorporating nisin against Listeria monocytogenes. J. Agric. Food Chem. 58: 1884-1891.

Masuda, T.; Ueno, Y. & Kitabatake, N. (2001). Sweetness and enzymatic activity of lysozyme. J. Agric. Food Chem. 49: 4937-4941.

Masuda, T.; Ide, N. & Kitabatake, N. (2005a). Structure-sweetness relationship in egg white lysozyme: role of lysine and arginine on the elicitation of lysozyme sweetness. Chem. Senses 30: 667-681.

Masuda, T.; Ide, N. & Kitabatake, N. (2005b). Effects of chemical modification of lysine residues on the sweetness of lysozyme. Chem. Senses 30: 253-264.

Matoba, T. & Doi, E. (1979). In vitro digestibility of succinylated protein by pepsin and pancreatic proteases. J. Food Sci. 44: 537-539.

Matsudomi, N.; Kato, A. & Kobayashi, K. (1982). Conformation and surface properties of deamidated gluten. Agric. Biol. Chem. 46: 1583-1586.

Matsudomi, N.; Sasaki, T.; Kato, A. & Kobayashi, K. (1985). Conformational changes and functional properties of acid-modified soy protein. Agric. Biol. Chem. 49: 1251-1256.

Maurer-Stroh, S.; Debulpaep, M.; Kuemmerer, N.; Lopez de la Paz, M.; Martins, I.C.; Reumers, J.; Morris, K.L.; Copland, A.; Serpell, L.C.; Serrano, L.; Schymkowitz, J.W.H. & Rousseau, F. (2010). Exploring the sequence determinants of amyloid structure using position-specific scoring matrices. Nature Methods 7: 237-242.

Marquardt, T. & Helenius, A. (1992). Misfolding and aggregation of newly synthesized proteins in the endoplasmatic reticulum. J. Cell Biol. 117: 505-513.

McClellan, S.J. & Franses, E.I. (2003). Effect of concentration and denaturation on adsorption and surface tension of bovine serum albumin. Colloids Surf, B Biointerfaces 28: 63-75.

McClements, J. (2005). Food emulsions: principles, practice, and techniques. 2nd Ed. Boca Raton, FL: CRC Press.

McMahon, D.J.; Adams, S.L. & McManus, W.R. (2009). Hardening of high-protein nutrition bars and sugar/polyol-protein phase separation. J. Food Sci. 74: E312-E321.

Means, G.E. & Feeney, R.E. (1971). Chemical modifications of proteins. Holden-Day, San Francisco.

Meinders, M.B.J.; van den Bosch, G.G.M. & de Jongh, H.H.J. (2001). Adsorption properties of proteins at and near the air/water interface from IRRAS spectra of protein solutions. Eur. Biophys. J. 30: 256-267.

Melnychyn, P. & Stapley, R. (1973). Acylated protein from coffee whitener formulations. U.S. Patent 3,764,711.

Mendis, E.; Rajapakse, N. & Kim, S. (2005). Antioxidant properties of a radical-scavenging peptide purified from enzymatically prepared fish skin gelatin hydrolysate. J. Agric. Food Chem. 53: 581-587.

Metcalfe, D.D.; Astwood, J.D.; Townsend, R.; Sampson, H.A.; Taylor, S.L. & Fuchs, R.L. (1996). Assessment of the allergenic potential of foods derived from genetically engineered crop plants. Crit. Rev. Fd. Sci. Nutr. 36: S165-S186.

Migneault, I.; Dartiguenave, C.; Bertrand, M.J. & Waldron, K.C. (2004). Gluteraldehyde: behavior in aqueous solution, reaction with proteins, and application to enzyme crosslinking. BioTechniques 37: 790-802.

Mine, Y. (1992). Sulfhydryl groups changes in heat-induced soluble egg white aggregates in relation to molecular size. J. Food Sci. 57: 254-255.

Ming, D. & Hellekant, G. (1994). Brazzein, a new high-potency thermostable sweet protein from Pentadiplandra brazzeana B. FEBS Lett. 355: 106-108.

Mirmoghatadaie, L.; Kadivar, M. & Shahedi, M. (2009). Effects of succinylation and deamidation on functional properties of oat protein isolate. Food Chem. 114: 127-131.

Moore, S. & Stein, W.H. (1948). Photometric ninhydrin method for use in the chromatography of amino acids. J. Biol. Chem. 176: 367-388.

Morel, B.; Varela, L. & Conejero-Lara, F. (2010). The thermodynamic stability of amyloid fibrils studied by differential scanning calorimetry. J. Phys. Chem. B 114: 4010-4019.

Morris, J.A. & Cagan, R.H. (1972). Purification of monellin, the sweet principle of Dioscoreophyllum cumminsii. Biochem. Biophys. Acta 261: 114-122.

Mossine, V.V.; Glinsky, G.V. & Feather, M.S. (1994). The preparation and characterization of some Amadori compounds (1-amino-1-deoxy-D-fructose derivatives) derived from a series of aliphatic omega-amino acids. Carbohydr. Res. 262: 257-270.

Mottram, D.S.; Wedzicha, B.L. & Dodson, A.T. (2002). Acrylamide is formed in the Maillard reaction. Nature 419: 448.

Moure, A.; Sineiro, J.; Domínguez, H. & Parajó, J.C. (2006). Functionality of oilseed protein products: a review. Food Res. Int. 39: 945-963.

Nakai, S. (1983). Structure-function relationships of food proteins with an emphasis on the importance of protein hydrophobicity. J. Agric. Food Chem. 31: 676-683.

Nakai, S. & Li-Chan, E. (1988). Hydrophobic interaction in food systems. CRC Press: Boca Raton, FL.

Narayana, K. & Rao, N. (1991). Effect of acetylation and succinylation on the physicochemical properties of winged bean (Psophocarpus tetragonolobus) proteins. J. Agric. Food Chem. 39: 259-261.

Niestijl Jansen J.J.; Kardinaal, A.F.M.; Huijbers, G.; Vlieg-Boerstra, B.J.; Martens, B.P.M. & Ockhuizen, T. (1994). Prevalence of food allergy and intolerance in the adult Dutch population. J. Allergy Clin. Immunol. 93: 446-456.

Nimni, M.E., Cheung, D.; Strates, B.; Kodama, M. & Sheikh, K. (1987). Chemically modified collagen: a natural biomaterial for tissue replacement. J. Biomed. Mater. Res. 21: 741771.

Nizet, V. (2006). Antimicrobial peptide resistance mechanisms of human bacterial pathogens. Curr. Issues Mol. Biol. 8: 11-26.

Ohta, K.; Masuda, T.; Ide, N. & Kitabatake, N. (2008). Critical molecular regions for elicitation of the sweetness of sweet-tasting protein, thaumatin I. FEBS J. 275: 36443652.

ÓKane, F.E.; Happe, R.P.; Vereijken, J.M.; Gruppen, H. & van Boekel, M.A.J.S. (2004). Heatinduced gelation of pea legumin: comparison with soy glycinin. J. Agric. Food Chem. 52: 5071-5078.

Okuyama, T. & Satake, K. (1960). On the preparation and properties of 2,4,6-trinitrophenylamino acids and -peptides. J. Biochem. 47: 454-466.

Onuchic, J.N.; Luthey-Schulten, Z. & Wolynes, P.G. (1997). Theory of protein folding: the energy landscape perspective. Ann. Rev. Phys. Chem. 48: 545-600.

Owusu Apenten, R.K.; Chee, C. & Hwee, O.P. (2003). Evaluation of a sulphydryl-disulphide exchange index (SEI) for whey proteins - beta-lactoglobulin and bovine serum albumin. Food Chem. 83: 541-545.

Pabit, S.A.; Roder, H. & Hagen, S.J. (2004). Internal friction controls the speed of protein folding from a compact configuration. Biochemistry 43: 12532-12538.

Paik, W. & Kim, S. (1972). Effect of methylation on susceptibility of proteins to proteolytic enzymes. Biochemistry 11: 2589-2594.

Panasiuk, R.; Amarowicz, R.; Kostyra, H. & Sijtsma, L. (1998). Determination of α-amino nitrogen in pea protein hydrolysates: a comparison of three analytical methods. Food Chem. 62: 363-367.

Paparcone, R. & Buehler, M.J. (2011). Failure of $A\beta(1-40)$ amyloid fibrils under tensile loading. Biomaterials 32: 3367-3374.

Paparcone, R.; Keten, S. & Buehler, M.J. (2010). Atomistic simulation of nanomechanical properties of Alzheimer's $A\beta(1-40)$ amyloid fibrils under compressive and tensile loading. J. Biomechanics 43: 1196-1201.

Park, H.J.; Chinnan, M.S. & Shewfelt, R.L. (1994). Edible corn-zein film coatings to extend storage life of tomatoes. J. Food Process Preserv. 18: 317-331.

Park, H.J. (1999). Development of advanced edible coatings for fruits. Trends Food Sci. Technol. 10: 254-260.

Patel, K. & Borchardt, R.T. (1990a). Chemical pathways of peptide degradation. II. Kinetics of deamidation of an asparaginyl residue in a model hexapeptide. Pharm. Res. 7: 703-711.

Patel, K. & Borchardt, R.T. (1990b). Chemical pathways of peptide degradation. III. Effect of primary sequence on the pathways of deamidation of asparaginyl residues in hexapeptides. Pharm. Res. 7: 787-793.

Patil, S.M.; Mehta, A.; Jha, S. & Alexandrescu, A.T. (2011). Heterogeneous amylin fibril growth mechanisms imaged by total internal reflection fluorescence microscopy. Biochemistry 50: 2808-2819.

Payne, P.I.; Holt, L.M.; Jackson, E.A. & Law, C.N. (1984). Wheat storage proteins: their genetics and their potential for manipulation by plant breeding. Phil. Trans. R. Soc. Lond. B 304: 359-371.

Pedroche, J.; Yust, M.M.; Lqari, H.; Giron-Calle, J.; Alaiz, M.; Vioque, J. & Millan, F. (2004). Brassica carinata protein isolates: chemical composition, protein characterization and improvement of functional properties by protein hydrolysis. Food Chem. 88: 337-346.

Pellegrino, L.; van Boekel, M.A.J.S.; Gruppen, H.; Resmini, P. & Pagani, M.A. (1999). Heatinduced aggregation and covalent linkages in β-casein model systems. Int. Dairy J. 9: 255-260.

Pereda, M.; Aranguren, M.I. & Marcovich, N.E. (2008). Characterization of chitosan/caseinate films. J. Appl. Polymer Sci. 107: 1080-1090.

Pereda, M.; Aranguren, M.I. & Marcovich, N.E. (2009). Water vapor absorption and permeability of films based on chitosan and sodium caseinate. J. Appl. Polym. Sci. 111: 2777-2784.

Petersen, A.; Schramm, G.; Schlaak, M. & Becker, W.M. (1998). Post-translational modifications influence IgE reactivity to the major allergen Phl p 1 of timothy grass pollen. Clin. Exp. Allergy 28: 315-321.

Pétra, P.H. (1971). Modification of carboxyl groups in bovine carboxypeptidase A. I. Inactivation of the enzyme by *N*-ethyl-5-phenylisoxazolium-3'-sulfonate (Woodward's reagent K). Biochemistry 10: 3163-3170.

Plaxco, K.W.; Simons, K.T.; Ruczinski, I. & Baker, D. (2000). Topology, stability, sequence, and length: defining the determinants of two-state protein folding kinetics. Biochemistry 39: 11177-11183.

Plaxco, K.W.; Simons, K.T. & Baker, D. (1998). Contact order, transition state placement and the refolding rates of single protein domains. J. Mol. Biol. 277: 985-994.

Plotkin, S.S. & Onuchic, J.N. (2002). Understanding protein folding with energy landscape theory. Part I: Basic concepts. Quart. Rev. Biophys. 35: 111-167.

Pogaku, R.; Eng Seng, C.; Boonbeng, L. & Kallu, U. (2007). Whey protein isolate-starch system - a critical review. Intl. J. Food Eng. 3: 104-119.

Ponce, A.G.; Roura, S.I.; Del Valle, C.E. & Moreira, M. (2008). Antimicrobial and antioxidant activities of edible coatings enriched with natural plant extracts: in vitro and in vivo studies. Postharvest. Biol. Technol. 49: 294-300.

Promeyrat, A.; Bax, M.L.; Traoré, S.; Abry, L.; Santé-Lhoutellier, V. & Gatellier, Ph. (2010). Changed dynamics in myofibrillar protein aggregation as a consequence of heating time and temperature. Meat Sci. 85: 625-631.

Qiu, L. & Hagen, S.J. (2004). A limiting speed for protein folding at low solvent viscosity. J. Am. Chem. Soc. 126: 3398-3399.

Quintanilla-Guerrero, F.; Duarte-Vázquez, M.A.; Tinoco, R.; Gómez-Suárez, M.; GarcíaAlmendárez, B.E.; Vazquez-Duhalt, R. & Regalado, C. (2008). Chemical modification of turnip peroxidase with methoxypolyethylene glycol enhances activity and stability for phenol removal using the immobilized enzyme. J. Agric. Food Chem. 56: 8058-8065.

Quintas, A.; Saraiva M.J. & Brito, R.M. (1997). The amyloidogenic potential of transthyretin variants correlates with their tendency to aggregate in solution. FEBS Lett. 418: 297300.

Quiocho, F.A. & Richards, F.M. (1966). The enzymic behaviour of carboxypeptidase-A in the solid state. Biochemistry 5: 4062-4076.

Rahma, E.H. & Narasinga Rao (1983). Effect of acetylation and succinylation of cottonseed flour on its functional properties. J. Agric. Food Chem. 31: 351-355.

Ramírez-Alvarado, M.; Merkel, J.S. & Regan, L. (2000). A systematic exploration of the influence of protein stability on amyloid fibril formation in vitro. Proc. Natl. Acad. Sci. U.S.A. 97: 8979-8984.

Ramírez-Jiménez, A.; Guerra-Hernández, E. & García-Villanova, B. (2000). Browning indicators in bread. J. Agric. Food Chem. 48: 4176-4181.

Richard, F.M. & Knowles, J.R. (1968). Gluteraldehyde as a protein cross-linkage reagent. J. Mol. Biol. 37: 231-233.

Riha, W.E.; Izzo, H.V.; Zhang, J. & Ho, C.T. (1996). Nonenzymatic deamidation of food proteins. Crit. Rev. Food Sci. Nutr. 6: 225-255.

Robinson, A.B. & Rudd, C. (1974). Deamidation of glutaminyl and asparaginyl residues in peptides and proteins. Curr. Top. Cell. Regul. 8: 247-295.

Rodríguez Patino, J.M.; Carrera Sánchez, C. & Rodríguez Niño, M.R. (2008). Implications of interfacial characteristics of food foaming agents in foam formulations. Adv. Colloid Interf. Sci. 140: 95-113.

Rondeau, P.; Navarra, G.; Cacciabaudo, F.; Leone, M.; Bourdon, E. & Militello, V. (2010). Thermal aggregation of glycated bovine serum albumin. Biochim. Biophys. Acta 1804: 789-798.

Rondeau, P.; Armenta, S.; Caillens, H.; Chesne, S. & Bourdon, E. (2007). Assessment of temperature effects on β-aggregation of native and glycated albumin by FTIR spectroscopy and PAGE: relations between structural changes and antioxidant properties. Arch. Biochem. Biophys. 460: 141-150.

Röper, H.; Röper, S.; Heyns, K. & Meyer, B. (1983). N.M.R. spectroscopy of N-(1-deoxy-D-fructos-1-yl)-L amino acids ("fructose-amino acids"). Carbohydr. Res. 116: 183-195.

Rosén, J. & Hellenäs, K.E. (2002). Analysis of acrylamide in cooked foods by liquid chromatography tandem mass spectrometry. Analyst 127: 880-882.

Roth, M. (1971). Fluorescence reaction for amino acids. Anal. Chem. 43: 880-882.

Rudd, P.M.; Colominas, C.; Royle, L.; Murphy, N.; Hart, E.; Merry, A.H.; Hebestreit, H.F. & Dwek, R.A. (2001). A high-performance liquid chromatography based strategy for rapid, sensitive sequencing of N-linked oligosaccharide modifications to proteins in sodium dodecyl sulphate polyacrylamide electrophoresis gel bands. Proteomics 1: 285-294.

Rudén, C. (2004). Acrylamide and cancer risk - expert risk assessments and the public debate. Food Chem. Tox. 42: 335-349.

Sala, G.; de Wijk, R.A.; van de Velde, F. & van Aken G.A. (2008). Matrix properties affect the sensory perception of emulsion-filled gels. Food Hydrocolloids 22: 353-363.

Salgó, A.; Granzler, K. & Jecsai, J. (1984). Simple enzymatic methods for prediction of plant protein digestibility. In: R. Lásztity, M. Hidvé. Proc. Int. Assoc. Cereal Chem. Symp. (pp. 311-321). Budapest: Akadémiai Kiadó.

Salunkhe, D.K.; Jadhav, S.J.; Kadam, S.S. & Chavan, J.K. (1982). Chemical, biochemical, and biological significance of polyphenols in cereals and legumes. Crit. Rev. Food Sci. Nutr. 17: 277-305.

Samejima, K.; Dairman, W. & Underfriend, S. (1971). Condensation of ninhydrin with aldehydes and primary amines to yield highly fluorescent ternary products. Anal. Biochem. 42: 222-247.

Sanchez-Ruiz, J.M. (2011). Probing free-energy surfaces with differential scanning calorimetry. Annu. Rev. Phys. Chem. 62: 231-255.

Sancho, A.I.;. Rigby, N.M.; Zuidmeer, L.; Asero, R.; Mistrello, G.; Amato, S.; GonzálezMancebo, E.; Fernández-Rivas, M.; van Ree, R. & Mills, E.N.C. (2005). The effect of thermal processing on the IgE reactivity of the non-specific lipid transfer protein from apple, Mal d 3. Allergy 60: 1262-1268.

Sanger, F. (1945). The free amino groups of insulin. Biochem. J. 39: 507-515.

Santé-Lhoutellier, V.; Astruc, T.; Marinova, P.; Greve, E. & Gatellier, P. (2008). Effect of meat cooking on physicochemical state and in vitro digestibility of myofibrillar proteins. J. Agric. Food Chem. 56: 1488-1494.

Satake, K.; Okuyama, T.; Ohashi, M. & Shinoda, T. (1960). The spectrophotometric determination of amine, amino acid and peptide with 2,4,6-trinitrobenzene-1sulfonic acid. J. Biochem. 47: 654.

Sawyer, W.H. (1968). Heat denaturation of bovine β-lactoglobulins and relevance of disulfide aggregation. J. Dairy Sci. 51: 323-329.

Schilling, E.D.; Burchill, P.I. & Clayton, R.A. (1963). Anomalous reactions of ninhydrin. Anal. Biochem. 5: 1-6.

Schwenke, K.D.; Knopfe, C.; Mikheeva, L.M. & Grinberg, V.Y. (1998). Structural changes of legumin from Faba beans (Vicia faba L.) by succinylation. J. Agric. Food Chem. 46: 2080-2086.

Sheih, I.C.; Wu, T.K. & Fang, T.J. (2009). Antioxidant properties of a new antioxidative peptide from algae protein waste hydrolysate in different oxidation systems. Bioresour. Technol. 100: 3419-3425.

Shental-Bechor, D. & Levy, Y. (2008). Effect of glycosylation on protein folding: a close look at thermodynamic stabilization. Proc. Natl. Acad. Sci. U.S.A. 105: 8256-8261.

Shih, F.F. (1990). Deamidation during treatment of soy protein with protease. J. Food Sci. 55: 127-129.

Shimada, K. & Cheftel, J.C. (1989). Sulfhydryl group/disulfide bond interchange reactions during heat-induced gelation of whey protein isolate. J. Agric. Food Chem. 3: 161168.

Shirahama, H.; Lyklema, J. & Norde, W. (1990). Comparative protein adsorption in model systems. J. Colloid Interf. Sci. 139: 177-187.

Siepen, J.A. & Westhead, D.R. (2002). The fibril_one on-line database: mutations, experimental conditions, and trends associated with amyloid fibril formation. Protein Sci. 11: 1862-1866.

Silva Freitas, D. & Abrahão-Neto, J. (2010). Batch purification of high-purity lysozyme from egg white and characterization of the enzyme modified by PEGylation. Pharm. Biol. 48: 554-562.

Sinha, U. & Brewer, J.M. (1985). A spectrophotometric method for quantitation of carboxyl group modification of proteins using Woodward's reagent K. Anal. Biochem. 151: 327-333.

Siu, M. & Thompson, L.U. (1982). In vitro and in vivo digestibilities of succinylated cheese protein concentrates. J. Agric. Food Chem. 30: 743-747.

Skraup, Z.H. & Kaas, K. (1906). Über die Einwirkung von salpetriger Säure auf Ovalbumin. Justus Liebigs. Ann. Chemie 351: 379-389.

Smythe, C.V. (1936). The reaction of iodoacetate and of iodoacetamide with various sulfhydryl groups, with urease, and with yeast preparations. J. Biol. Chem. 114: 601612.

Soderling, T.R. (1975). Regulation of glycogen synthetase. Specificity and stoichiometry of phosphorylation of the skeletal muscle enzyme by cyclic 3′: 5′-AMP-dependent protein kinase. J. Biol. Chem. 250: 5407-5412.

Song, Y.; Azakami, H.; Hamasu, M. & Kato, A. (2001). In vivo glycosylation suppresses the aggregation of amyloidogenic hen egg white lysozymes expressed in yeast. FEBS Lett. 491: 63-66.

Sorci, M.; Silkworth, W.; Gehan, T. & Belfort, G. (2011). Evaluating nuclei concentration in amyloid fibrillation reactions using back-calculation approach. PLoS One 6: e20072.

Stadler, R.H.; Blank, I.; Varga, N.; Robert, F.; Hau, J.; Guy, P.A.; Robert, M.-C. & Riediker, S. (2002). Acrylamide from Maillard reaction products. Nature 419: 449.

Straub, J.E. & Thirumalai, D. (2011). Toward a molecular theory of early and late events in monomer to amyloid fibril formation. Annu. Rev. Phys. Chem. 62: 437-463.

Sun, Y.; Hayakawa, S. & Izumori, K. (2004). Modification of ovalbumin with a rare ketohexose through the Maillard reaction: effect on protein structure and gel properties. J. Agric. Food Chem. 52: 1293-1299.

Sun, Y. & Hayakawa, S. (2002). Heat-induced gels of egg white/ovalbumins from five avian species: thermal aggregation, molecular forces involved, and rheological properties. J. Agric. Food Chem. 50: 1636-1642.

Suttiprasit, P.; Krisdhasima, V. & McGuire, J. (1992). The surface activity of α-lactalbumin, β lactoglobulin and bovine serum albumin. I. Surface tension measurements with single-component and mixed solutions. J. Colloid Interface Sci. 154: 316-326.

Tareke, E.; Rydberg, P.; Karlsson, P.; Eriksson, S. & Törnqvist, M. (2002). Analysis of acrylamide, a carcinogen formed in heated foodstuffs. J. Agric. Food Chem. 50: 49985006.

Tawfik, D.S. (2002). Chapter 63: Side chain selective chemical modifications of proteins. In: The Protein Protocols Handbook, 2nd Edition, Edited by J.M. Walker, Humana Press Inc. Totowa, NJ, pp 465-467.

Teplow, D.B.; Lazo, N.D.; Bitan, G.; Bernstein, S.; Wyttenbach, T.; Bowers, M.T.; Baumketner, A.; Shea, J.E.; Urbanc, B.; Cruz, L.; Borreguero, J. & Stanley, H.E. (2006). Elucidating amyloid-β protein folding and assembly: a multidisciplinary approach. Acc. Chem. Res. 39: 635-645.

Thomas, M.E.C.; Scher, J.; Desobry-Banon, S. & Desobry, S. (2004). Milk powders ageing: effect on physical and functional properties. Crit. Rev. Food Sci. Nutr. 44: 297-322.

Tyler-Cross, R. & Schirch, V. (1991). Effects of amino acid sequence, buffers, and ionic strength on the rate and mechanism of deamidation of asparagines residues in small peptides. J. Biol. Chem. 266: 22549-22556.

Van den Berg, L.; Carolas, A.L.; van Vliet, T.; van der Linden, E.; van Boekel, M.A.J.S. & van de Velde, F. (2008). Energy storage controls crumbly perception in whey proteins/polysaccharide mixed gels. Food Hydrocolloids 22: 1404-1417.

Van der Wel, H. & Loeve, K. (1972). Isolation and characterization of thaumatin I and II, the sweet-tasting proteins from Thaumatococcus danielli Benth. Eur. J. Biochem. 31: 221225.

Van Teeffelen, A.M.M.; Broersen, K. & de Jongh, H.H.J. (2005). Glucosylation of β lactoglobulin lowers the heat capacity change of unfolding; a unique way to affect protein thermodynamics. Protein Sci. 14: 2187-2194.

van Vliet, T. & Walstra, P. (1995). Large deformation and fracture behaviour of gels. Faraday Discussions 101: 359-370.

Veerman, C.; Ruis, H.; Sagis, L.M.C.; van der Linden, E. (2002). Effect of electrostatic interactions on the percolation concentration of fibrillar β-lactoglobulin gels. Biomacromolecules 3: 869-873.

Visschers, R.W. & de Jongh, H.H.J. (2005). Disulphide bond formation in food protein aggregation and gelation. Biotechnol. Adv. 23: 75-80.

Vioque, J.; Sánchez-Vioque, R.; Clemente, A.; Pedroche, J.; Bautista, J. & Millán, F.J. (1999). Production and characterization of an extensive rapeseed protein hydrolysate. J. Am. Oil Chem. Soc. 76: 819-823.

Wada, R. & Kitabatake, N. (2001). β-Lactoglobulin A with N-ethylmaleimide-modified sulfhydryl residue, polymerized through intermolecular disulfide bridge on heating in the presence of dithiothreitol. J. Agric. Food Chem. 49: 4971-4976.

Walder, R. & Schwartz, D.K. (2010). Single molecule observations of multiple protein populations at the oil-water interface. Langmuir 26: 13364-13367.

Walsh, D.M.; Lomakin, A.; Benedek, G.B.; Condron, M.M. & Teplow, D.B. (1997). Amyloid β-protein fibrillogenesis. Detection of a protofibrillar intermediate. J. Biol. Chem. 272: 22364-22372.

Wanasundara, P.K.J.P.D. & Shahidi, F. (1997). Functional properties of acylated flax protein isolates. J. Agric. Food Chem. 45: 2431-2441.

Wang, C.; Eufemi, M.; Turano, C. & Giartosio, A. (1996). Influence of the carbohydrate moiety on the stability of glycoproteins. Biochemistry 35: 7299-7307.

Waniska, R.D. & Kinsella, J.E. (1979). Foaming properties of proteins: evaluation of a column aeration apparatus using ovalbumin. J. Food Sci. 44: 1398-1411.

Waniska, R.D. & Kinsella, J.E. (1985). Surface properties of β-lactoglobulin: adsorption and rearrangement during film formation. J. Agric. Food Chem. 33: 1143-1148.

Weijers, M.; Broersen, K.; Barneveld, P.A.; Cohen Stuart, M.A.; Hamer, R.J.; de Jongh, H.H. & Visschers, R.W. (2008). Net charge affects morphology and visual properties of ovalbumin aggregates. Biomacromolecules 9: 3165-3172.

Wierenga, P.A.; Kosters, H.; Egmond, M.R.; Voragen, A.G.J. & de Jongh, H.H.J. (2006). Importance of physical vs. chemical interactions in surface shear rheology. Adv. Colloid Interface Sci. 119: 131-139.

Wierenga, P.A.; Meinders, M.B.J.; Egmond, M.R.; Voragen, F.A.G.J. & de Jongh, H.H.J. (2003). Protein exposed hydrophobicity reduces the kinetic barrier for adsorption of ovalbumin to the air-water interface. Langmuir 19: 8964-8970.

Wierenga, P.A.; Meinders, M.B.; Egmond, M.R.; Voragen, A.G. & de Jongh, H.H. (2005). Quantitative description of the relation between protein net charge and protein adsorption to air-water interface. J. Phys. Chem. B 109: 16946-16952.

Wilde, P.J. (2000). Interfaces: their role in foam and emulsion behaviour. Curr. Opin. Colloid Interface Sci. 5: 176-181.

Wilde, P.J.; Mackie, A.; Husband, F.; Gunning, P. & Morris, V. (2004). Proteins and emulsifiers at liquid interfaces. Adv. Colloid Interface Sci. 108-109: 63-71.

Wimley, W.C. (2010). Describing the mechanism of antimicrobial peptide action with the interfacial activity model. ACS Chem. Biol. 5: 905-917.

Wolynes, P.G. (2005). Recent successes of the energy landscape theory of protein folding and function. Quart. Rev. Biophys. 38: 405-410.

Woodward, R.B. & Olofson, R.A. (1961). The reaction of isoxazolium salts with bases. J. Am. Chem. Soc. 83: 1007-1009.

Woodward, R.B.; Olofson, R.A. & Mayer, H. (1961). A new synthesis of peptides. J. Am. Chem. Soc. 83: 1010-1012.

Wright, H.T. & Urry, D.W. (1991). Nonenzymatic deamidation of asparaginyl and glutaminyl residues in proteins. Crit. Rev. Biochem. Mol. Biol. 26: 41-52.

Xu, L.J.; Sheldon, B.W.; Larick, D.K. & Carawan, R.E. (2002). Recovery and utilization of useful by-products from egg processing wastewater by electrocoagulation. Poult. Sci. 81: 785-792.

Xu, Z.; Paparcone, R. & Buehler, M.J. (2010). Alzheimer's $A\beta(1-40)$ amyloid fibrils feature size-dependent mechanical properties. Biophys. J. 96: 2053-2062.

Yemm, E.W. & Cocking, E.C. (1955). The determination of amino acids with ninhydrin. Analyst 80: 209-213.

Zamora, R. & Hildalgo, F.J. (2006). Coordinate contribution of lipid oxidation and Maillard reaction to the nonenzymatic food browning. Crit. Rev. Food Sci. Nutr. 45: 49-59.

Zavodsky, M.; Chen, C.W.; Huang, J.K.; Zolkiewski, M.; Wen, L. & Krishnamoorthi, R. (2001). Disulfide bond effects on protein stability: designed variants of Cucurbita maxima trypsin inhibitor-V. Protein Sci. 10: 149-160.

Zhang, J.; Lee, T.C. & Ho, C.-T. (1993). Comparative study on kinetics of nonenzymatic deamidation of soy protein and egg white lysozyme. J. Agric. Food Chem. 41: 22862290.

Zhao, Y.; Ma, C.-Y.; Yuen, S.-N. & Phillips, D.L. (2004a). Study of succinylated food proteins by raman spectroscopy. J. Agric. Food Chem. 52: 1815-1823.

Zhao, Y.; Ma, C.-Y.; Yuen, S.-N. & Phillips, D.L. (2004b). Study of acetylated food proteins by raman spectroscopy. J. Food Sci. 69: 206-213.

Zhou, P.; Liu, X. & Labuza, T.P. (2008). Effects of moisture-induced whey protein aggregation on protein conformation, the state of water molecules, and the microstructure and texture of high-protein-containing matrix. J. Agric. Food Chem. 56: 4535-4540.

Zidar, J. & Merzel, F. (2011). Probing amyloid-beta fibril stability by increasing ionic strengths. J. Phys. Chem. B 115: 2075-2081.

Zinoviadou, K.G.; Koutsoumanis, K.P. & Biliaderis, C.G. (2009). Physico-chemical properties of whey protein isolate films containing oregano oil and their antimicrobial action against spoilage flora of fresh beef. Meat Sci. 82: 338-345.

Chapter-6

Design, Intensification, and Modeling of Rational Asymmetric Catalysts

1. Introduction: An Overview

The advancement of novel catalysts for diverse chemical processes necessitates a fundamental comprehension at the molecular level of their mechanisms, alongside expertise in surface science, chemistry, materials science, process modeling, and process systems engineering. Catalyst design and kinetic modeling have traditionally relied on chemical intuition, which encompasses a substantial empirical database and qualitative principles of chemical reaction engineering and surface science. Recently, first-principle kinetic modeling has emerged as a crucial instrument for examining catalytic processes and catalyst architectures, facilitating enhanced operational advantages through the integration of first-principle guided exploration with experimental data. Nanocatalysis design constitutes the foundational elements of micro-level controlled reaction engineering, with size and shape-dependent material properties serving as the critical enabling factors of this new technology. The qualitative difference pertains to the material qualities that alter at the 1-10 nanometer scale. Significant advancements in nanomaterial production capabilities, with enhanced control over size and form, have been reported in the past two decades; yet, the criteria for rational catalyst design remain underdeveloped. The examination of reactions introduces inherent complexity in the development of nanomaterial topology for catalyst design and underscores the importance of a systematic multiscale simulation approach for design and prediction.In the field of heterogeneous catalysis, desired topological improvements on the basis of experimental information is well known and tedious [1]. Advances in surface science enable data manipulation of individual atoms on the catalyst surface with experiments provides initial guess towards systematic design. Number of model based catalyst design strategies were reported in open literature like use of qualitative reasoning and knowledge-based systems [2,3], computational models [4-6] and detailed microkinetic modeling for catalytic systems [7]. In order to get the job done, compositional analysis of physical, chemical, and/or biological properties and validation of appropriate molecular structure (robust forward model linked with material description and knowledge extraction from experiment). Reaction modeling systems, optimization, and artificial intelligence-based tools were widely developed in an automated environment to eliminate the necessity for reaction engineering skills in constructing robust kinetic models. These instruments provide thorough examination of various reaction mechanisms based on experimental data. Katare et al. [6] established a kinetic model for propane aromatization with an H-ZSM-5 zeolite catalyst, providing a proof-of-concept in their research. Similarly, organometalic catalyst has interests in the broad areas of homogeneous catalysis, and catalyst modeling and mechanism elucidation. The main thrust of the field is in catalyst design and investigation of mechanistic aspects includes organic synthetic, organometallic and inorganic chemistry, molecular modeling and reaction monitoring. Fundamental research provide a theoretical basis and looks at the catalyst modeling using both experimental and theoretical techniques, providing information on active species and likely transition states. Ultimately helps in designing real experimental systems to be modeled with reliability and confidence. Mechanistic investigations using spectroscopic and kinetic methods help elucidate catalytic pathways and possible catalyst decomposition routes. Assessment and optimization of catalyst performance factors effecting catalyst behavior are important. The study of heterocyclic carbenes and synthesis of mixed carbene-donor ligands (transition metal complexes) were under extensive consideration. The structure, stoichiometric reaction behaviour and catalytic potential of the new complexes are the key goals to elaborate ways of manipulating the chemistry of the complexes in catalysis. The synthetic work is integrated with theoretical analyses of orbital behavior and its impact on the energetics of decomposition reactions. The synthesis of novel polydentate ligands and their transition metal complexes presents opportunities for the development of new catalysts for alkene polymerization, copolymerization, and oligomerization. Ligands containing phosphorus, nitrogen, and/or oxygen donors constitute the primary categories of ligands, and their selection is contingent upon the metal under examination. Consequently, the technique for catalyst discovery aims to identify an appropriate strategy grounded in the conceptual frameworks of surface mechanisms, molecular chemistry, biochemistry, and solid-state chemistry.

2. Reaction modeling suite

The model development process requires swift evaluation of reaction mechanistic hypotheses that may elucidate the experimental data. A number of user-driven tools were provided that assist knowledge archiving and retrieval for automated reaction kinetic modeling. These modeling packages formulate reaction mechanisms and immediately generate kinetic models. The architectural principles underlying complex mathematical modeling are illustrated in picture 1. Modeling can yield reaction energies (ΔH_r), energy barriers (ΔE_a), the positions of intermediates, and it can investigate catalyst composition, toxins, the effects of promoters, and the efficiency of catalysts by alternate pathways. In the design of homogeneous catalysts, the regulation of activity, ligand type, co-catalyst concentration, molecular weight, and tacticity (which is contingent upon equilibrium) are also significant factors. The automated kinetic models development software tools of reaction networking, parameter optimization and overall kinetic modeling etc., are available in bulk like Reactor, Reaction Modeling Suite (RMS), KINAL A, Aspen Custom Modeler, gPROMS, Presto Kinetics, GREGPAK, Matlab, MLAB, ParaMetra, Scientist, Eurokin, polymath, Chemkin, Mitsubishi, MKM, IBM CKS, NetGen, XMG, Dynetica, Forcite, COMPASS, Sorption, CASTEP, DMol [3], Gepasi, Athena Visual Studio parameter estimation tool, E-cell, DBsolve,

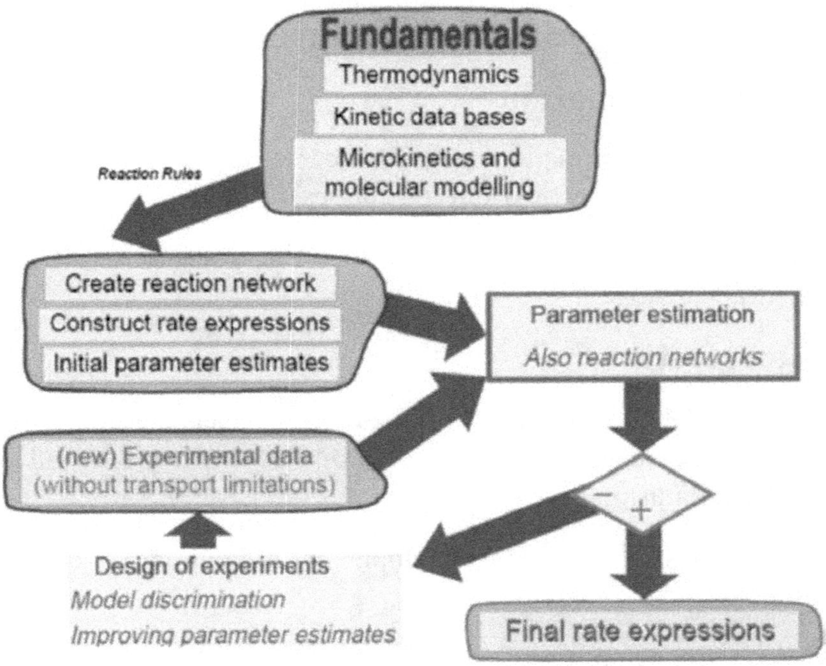

Fig. 1. Inventory of model building process

etc. For catalyst design Katare et al. developed a predictive model for surface reactions studies and claims to be rational and robust [6]. The essential steps in this model construction include the formulation of the most straightforward plausible reaction mechanism, the conversion of the reaction mechanism into a mathematically manageable computational model, the resolution of the model to estimate parameters considering high-throughput and/or limited experimental data, the refinement of the model to enhance data alignment by modifying the mechanism, and the proposal of new experiments to differentiate between multiple models. These software products required cutting-edge functionality for certain reasons [6]:

Generation and networking of reaction mechanisms

Accurate representation of the large number of elementary reactions, overall reactions and species.

- Robust compiler that fallows the generic reaction rules to generate network generator.
- Pruning the reaction mechanism to get it in simpler and sophisticated.
- Reduce the number of thermodynamic and/or kinetic parameters to optimize based on thermokinetic and experimental data manipulation.

Kinetic parameter estimation

- Robust solvers that can handle large number of differential-algebraic equations and validate parameter estimation techniques.
- Evaluation of the multiple solutions for the parameters that explain the experimental data also.

Results and forecasting

- Discrepancies identification and optimization between the key features of model and experimental data that leads to design of experiment and questions on generated mechanism.
- Explain robustness of the model development with boundaries and shows statical analysis.

The most critical and important task in building a robust model is development of kinetic model and its also the most-time-consuming (excepting experimental or pilotplant campaigns). Without a satisfactory reaction mechanism and kinetics, a model may be applicable to narrow ranges of conditions and dangerous to use for predictions. The open literature information may not exactly match your conditions or catalyst formulation, therefore, formulation or selection of reaction-mechanism and rate expressions is best suggested. Establishing the reaction kinetics involves selection of rate expressions, and determination of rate parameters. Rate expressions involves four parameters; preexponential factor, activation energies, reaction order to each component and adsorption constants. The pre-exponential factors must be determined from the actual experimental data for catalyst or reaction conditions. This chapter give pathway to model information for catalyst design and intensification from state of the art reaction kinetics data generation, to the developing of computer-based knowledge organization for catalyst development, with brief review said techniques and models.

3. Mechanism of Thermodynamical consistency

In mechanism development and rate parameter optimization, thermodynamic consistency at both the enthalpic and entropic levels was often overlooked. An inconsistency in enthalpies gives incorrect solutions to the energy conservation equation, which lead towards ambiguous predictions of heat exchange, conversion, selectivity, etc.; while entropic inconsistency explains fundamental inconsistency (i.e. the pre-exponential factors). As there is no specific criterion for thermodynamic consistency evaluation and it distorts underlying equilibrium constant, only few published mechanisms are thermodynamically consistent [912]. In general, for any i-th reaction in a, the following equations form the basis of the enthalpic and entropic constraints [13]. For thermodynamically consistent mechanism the ratio of equilibrium constants (determined by gas phase thermochemistry) should be closer to 1 and for rigorous account of temperature variations data its necessary.

$$\Delta H_i = E_i^f - E_i^b = \sum_{j=1}^{L} c_{ij}\left(E_j^f - E_j^f\right), i = L+1, I$$

$$A_i^f / A_i^b = e^{\Delta S_i / R}$$

$$\frac{\Delta S_i}{R} = \ln\left(A_i^f / A_i^b\right) = \ln \prod_{j=1}^{L} \left(A_j^f / A_j^b\right)^{c_{ij}}, i = L+1, I,$$

where f stands for the forward, b for backward reaction steps, A is the pre-exponential factor, E is the activation energy, R is the universal gas constant, ΔH is enthalpy change and ΔS is entropy change of the reaction, respectively. For a linearly independent set of reactions network, if L is the size of reaction mechanism (unique in size) having I number of reactions, then L ≤ I. The linear dependent reactions constraints can be expressed as c_{ij}, those are the coefficients of i-th linearly decomposed reaction onto the reaction basis.

4. Computer-assisted catalyst design

Computer-aided catalyst design and intensification represents a cohesive methodology in catalysis research and development, since chemical engineering is presently experiencing a resurgence under the term process intensification. The notion of process intensification in catalysis is novel and not firmly established within the catalyst design sector. It is especially appropriate for the design and development of novel catalysts. The integration of suitably designed micro-structured catalysts with precise control over the reaction's surface chemistry is shown. The modeling of processes at the active sites of catalysts is occasionally conducted using density functional theory, which will not be addressed here.

5. Reaction modeling suite

The Reaction Modeling Suite (RMS) was created at Purdue University, USA [6]. The quantitative kinetic characterization of a reaction mechanism necessitates the formulation of both differential and algebraic equations for each reaction component. RMS may formulate reaction mechanisms that generally encompass over 10 differential equations for a basic reaction, and for intricate systems, it may require 100 or more differential equations. Every system of differential equations comprises several kinetic parameters that must be derived from experimental data through regression analysis. Developing code and solving the relevant differential equations is a laborious and time-intensive endeavor, with a significant potential for errors due to its complexity. The essential steps in this process include the formulation of a reaction mechanism based on chemical principles, the translation of this mechanism into a manageable mathematical model using computer language, the integration of a set of differential and algebraic equations with appropriate initial conditions, the estimation of optimal model parameters, and ultimately, the assessment of the reliability of these parameters. Each of the aforementioned steps demands substantial skill, time, and effort. Furthermore, numerous potential kinetic models usually require evaluation, necessitating the repetition of model creation and solving tasks for each new physical model. RMS provides the necessary tools that enable chemical professionals to concentrate on chemistry without being too encumbered by the execution of requisite numerical procedures. Purdue University has successfully created a suite of systems tools, known as the Reaction Modeling Suite (RMS), for the quick construction of intricate kinetic models and the evaluation of these models using extensive and varied experimental data. These system tools are crucial for ensuring that model development and assessment align with the rapid data generation from high-throughput experimentation [6]. In RMS, the Chemistry Compiler processes a collection of chemistry rules (i.e., any kinetic model is characterized by a set of reaction rules), verifies the chemical consistency of the rule set, and then produces the corresponding kinetic expressions. Subsequently, an Equation Generator formulates the kinetic expressions into suitable differential and algebraic equations that may be solved numerically. Parameter Optimization is the subsequent phase in which the optimal parameters for a specific model required to characterize the experimental data were identified, particularly challenging due to the extensive, nonlinear, coupled sets of DAEs linked to intricate kinetic mechanisms. A Statistical Analyzer has been created to objectively evaluate the reliability of model parameters, incorporating nonlinear statistics. The RMS tools provide the nearly automated investigation of various reaction mechanisms and kinetics, aiding in catalyst design.

6. Enhancement and optimization of catalysts

The following is an overview of the catalyst modeling and optimization features necessary in developing attractive suite. Catalysts can be modeled in several different ways, with increasing levels of complexity by considering pellet balances, pellet factor effectiveness factor using is basic activity profiles, etc. With out pellet balances optimization is a very basic modelling level, it is not necessary to make any extra calculations due to the presence of a catalyst by expressing reaction kinetics in terms of composition of the catalyst. The catalyst composition can then be made an optimization variable. In the case of plug flow reactors, the composition can be varied along the length of the reactor, and viewed using the catalyst profile. Pellet factor calculation involves balances between component concentrations in the bulk, and within the catalyst pellet. The activity distribution inside the pellet is represented by a "Dirac-delta" function (infinite concentration at a single location). Although this cannot be achieved in practice, the Dirac-delta function can be closely approximated by a narrow step distribution of active material around the optimal location of the delta function without noticeable loss in performance.

Catalyst effectiveness factor calculations may involve Thiele modulus and effectiveness factor for the pellet. The effectiveness factor is a function of temperature, pressure and catalyst properties such as shape, size and diffusivity of components. When using this method the catalyst properties become optimizable parameters, and optimized profiles can be obtained. This method is sub-divided into basic activity profile as represented below:

- Uniform - The active material is uniformly distributed through the pellet.
- Egg-shell - The active material is concentrated at the pellet surface, and rarefied towards the centre.
- Egg-yolk - The active material is concentrated at the pellet centre, and rarefied towards the surface.
- Peaked - The active material is concentrated at a location partway between the centre and surface, and rarefied towards the centre and surface.

- Layered - The active material is distributed in a layer of thickness T , and location L from the pellet centre, where T and L are expressed in fractions of the pellet diameter.

The activity profiles are represented as infinitely variable profiles and by definition, when using the effectiveness factor method, the intrapellet mass balances are solved to determine the conentration profile throughout the catalyst pellet. There are also three further levels of complexity which can be selected to improve the model at the expense of computational time:

- Interphase mass balances causes the concentration profile between the pellet surface and the bulk material to be calculated.
- Interphase heat balances calculates the temperature profile between the pellet surface and the bulk material
- Intrapellet heat balances calculates the temperature profile within the catalyst pellet

Catalyst deactivation is defined as the steady degradation of the performance of the catalyst over time. Deactivation can be modeled by describing a deactivation correlation which relates the catalyst performance to its age. The deactivation factor varies from 1.0 (fresh catalyst) to 0.0 (fully depleted catalyst). If catalyst deactivation is significant it is important to consider its effect when optimising the catalyst properties. Properties which are optimal when the catalyst is fresh may become highly non-optimal by the time catalyst has degraded. By considering catalyst deactivation during optimization it is possible to design a catalyst that, while non-optimal initially, becomes optimal when its performance is averaged out over its lifespan. An additional factor which may be considered is the possibility of modifying the reactor temperature over time, to help minimize the impact of the catalyst deactivation.

Center of Process Integration, University of Manchester, United Kingdom provides a unique software suite have all above features named "REACTOR". Figure 2,3 and 4 shows its user interface for optimizing catalyst properties. It also develops rate expressions and optimized the reactor at desired operating conditions. Heterogeneous catalyst's fluid-particle interface is modeled in mass and energy balance below.

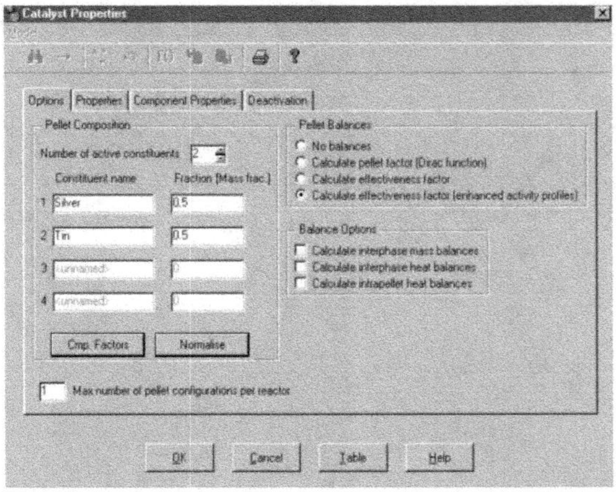

Fig. 2. Catalyst properties options in REACTOR

Mass balance:

$$k_{g,i}\left(C_{s,i} - C_{j,i}\right) = -D_{e,i}\frac{dC_i}{dz} = \eta \sum \left(V_{i,j} - R_{s,j}\right)\rho p S' \tag{1}$$

Energy balance:

$$h\left(T_s - T_g\right) = -\lambda\frac{dT}{dz} = \eta \sum \left(\Delta H_j - R_{s,j}\right)\rho p S' \tag{2}$$

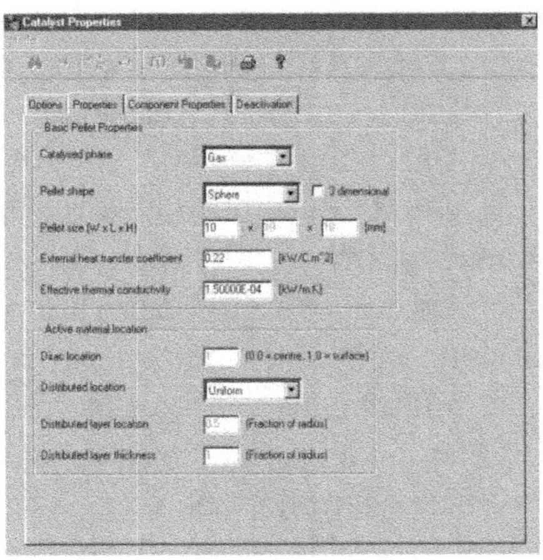

Fig. 3. Catalyst physical properties options in REACTOR

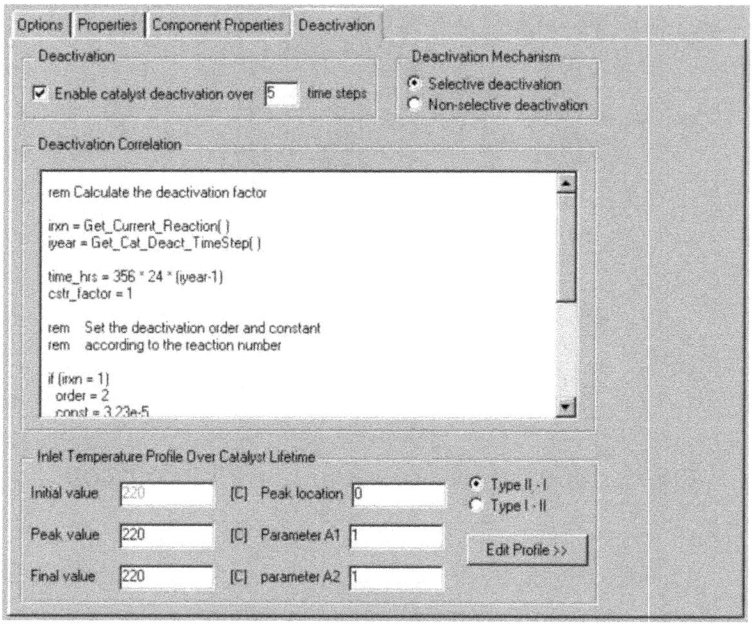

Fig. 4. Catalyst deactivation modeling window in REACTOR

7. Design and modeling of microchannels

Microchannel has distinct identical channels and shape with characteristic dimensions in the order of micrometers and surface area-to-volume ratios ca. 50-100 times higher than those of their conventional counterparts. Due to the presence of high surface areas, submillimeter dimensions and metallic substrates usage, heat can be distributed quickly and uniformly distributed along the channel walls of the catalyst. High heat integration opertunities are available to drive endothermic and exothermic reactions independently and/or in parallel. The representative units can be modeled by incorporating twodimensional continuity, momentum, energy conservation and species mass transport equations for the fluid and catalytic wall phase, and energy conservation equation for the metallic/catalytic wall phase. Simultaneous solution of these equations may carry out using finite elements methods. The model is then used to figure out the effects of various configurationally parameters such as wall thickness, type of wall material and the presence of micro-baffles on temperature distribution. The results forecast the parameters affects the degree of heat flow between channels and provide useful insights for the design of microchannels/catalysts.

8. Thermoanalytical methodology

Thermoanalytical approaches were regarded as transient response methods that correlate the distinctive features of solid catalysts with temperature-programmed heating. Exchanges of matter and energy on the surfaces of solid catalysts facilitate the observation of physical or chemical reactions. The response recorded at specific temperatures (thermogram) indicates the characteristics of the system and appropriate experimental settings. This analysis serves as a mechanism for both quantitative and qualitative assessment of the impact of various activity components. Table 1 enumerates the primary thermoanalytical techniques. A variety of approaches are available in the free literature for extracting kinetic information from thermoanalytical data [14]. In heterogeneous catalysis, these techniques serve as potent instruments for examining the effects of composition, preparation process, and pretreatment on the reactivity of the surface or bulk with gases.

Table 1. Thermal analysis techniques and their uses

Thermal analysis techniques	Characteristic factor monitored
TPx Temperature-programmed reaction Temperature-programmed desorption (TPD)	Gas composition at the reactor outlet Characterisation of adsorptive properties of materials, surface acidity, total adsorption and desorption capacity with temperature, temperatures of rate maxima, metal surface area and dispersion, surface energetic heterogeneity, binding states and energies of adsorbed molecules, mechanism and kinetics of adsorption and desorption
Temperature-programmed reduction (TPR)	Characterisation of redox properties of materials, consumption of reducing agent with temperature range, temperatures of rate maxima, valence states of metal atoms in zeolites and metal oxides, metal oxide and support interaction, indication of alloy formation in bimetallic catalysts
Temperature-programmed oxidation (TPO)	Characterisation of coke species in deactivated catalysts and redox properties of metal oxides, total coke content in deactivated catalysts, mechanism and kinetics of oxidation reactions, deactivation kinetic mechanism
Thermogravimetry (TG)	Weight of sample

Differential thermogravimetry (DTG)	Rate of weight changes
Thermomechanical analysis (TMA), Dilatometry	Specific volume of solid sample
Thermomagnetic analysis (TMA)	Magnetic susceptibility
DMC Differential microcalorimetry	Enthalpy difference between sample and reference

The suitable model can account for heterogeneity in terms of chemisorption by introducing a sufficient number of different adsorption states. The readsorption formulated and tested against experimental data by describing the intrinsic dynamics of an adsorption state as a quasiequilibrium adsorption/desorption between the gas phase and the surface. Conceptually it is potentially useful for the elucidation of fundamental reaction mechanistic information for design and intensification of catalyst. The transient response methods place similar requirements on experimentation as conventional methods of kinetic analysis. In slow transient input, all or part of the elementary reactions may proceed in quasiequilibrium and their individual rates then can not properly elucidated. In general this is based on nonlinear regression analysis, and omits assumptions of pseudo-steady state.

In our extensive studies of catalyst intensification we conduct series of kinetic analysis for TPx experiments in phenomenological way [15-17]. Modelling of TPD data is based on adsorption and desorption theories, while TPR and TPO call for gas-solid surface reaction mechanisms, which may include topochemical characteristics. Further the rate of surface reaction divided into intrinsic rate of reaction per area of reaction interface and change in the reaction interface in reaction [18]. In general kinetic analysis of thermoanalytical data where there is a linear temperature rise (β) represented in the rate equation, where, rate of reaction is proportional to Arrhenius temperature dependence, A is pre-exponential factor, E is activation energy, and these are function of degree of conversion, f(a).

$$\beta \frac{d\alpha}{dT} = A\exp\left[\frac{-E}{RT}\right] f(\alpha) \quad (3)$$

The transforming eq. 3 by taking natural logarithms on both sides gives

$$\ln\left[\beta\frac{d\alpha}{dT}\right] = \ln\left(Af(\alpha)\right) - \frac{E}{RT} \quad (4)$$

The plot of ln ($\beta d\alpha/dT$) versus 1/T results straight line at a selected conversion level, the slope of the line gives -E/R at certain conversion level. This also did not explain reaction mechanism but by varing activation energy E = E(α) results from the Friedman analysis explains that eq. 3 with a normal physico-chemical interpretation will not describes the reaction kinetics.

9. Development and enhancement of a Pt-Sn-based catalyst for alkane dehydrogenation

Regulating chemical reaction pathways at the molecular level is undeniably the foremost scientific difficulty in achieving fully sustainable, thermodynamically efficient chemical processes. Integrated molecular reaction control is a significant advancement in the design of sustainable and efficient chemical processes. A plethora of foundational studies on this subject has been documented in the literature, and the primary task currently is to progress them interdisciplinarily towards industrial catalysis. The catalyst architecture addressed the molecular alignment, collision geometry, and selectivity of the desired product. This example addresses the design and enhancement of Pt-Sn based catalysts for the dehydrogenation of light alkanes to light alkenes. The constraints of this process are intricate due to endothermic limitations and endothermicity.Bringing more molecules at the energy levels exceeding the activation energy threshold by conductive heating but it offers only a macroscopic control over the reaction and is thermodynamically inefficient. There are two broader approaches to control stereochemistry of reaction. First consists of immobilizing the molecules during reaction in confined nanospaces by imposing "hard walls" of structures or of other molecules, or to force reacting species to assume certain position inside the offered structure. The second approach consists in acting upon moving molecules with spatially oriented external fields. Methods for controlling molecular alignment and orientation via nano-structural confinement invite shape-selective and imprinted catalyst. Shape-selective zeolites (having distinct sized nanocage) discriminate between reactants and products and between tran-sition states molecules within the pores [19, 20].

Pt-Sn-based catalysts supported on amorphous supports or zeolites, for propane dehydrogenation have been discussed in many studies [15-29]. The selection of Sn as a promoter has been explained in terms of geometric and/or electronic effects [30]. In the geometric effect, Sn decreases the size of platinum ensembles, which reduces hydrogenolysis and coking reactions. Sn also modifies the electronic density of Pt, either by positive charge transfer from the Snn+ species or the different electronic structures in Pt-Sn alloys [31]. The deactivation of Pt-based catalyst during propane dehydrogenation is fatal, and mainly due to the aggregation of Pt particles and carbon deposition [31, 32]. However, the mechanistic understanding is still controversial and the reaction results were not conclusive, especially on stability and selectivity. In early days much focus has been given to Pt-Sn-based catalysts supported on Al_2O_3 for propane dehydrogenation. But the lifetime of these catalysts is too short. Therefore, there is a growing interest in developing zeolite supported catalysts, e.g. Pt-Sn/ZSM-5 zeolite for propane dehydrogenation, because of longer activation times and higher conversions [15-17]. Many studies have improved the performance of Pt-Sn/ZSM-5 catalysts by adding a more metallic promoter (particularly from alkali or alkaline earth metals) like Na, Fe, Zn, etc. [33, 34] and by increasing the Si/Al ratio of the ZSM-5 support [16, 34]. However, ZSM-5 supported catalysts were detrimentally affected by frequent regeneration with steam [35] and have lower propylene selectivity [16]. ZSM-5 supported catalysts can further take part in cracking and their large pore size fails to create shape selectivity effect for propylene. Moreover comparing the effects of supports and additional promoters, it was observed that the performance of propane dehydrogenation reaction significantly dependent on support. In particular it was noted that the control of intermediates products to desired product over the bi-functional catalyst was significantly support dependent.

Detailed TPx experimentation and kinetic studies with continuous experimental validity took us towards the innovative design of Pt-Sn-based zeolite supported catalyst. In this case, competitive reactions present a picture of a nano-snooker game. Unfortunately, in practicing industrial reactors has very limited degree of control of molecular-level events, therefore the catalyst should be robust enough. Extensive analysis suggests SAPO-34 as support (almost inert to dehydrogenation), with good shape selective opportunities and hydrothermally stable. The silico-aluminophosphate zeolite SAPO-34 is a microporous sieve with a chabasite-like structure (structural code CHA), has an extremely good shape selective effect for propylene [36-38]. Their Bronsted acid sites are responsible for their activities and based on two distinct hydroxyls (OH -bridges to Al and Si) on the structural framework. The structure, acidity and catalytic properties of SAPO-34 depend on the number and distribution of Si in the framework [39, 40]. Its Si based acid sites make it resistant to hydrothermal treatment and gives superiority over ZSM-5, where dealumination occur frequently [33,34]. These observations suggest that the weak acidic support SAPO-34, which can sophisticatedly control the stereo-chemistry and gives shape selectivity effect for propylene in addition to being a bi-functional catalyst, $Pt - Sn/SAPO - 34$.

The impact of the support on the reduction characteristics of the Pt-Sn/SAPO-34 catalyst was examined using H_2-TPR, with results presented in Fig. 5. The reduction peaks at elevated temperatures (exceeding 450 °C) suggest that the platinum exhibited strong interaction with the support in the presence of tin [5,6,14,34]. The peak height and width augment with increased Sn loading between 400 to 600°C, likely attributable to joint reduction and significant interaction between Pt and Sn on the SAPO-34 [24]. At elevated temperatures, Sno species are formed. The oxidation state of Sn(+2) in a bimetallic Pt-Sn catalyst is significant, whereas the creation of a Pt-Sn alloy leads to irreversible deactivation [16, 24, 30]. The platinum linked to tin oxide is thought to be active for the dehydrogenation reaction. The decrease peaks at elevated temperatures revealed that the Pt exhibited a robust interaction with the SAPO-34 compared to other supports. The H_2 reduction results from Pt-Sn/SAPO-34 exhibited distinct reduction peaks, confirming the synergistic reduction and significant interaction between Pt and Sn on the SAPO-34.

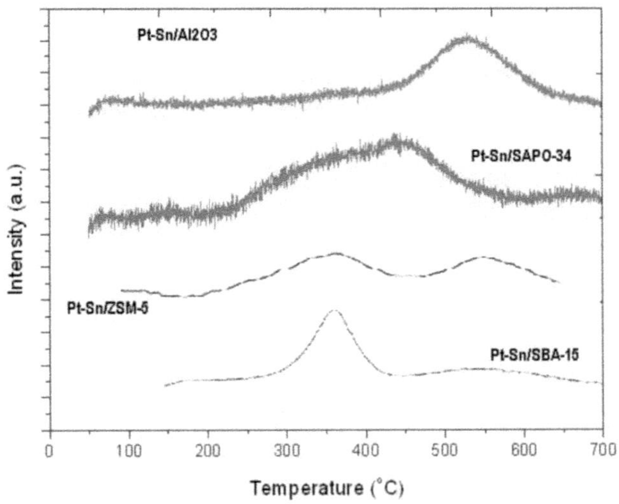

Fig. 5. H_2**-TPR profiles of PtSn based different support catalysts of identical Pt and Sn composition.**

Coke formation during propane dehydrogenation is the inherent factors that adversely affect catalyst performance. The amount of coke formed over different catalysts (after 8 hr reaction) was analyzed by TPO. The results are shown in Fig. 6. The typical TPO profile shows two successive peaks, indicating coke deposited on the metallic sites and support, while the peak height represents the intensity of deposition. The peaks between 300 – 400 °C and 550 – 650°C corresponds to coke deposited on metallic site and support, respectively. It was noted that the strong electronic attachment over SAPO-34 facilitated the transfer of carbon deposits on metal sites to support. It's only possible when the metals are well interacted with the support. Moreover Sn modifies Pt ensembles by decreasing surface area and reduces carbon-metal bonds/interaction, and allows the intermediates to mobile easily to support [41].

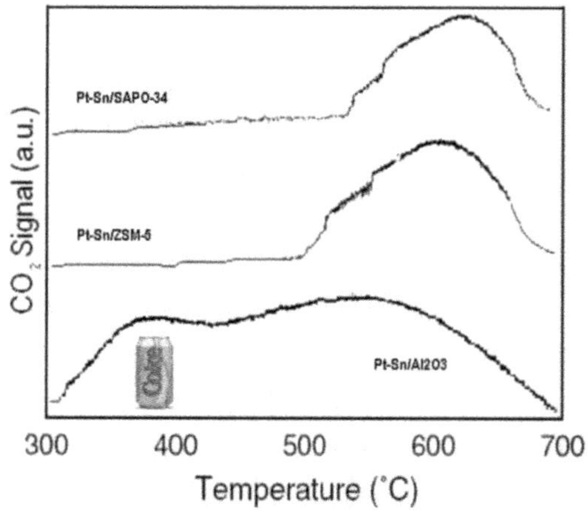

Fig. 6. TPO profiles of coke deposits on different supported PtSn-based catalysts after 8 hr propane dehydrogenation reaction under identical operating conditions.

The superiority of novel support SAPO-34 performance was experimentally verified (results are shown in Fig. 7 and 8), and it was observed that Pt-Sn/SAPO-34 catalyst controls the stereo-chemistry of propane dehydrogenation reaction much better than PtSn -based catalysts supported on ZSM-5 and Al_2O_3.

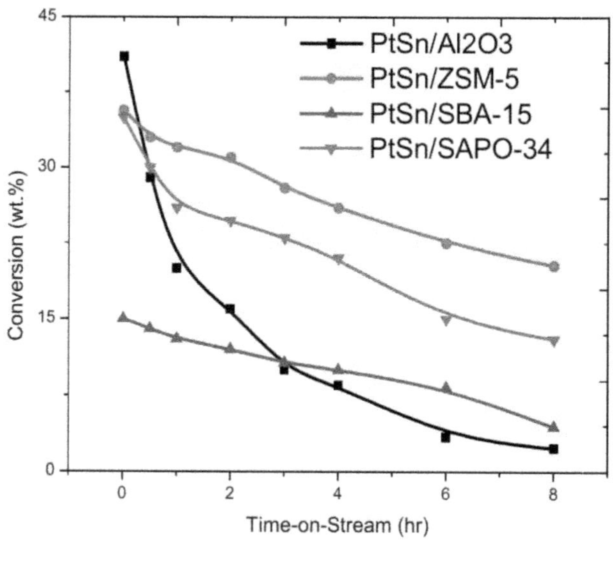

Fig. 7. Catalysts activity at 586 °C **for propane conversion**

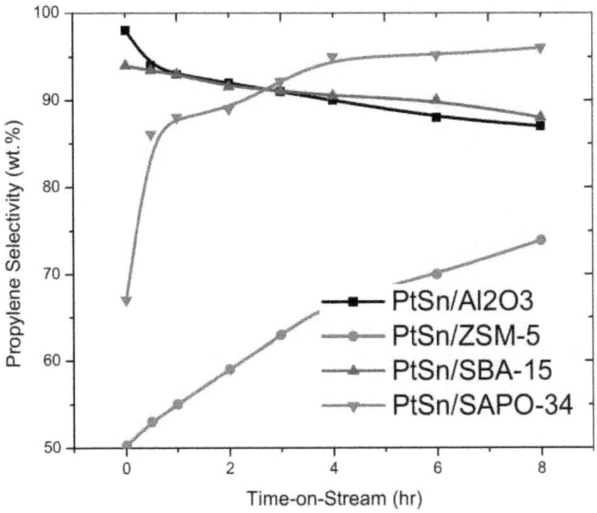

Fig. 8. Catalysts selectivity of propylene over different supported PtSn-based catalysts

10. Concise Summary

Stringent time and budget limitations compel catalysis research to consistently minimize experimental endeavors throughout catalyst creation and to promote the implementation of model-based strategies. Nonetheless, the endeavor to establish an advanced model or criteria in conjunction with innovative TPx analysis for the design and enhancement of catalysts remains substantial. To address bottlenecks in catalyst modeling, significant effort must be dedicated to the systematization of kinetic models, formalization of reaction pathways, and the creation of knowledge-based software tools. This type of contribution is asserted by numerous software developers, while TPx analysis is essential for achieving a robust catalyst design. This will demonstrate how informal textual or metal-support information delineates criteria for catalysts intended for a certain application. A stochastic optimization method can be utilized and validated through testing to identify the ideal design amidst these uncertainties. This study presents the ideal characteristics of a catalyst for the direct dehydrogenation of propane to propylene, while modeling the uncertainties related to the reactions and their parameters to develop a sustainable catalyst.

11. References

[1] Jones, C. W.; Tsuji, C.; Davis, M. E. Organic-Functionalized Molecular Sieves as ShapeSelective Catalysts. Nature 1998, 393, 52-54.

[2] Banares-Alcantara, R.; Westerberg, A. W.; Ko, E. I.; Rychener, M. D. Decadesa Hybrid Expert System for Catalyst Selection-I. Expert System Consideration. Comput. Chem. Eng. 1987, 11 (3), 265-277.

[3] Banares-Alcantara, R.; Ko, E. I.; Westerberg, A. W.; Rychener, M. D., Decadesa Hybrid Expert System for Catalyst Selection-Ii. Final Architecture and Results. Comput. Chem. Eng. 1988, 12 (9-10), 923-938.

[4] Van Santen, R. A. Theory, Spectroscopy and Kinetics of Zeolite Catalyzed Reactions. Catal. Today 1999, 50, 511-515.

[5] Jacobsen, J. H. C.; Dahl, S.; Clausen, B. S.; Bahn, S.; Logadottir, A.; Norskov, J. K. Catalyst Design by Interpolation in the Periodic Table: Bimetallic Ammonia Synthesis Catalysts. J. Am. Chem. Soc. 2001, 123, 8404-8405.

[6] Santhoji Katare, James M. Caruthers, W. Nicholas Delgass, and Venkat Venkatasubramanian, An Intelligent System for Reaction Kinetic Modeling and Catalyst Design, Ind. Eng. Chem. Res. 2004, 43, 3484-3512.

[7] Dumesic, J. A.; Rudd, D. F.; Aparicio, L. M.; Rekoske, J. E.; Trevino, A. A. The Microkinetics of Heterogeneous Catalysis; American Chemical Society: Washington, DC, 1993; p 316.

[8] J.M. Caruthers, J.A. Lauterbach, K.T. Thomson, V. Venkatasubramanian, C.M. Snively, A. Bhan, S. Katare, and G. Oskarsdottir, Catalyst design: knowledge extraction from high-throughput experimentation, Journal of Catalysis 216 (2003) 98-109.

[9] Dumesic, J. A.; Rudd, D. F.; Aparicio, L. M.; Rekoske, J. E.; Trevino, A. A., The Microkinetics of Heterogeneous Catalysis, American Chemical Society: Washington, D.C., 1993.

[10] Stoltze, P., Prog. Surf. Sci., 2000, 65, 65-150.

[11] Schmitz, G., J. Chem. Phys., 2000, 112, 10714-10717.

[12] Aghalayam, P.; Park, Y. K.; Vlachos, D. G., AIChE J., 2000, 46, 2017-2029, Chem. Eng. Progress, Oct. issue, 2000.

[13] Ashish B. Mhadeshwar, Hai Wang, and Dionisios G. Vlachos, THERMODYNAMIC CONSISTENCY IN MICROKINETIC DEVELOPMENT OF SURFACE REACTION MECHANISMS, Prepr. Pap.-Am. Chem. Soc., Div. Fuel Chem. 2003, 48(2), 514.

[14] Sestak, J., Satava V., Wendlandt, W. W., The study of heterogeneous processes by thermal analysis, Thermochim. Acta 7 (1973) 333-556.

[15] Zeeshan Nawaz and Fei Wei, Pt-Sn-based catalyst's intensification using Al2O3-SAPO34 as a support for propane dehydrogenation to propylene, Journal of Industrial and Engineering Chemistry, 17 (2011) 389-393.

[16] Zeeshan Nawaz, Tang Xiaoping, Qiang Zhang, Wang Dezheng and Fei Wei, A highly selective Pt – Sn/SAPO – 34 catalyst for propane dehydrogenation to propylene, Catalysis Communications, 10 (2009) 1925-1930.

[17] Zeeshan Nawaz, Xiaoping Tang, Yao Wang and Fei Wei, Parametric characterization and influence of Tin on the performance of Pt – Sn/SAPO – 34 catalyst for selective propane dehydrogenation to propylene, Industrial & Engineering Chemistry Research, 49 (2010) 1274-1280.

[18] Haber, J., Kinetics and mechanism of the reduction of VI-b transition metal oxides and their oxysalts, J. Less-Comm. Met. 54 (1977) 243-261.

[19] M.E. Davis, Katz A, W.R. Ahmad, Rational catalyst design via imprinted nanostructured materials, Chem. Mater. 8 (1996) 1820-1839.

[20] M. Stöcker, Gas phase catalysis by zeolites, Microp. Mesop. Mater. 82 (2005) 257-292.

[21] H. Lieske, J. Volter, J. Catal. 90 (1984) 96-105.

[22] G.D. Angel, A. Bonilla, Y. Pena, J. Navarrete, J.L.G. Fierro, D.R. Acosta, J. Catal., 219 (2003) 63-73.

[23] T. Waku, J.A. Biscardi, E. Iglesia, Chem. Commun., (2003) 1764-1772.

[24] F.B. Passos, M. Schmal, M.A. Vannice, J. Catal. 160 (1996) 106-117.

[25] P.L.D. Cola, R. Gläser, J. Weitkamp, Appl. Catal. A. Gen. 306 (2006) 85-97.

[26] M.S. Kumar, D. Chen, J.C. Walmsley, A. Holmen, Catal. Commun. 9 (2008) 747-750.

[27] S.M. Stagg, C.A. Quenrini, W.E. Alvarez, D.E. Resasco, J. Catal. 168 (1997) 75-94.

[28] Y. Zhang, Y. Zhou, A. Qiu, Y. Wang, Y. Xu, P. Wu, Catal. Comm. 7 (2006) 860-866.

[29] G. Aguilar-Rios, P. Salas, M.A. Valenzuela, H. Armendariz, J.A. Wang, Salmones, Catal. Lett. 60 (1999) 21-25.

[30] J. Llorca, N. Homs, J. Leon, J. Sales, J.L.G. Fierro, P. Ramirez, Appl. Catal. A. Gen. 189 (1999) 77-86.

[31] C. Kappenstein, M. Guierin, K. Liaziar, K. Matusek, Z. Paial, J. Chem. Faraday Trans. 94 (1998) 2463-2474.

[32] M.S. Kumar, D. Chen, A. Holmen, J.C. Walmsley, Catal. Today. 142 (2009) 17-23.

[33] Y. Zhang, Y. Zhou, A. Qiu, Y. Wang, Y. Xu, P. Wu, Acta Phys.-Chim. Sin., 22 (2006) 672-678.

[34] Y. Zhang, Y. Zhou, H. Liu, Y. Wang, Y. Xu, P. Wu, Appl. Catal. A. Gen. 333 (2007) 202210.

[35] Y. Zhang, Y. Zhou, K. Yang, Y. Li, Y. Wang, Y. Xu, P. Wu, Micropor. Mesopor. Mater. 96 (2006) 245-254.

[36] M.A. Anderson, B. Sulikowski, B.J. Barrie, J. Klinowski, J. Phys. Chem. 94 (1990) 27302734.

[37] Y. Xu, C.P. Grey, J.M. Thomas, A.K. Cheetham, Catal. Lett. 4 (1990) 251-260.

[38] J. Chen, P.A.Wright, J.M. Thomas, S. Natarajan, L. Marchese, S.M. Bradley, G. Sankar, C.R.A. Catlow, P.L. Gai-Boyes, R.P. Townsend, C.M. Lok, J. Phys. Chem. 98 (1994) 10216-10224.

[39] G. Sastre, D.W. Lewis, C. Richard, A. Catlow, J. Phys. Chem. B 101 (1997) 5249-5262.

[40] E. Dumitriu, A. Azzouz, V. Hulea, Micropor. Mesopor. Mater. 10 (1997) 1-8.

[41] F.N. Thomas, V. Hannelore, A.L. Johannes, J. Catal., 157 (1995) 388-395.

Chapter-7

Jacobsen's Catalyst: Preparation, Catalytic Properties & Recycling Capabilities

1. Introduction: An Overview

Metal salen complexes, exemplified by Jacobsen's catalyst, have garnered significant attention in recent decades due to their distinctive catalytic properties, particularly as olefin epoxidation catalysts in conjunction with terminal oxidants such as iodosylbenzene (PhIO), sodium hypochlorite (NaOCl), and meta-chloroperoxybenzoic acid (m-CPBA) (Gladasi, 2007). The salen ligands of Jacobsen's catalyst coordinate with manganese ions via four atoms: two nitrogen and two oxygen atoms. This tetradentate-binding motif resembles the porphyrin structure found in heme-based oxidative enzymes (Canali & Sherrington, 1999). Nevertheless, salen derivatives are more readily produced than porphyrins, and their structures can be more easily modified to establish an asymmetric environment surrounding the active metal site (Canali & Sherrington, 1999). However, this homogeneous catalyst cannot be separated from the reaction media and, subsequently, cannot be recycled. Moreover, it suffers deactivation in homogeneous phase by either formation of dimeric μ-oxo-manganese (IV) species, which are inactive in the olefin epoxidation or oxidative degradation of the salen ligand through the imine group (Figure 1) (Xia et al., 2005). The conventional ways to solve these problems are to immobilize Jacobsen-type catalysts onto solid supports. The last decade has witnessed an intense research effort to heterogenise Jacobsen-type catalysts, and in general chiral manganese(III) salen complexes, using several types of supports in order to make them recyclable as well as economical (Murzin et al., 2005). Reports on the heterogenization of Jacobsen-type catalysts have been centered on their covalent binding to organic polymers (Clapham et al., 2001) and on their encapsulation, entrapment, adsorption and covalent attachment to porous inorganic supports, such as zeolites, Si-MCM-41, Al-MCM-41 and clays (Cubillos, 2010; Dasa, 2006; Kureshy, 2006), and also on activated carbon (Mahata et al., 2007). Unfortunately, the catalytic activity of the recovered catalyst decreases during the catalytic tests of reuse. It is found that isolating Mn(III) salen complexes onto a solid support increases the catalyst stability (Baleizaõ & García, 2006) by suppressing the formation of inactive dimeric μ-oxo manganese(IV) species. However, the deactivation route by ligand oxidation cannot be avoided by anchoring the catalyst to a solid matrix, since it depends on the oxidation conditions. On the other hand, the immobilized catalysts usually lead to partial loss of activity and/or enantioselectivity as compared to their analogous homogeneous catalysts (Baleizaõ & García, 2006). It is well known that in homogeneous phase, the catalyst acquires an appropriate geometric configuration that promotes both, the oxygen transfer from oxo- Mn(V) active species to the double bond of olefin and chiral induction (Adam et al., 2000).

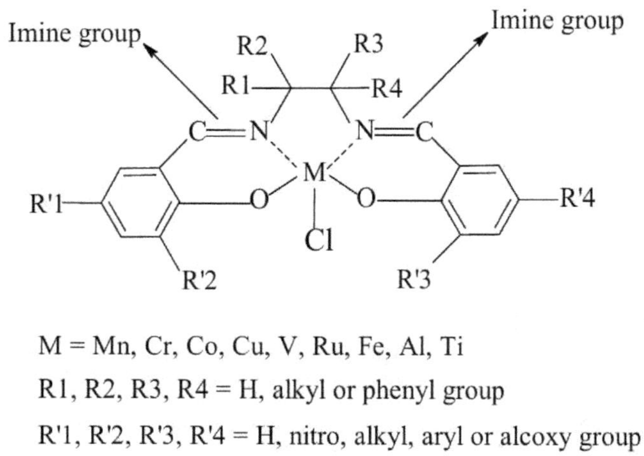

Fig. 1. Chemical structure of metal salen complexes.

In contrast, when the catalyst is not free to move due to the influence of the solid support, an inappropriate geometrical configuration can be obtained (Fan et al., 2002). In addition to that, structural modification of either the catalyst or the solid support during the immobilization process, generally lead to diminished catalytic activity (Mastrorilli & Nobile, 2004). The most common immobilization methods by covalent bond modify one or two tert-butyl groups of the salen ligand. It is known that the tert-butyl groups are very important, since they define the optimal trajectory of the incoming olefin towards the oxo-Mn(III) active species (Linde et al., 2005). In summary, compared with the homogeneous counterparts, some of the immobilized complexes often suffer from various disadvantages, such as poor activity, leaching of the active species into the reaction medium, and low substrate accessibility (Baleizaõ & García, 2006). Therefore, for industrial practical merit and academic interest of homogeneously catalyzed reactions, the development of an efficient strategy for catalyst recovery is still challenging.

An alternative convenient and economical strategy is to adjust the solubility of homogeneous catalysts by varying the reaction conditions, resulting in the direct separation of the catalysts during the reaction (Kureshy et al., 2001). Recently, I and my colleagues described that using dimethyldioxirane (DMD) as oxidizing agent for the epoxidation of R – (+)-limonene and cisethyl cinnamate with Jacobsen-type catalysts could facilitate product isolation and catalyst recovery by segregating the catalysts into a solid phase during the reaction (Cubillos et al., 2009). Limonene di-epoxide was the main product from limonene, whereas an epoxide was obtained with 78% e.e in the case of cis-ethyl cinnamate. Moreover, it was reported that catalyst stability towards the oxidative degradation could be enhanced in the reaction medium, using DMD as oxidizing agent (Cubillos et al., 2009). On the other hand, it was found that the stereogenic center of the pure enantiomerically Jacobsen's catalyst did not influence the catalytic activity for the epoxidation of R-(+)-limonene (Cubillos et al., 2010).

Here I report the preparation and the catalytic properties as well as recycling capabilities of pure enantiomerically Jacobsen's catalysts (R,R-Jacobsen and S,S-Jacobsen) and its racemic form for epoxidation of R -(+)-limonene and cis-ethyl cinnamate with in situ generated DMD as oxidant. In order to compare the activity, recovery and reuse of the catalysts with DMD, other oxidizing agents (NaOCl and *m*-CPBA) most commonly used, were employed for R -(+)-limonene epoxidation with R,R-Jacobsen as catalyst. The influence of catalyst optical configuration on catalytic performance for this reaction is also revised.

2. Experimental Design/Methods

The main reagents used for the synthesis of Jacobsen-type catalysts were 2,4-di-tertbutylphenol, tin(IV) chloride (99%, Aldrich ®), paraformaldehyde (99.5%, Aldrich ®), L-(+)tartaric acid (99.5%, Aldrich ®), D-(-)-tartaric acid (99 , Aldrich ®), cis/trans 1,2diaminocyclohexane (99%, Aldrich ®), potassium carbonate (99.995%, Aldrich ®), manganese acetate tetrahydrate (99%, Aldrich ®) and lithium chloride (99.99%, Aldrich ®). All these materials were directly used as received. The solvents were absolute ethanol, chloroform, acetone and dichloromethane. 3,5-di-tert-butyl salicylaldehyde was prepared from the formylation of 2,4-di-tert-butylphenol using paraformaldehyde as reactant and tin(IV) chloride as catalyst (Deng & Jacobsen, 1992). (R,R)-1,2-diammoniumcyclohexane and (S,S)-1,2-diammoniumcyclohexane mono-(+)-tartrate salt were prepared from the racemic resolution of cis/trans 1,2-diaminocyclohexane with L-(+)-tartaric acid and D-(-)-tartaric acid, respectively (Deng & Jacobsen, 1992). Other reagents included R-(+)-limonene (97%, Aldrich ®), ethyl phenylpropiolate (98%, Aldrich ®), Lindlar's catalyst (5wt. %

on calcium carbonate, Aldrich ®), Oxone ®($2KHSO_5 \cdot KHSO_4$, K_2SO_4, Aldrich ®), manganese(II) acetate tetrahydrate (99%, Aldrich ®), and sodium bicarbonate (Merck), aqueous sodium hypochlorite (NaOCl, 7% active chlorine basis, Carlo Erba), 4-phenylpyridine N -oxide (98%, Aldrich ®), meta-chloroperoxybenzoic acid (*m*-CPBA, 77%, Aldrich), 4 -methylmorpholine Noxide (97%, Aldrich ®). Cis-ethyl cinnamate was prepared from the partial hydrogenation of ethyl phenylpropiolate using the Lindlar's catalyst (Larrow & Jacobsen, 1994).

Different salen ligands R,R-Jacobsen, S,S-Jacobsen and racemic were prepared according to the methods previously described (Deng & Jacobsen, 1992), by condensing the appropriate diamine (3.5 mmol) and 3,5-di-tert-butyl salicylaldehyde (7.0 mmol), for 2 h , using ethanol as a solvent. Upon reaction, completion of a yellow-orange precipitate appeared. These precipitates were filtered and dried under vacuum. The obtained solids were characterized by FT-IR (Bahramian et al., 2006). The Jacobsen's catalysts R,R-Jacobsen, S,S-Jacobsen and Jacobsen racemic (Figure 2), were prepared as previously described (Deng & Jacobsen, 1992). Ethanolic solutions of 3.0 mmol of different ligands, and 3.3 mmol of manganese (II) acetate tetrahydrate were refluxed, for 1 h . During reflux, a change in the color of the solution from yellow-orange to dark-brown was observed. The manganese complexes were recrystallized in heptane and dried under vacuum. The FT-IR and DR UV-Vis spectral bands of all complexes were found to be identical to those reported (Bahramian, 2006; Chaube, 2005).

R,R-Jacobsen, S,S-Jacobsen and Jacobsen racemic catalysts were then used as catalysts for the epoxidation of R-(+)-limonene (1) and cis-ethyl cinnamate (2). In a standard procedure, 1.0 mmol of substrate, 1.2 mmol of sodium bicarbonate and 0.05 mmol of catalyst were dissolved in 4 mL of acetone. A buffer solution (aqueous $NaHCO_3$, 5%wt) was added to bring the pH in the range between 8.0 – 8.5 (mixture A). In another vessel, 2.0 mmol of $KHSO_5$ (Oxone ®) was dissolved in 4 mL of water (mixture B). While mixture A was being stirred, mixture B was slowly added, keeping the pH in the range 8.0 – 8.5 using $NaHCO_3$ solution (5%w/w aqueous). When mixture B was completely added, stirring was stopped and the formed solids separated by filtration and/or centrifugation. The liquid phase was extracted with dicloromethane in a separation funnel and analyzed by GC. The solid phase was washed with sufficient water up to reach a constant weight in the obtained residue. This dark brown residue (catalyst) was easily dissolved in acetone and thus ready for recycling. Also, the chemical identity of this residue was analyzed by FTIR.

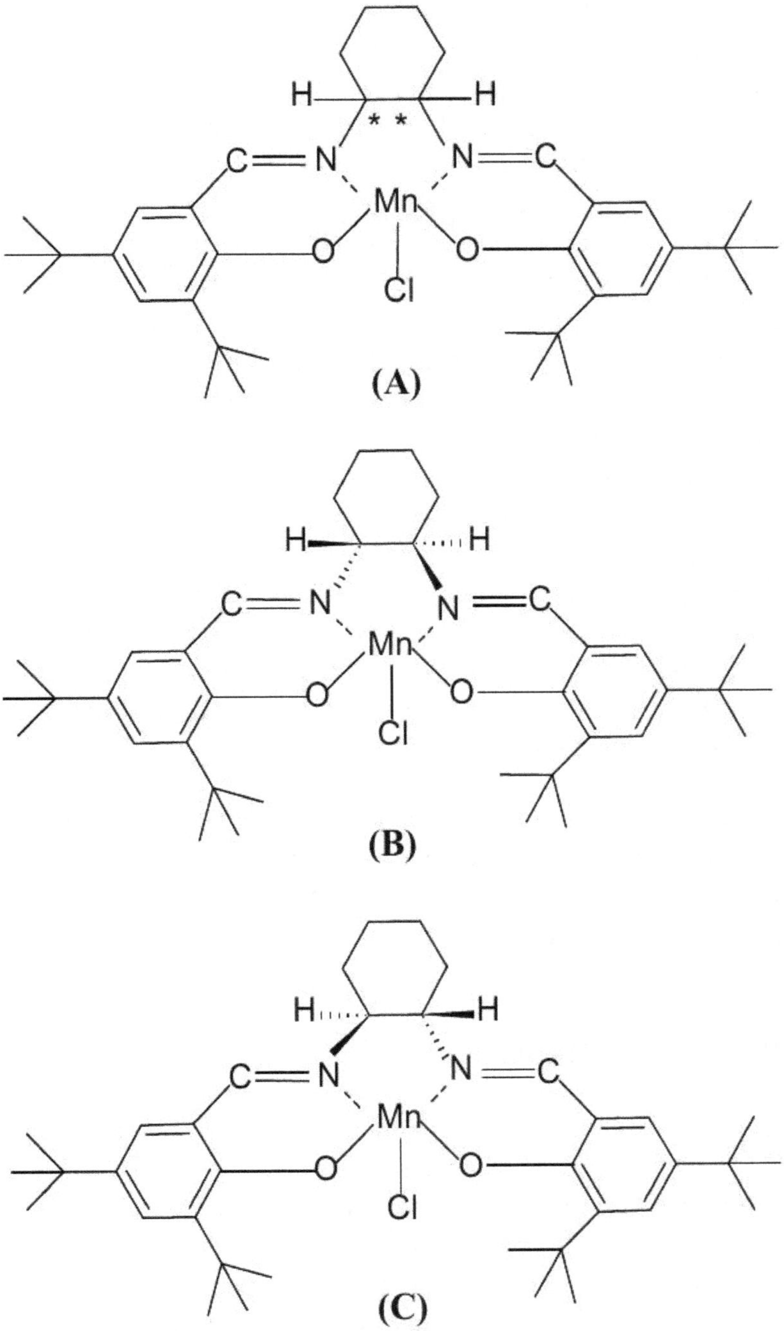

Fig. 2. Chemical structure of Jacobsen type catalysts. A: Racemic Jacobsen, B: R,R-Jacobsen, C: S,S-Jacobsen.

An Agilent Technologies 7890A gas chromatograph (GC), equipped with a DB-1 capillary column (50 m long, 0.32 mm ID and 1.20 mm film thickness) and a FID detector was used for the analysis of solvent purity, olefin and oxidation products. Ultra high pure helium was used as carrier gas (30 mL/min). The injection port temperature was kept at 300°C. For separation of R − (+)-limonene the column temperature was programmed between 80 and 140°C while for cis-ethyl cinnamate it was kept isothermal at 140°C. The area normalization method was used to determine conversion, selectivity and relative yield. The enantiomeric excess (ee) for the single epoxide derived from cis-ethyl cinnamate epoxidation was determined by GC using a chiral capillary column, i.e. Betadex GTA (60 m long, 250 mm ID and 0.25 mm film thickness). In this case, a commercially available 3-ethyl-phenylglycidate (cis/trans = 10/90, Aldrich) racemic mixture was used. In the case of the chiral epoxides derived from R-(+)-limonene, 1,2limonene oxide (97%, mixture of cis and trans, Aldrich) was used. Limonene diepoxide was prepared by oxidation of R-(+)-limonene (1.0 mmol) oxide using *m*-CPBA (5.0 mmol) as oxidizing agent and confirmed by GC-MS. The optical configuration was assigned by comparing the chromatogram of our products with those of cis-ethyl cinnamate isomers available in literature (Steiner et al., 2002). Also, isolated yields of the major epoxide product for either substrate were calculated. Thus, the major epoxide originating from the cis-ethyl cinnamate oxidation was purified by short-path distillation (110°C and 0.5 mmHg), while in the case of R-(+)-limonene, diepoxide was collected at 140°C and 0.5 mmHg .

In order to compare the catalytic activity, recovery and reuse of catalysts with DMD, other oxidizing agents such as NaOCl and *m*-CPBA were explored for R − (+)-limonene epoxidation according to reported conventional methods (Wang et al., 2006), using the same molar ratio of substrate/oxidant (0.5mmol/mmol), substrate/catalyst (20mmol/mmol), reaction time (30 min) and reaction temperature (25°C) as DMD.

3. Results and discussion

Figure 3 shows the FTIR spectra of the salen ligands and their corresponding catalysts. The salen ligands show a characteristic band around 1620 cm^{-1}, which is associated with the vibrations of the imine group (HC = N) (Bahramian et al., 2006). In the catalyst samples, this band is displaced towards lower wavelengths (1600 − 1590 cm^{-1}) as the first evidence of the formation of the organometallic complex. Additionally, characteristic bands at 1530 (C-O), 550(Mn − O) and 480 cm^{-1}(Mn − N) are also associated with the complexation of manganese by the salen ligand (Bahramian et al., 2006).

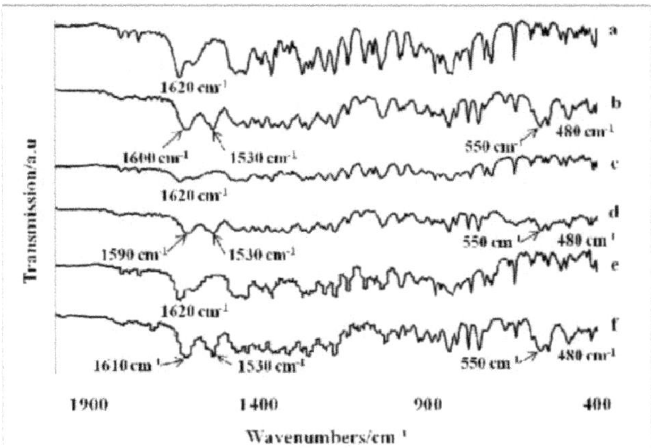

Fig. 3. FT-IR Spectra of (R,R) Jacobsen's ligand (a), (R,R) Jacobsen's catalyst (b), (S, S) Jacobsen's ligand (c), (S,S) Jacobsen's catalyst (d), Jacobsen's racemic ligand (e), Jacobsen's racemic catalyst (f).

Figure 4 shows DR UV-vis spectra of the salen ligands and their corresponding catalysts. The salen ligands exhibit absorption bands at 265 nm and 335 nm . These bands are attributed to Π → Π* transitions. The band at 265 nm has been assigned to the benzene ring and the one at 335 nm , to the imino groups (Chaube et al., 2005). The imino Π→Π* transitions in the Mn salen complexes is shifted to larger wavelengths due to metal coordination, confirming the formation of Mn(III) salen complex (Chaube et al., 2005).

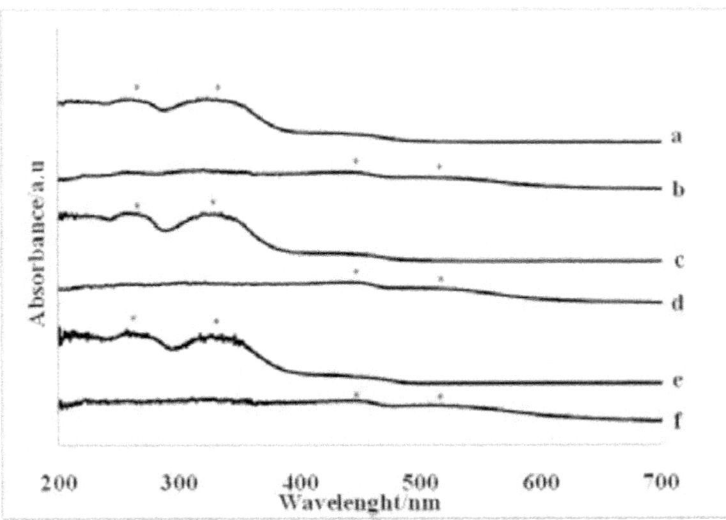

Fig. 4. DR UV-vis Spectra of (R,R) Jacobsen's ligand (a), (R,R) Jacobsen's catalyst (b), (S, S) Jacobsen's ligand (c), (S,S) Jacobsen's catalyst (d), Jacobsen's racemic ligand (e), Jacobsen's racemic catalyst (f).

UV-Vis and FT-IR spectra revealed that the salen ligands are unaffected and are not decomposed upon coordination of the organo-funtional groups with manganese (Chaube et al., 2005).

R,R-Jacobsen, S,S-Jacobsen and Jacobsen racemic catalysts were examined in the liquid phase oxidation reaction of cis-ethyl cinnamate and R-(+)-limonene using in situ generated DMD as oxidizing agent. Different oxidation products were obtained, depending on the substrate type. Figures 5 and 6 outline the test reactions. Di-epoxides appear as the main products from R-(+)-limonene oxidation (Figure 5), while a single epoxide ((2R,3R)-cis ethyl-3-phenylglycidate) was produced from cis-ethyl cinnamate oxidation (Figure 6). Similarly to the previously reported experiments of olefins oxidation (Cubillos, 2009, 2010), the Jacobsen type catalyst was easily separated from the obtained solid phase at the end of reaction.

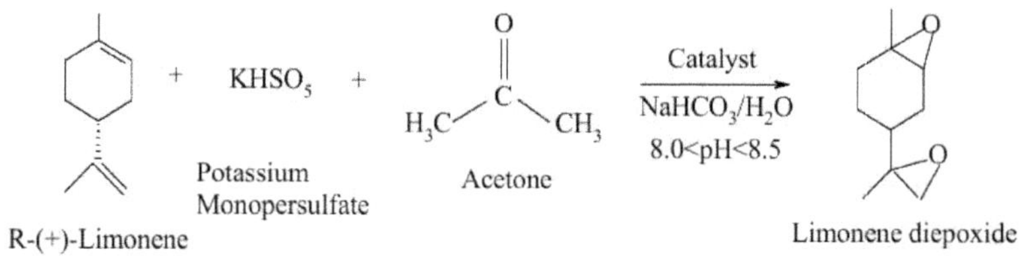

Fig. 5. Selective oxidation of R-(+)-limonene (1).

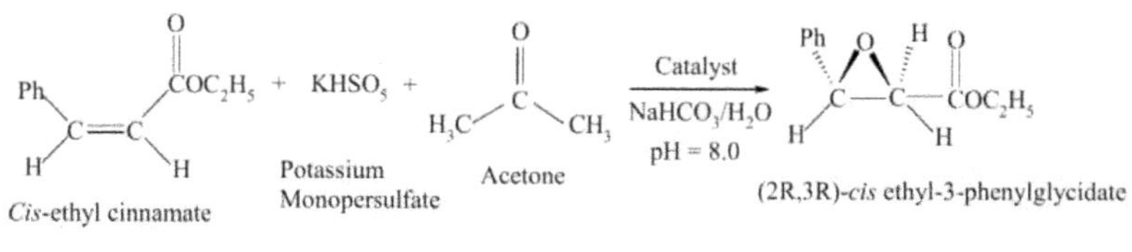

Fig. 6. Enantioselective oxidation of cis-ethyl cinnamate (2).

The results of catalytic activity and recyclability are collected in Tables 1 and 2. In the absence of catalyst, reaction was observed for R-(+)-limonene (conversion 53% and relative rates of reaction 0.145 mol/Lt $\times$ h). In this case, 79% selectivity to limonene diepoxide with 46% relative yield and 30% isolated yield was achieved. 1,2-limonene oxide was obtained as secondary product with very low selectivities (Table 1, entry 1,). In contrast, no reaction was observed using cis-ethyl cinnamate as substrate (Tables 1 and 2, entry 8). It is known that unfunctionalized olefins are more reactive than its functionalized counterparts when Jacobsen type catalysts are used (Porter & Skidmore, 2000).

Table 1. Results of catalytic activity and reutilization

Entry	Substrate	Catalyst	Conversion (%)	Epoxide selectivity (%)	Diepoxide selectivity (%)	Relative yield [h] (%)
1	(1) [a]	None	58	18 [e]	79g	46[i]
2	(1) [a]	I	100	5 [e]	93g	93 [i]
3	(1) [a]	II	100	3 [e]	93g	93 [i]
4	(1) [a]	III	100	4 [e]	958	95 [i]
5	(1) [b]	I	99	7 e	90g	89[i]
6	(1) [c]	I	98	5 [e]	87g	85 [i]
7	(1)[d]	I	98	8 [e]	90g	88 [i]
8	(2) [a]	None	0	-	-	0
9	(2) [a]	II	18	88 [f]	-	16j
10	(2) [a]	III	20	84 [f]	-	17j
11	(2) [a]	I	15	90 [f]	-	14j
12	(2) [b]	II	12	82 [f]	-	10j
13	(2) [c]	II	8	83 [f]	-	7j
14	(2) [d]	II	6	82 [f]	-	5j

[a] Reaction conditions: substrate = 1.0mmol; $KHSO_5$ = 2.0mmol (from Oxone ®), catalyst = 0.05mmol; acetone = 4.0 mL; water = 4.0 mL; reaction time = 30 min. [b] First reuse. [c] Second reuse. [d] Third reuse. [e] Selectivity to 1,2-limonene oxide. [f] Selectivity to ethyl-3-phenylglycidate (sum of cis and trans). g Sum of the four diepoxides of R-(+)-limonene. [h] Relative yield = Conversion × Selectivity/10,000. [i] Relative yield to the diepoxides. j Relative yield to ethyl-3-phenylglycidate (sum of cis and trans).

Similar yields (relative yields 93 – 95% and isolated yields 56 – 60%) to limonene diepoxide were reached with all catalysts, while in the absence of catalyst 46% and 30% respectively, were obtained. This finding suggests that the chiral center of enantiomerically pure (R, R and S,S-Jacobsen) appear to have little or no influence on the catalytic activity; rather the state of coordination given by the salen ligand to the manganese appears to be crucial. Here, the contribution of the catalyst is proven once again by an increase of both R -(+)-limonene conversion (or relative rates of reaction) and selectivity to limonene diepoxide (sum of the four diepoxides of R − (+)-limonene). None of the four diepoxides of R − (+)-limonene predominated. In contrast, the catalytic oxidation of cis-ethyl cinnamate offers the possibility.

Table 2. Results of catalytic activity and reaction rates

Entry	Substrate	Catalyst	Relative rate of reaction [e] (mol/ lt × h)	Enantiomeric excess (%)	Isolated yield (%)
1	(1) [a]	None	0.145	-	30 [h]
2	(1) [a]	I	0.25	-	56 [h]
3	(1) [a]	II	0.25	-	57 [h]
4	(1) [a]	III	0.25	-	60 [h]
5	(1) [b]	I	0.25	-	52 [h]
6	(1) [c]	I	0.245	-	53 [h]
7	(1) [d]	I	0.245	-	50[h]
8	(2) [a]	None	0	-	0
9	(2) [a]	II	0.045	78 [f]	11[i]
10	(2) [a]	III	0.050	55g	9 [i]
11	(2) [a]	I	0.038	0	7 [i]
12	(2) [b]	II	0.030	76 [f]	7 [i]
13	(2) [c]	II	0.020	74 [f]	5 [i]
14	(2) [d]	II	0.015	73 [f]	4 [i]

[a] Reaction conditions: substrate = 1.0mmol; $KHSO_5$ = 2.0mmol (from Oxone ®); catalyst = 0.05mmol; acetone = 4.0 mL; water = 4.0 mL. Reaction time = 30 min. [b] First reuse. [c] Second reuse. [d] Third reuse. [e] Relative rate of reaction = Conversion × initial mol/ reaction volume × reaction time. [f] Enantiomeric excess of (2R,3R)-cis-ethyl-3-phenylglycidate over (2S,3S)-cis-ethyl-3-phenylglycidate. g Enantiomeric excess of (2S,3S)-cis-ethyl-3-phenylglycidate over (2R,3R)-cis-ethyl-3-phenylglycidate. [h] Isolated yield to R-(+) limonene diepoxides. [i] Isolated yield to ethyl-3-phenylglycidate (sum of cis and trans). to epoxidize the unique C = C double bond located in its chemical structure. In order to perform enantioselective epoxidation, a pure enantiomerically catalyst is required, which can be reached either using R,R-Jacobsen or S,S-Jacobsen. In general, good selectivities to ethyl-3-phenylglycidate (sum of cis and trans, 84-90%) and good enantiomeric excesses to (2R,3R)-cis-ethyl-3-phenylglycidate (78%) were obtained with R,R-Jacobsen. In contrast, lower yields were reached (relative yields 16 and isolated yields 11). Additionally, it is worth to note that the product stereochemistry is strongly

dependent on the absolute configuration of catalyst. Thus, cis-ethylcinnmate with R,R-Jacobsen gives(2R,3R)-cis-ethyl-3-phenylglycidate (Table 2, entry 9), whereas R-(+)-limonene with S,S-Jacobsen gives (2S,2S)-cis-ethyl-3-phenylglycidate (Table 2, entry 10). This shows clearly the specificity of a pure enantiomeric catalyst for inducing the preferential formation of the observed product.

Table 3 shows the results of catalytic activity and catalyst recovery for R − (+)-limonene epoxidation using the R,R-Jacobsen catalyst and three oxidants: DMD, NaOCl and *m*-CPBA. For these experiments, Figure 7 shows the spectra of the fresh and used catalyst after reaction with either oxidant. As listed in Table 3, the largest conversion was obtained with DMD (100%), although different selectivities were obtained. Thus, 1,2-limonene oxide was the major product when NaOCl and *m*-CPBA were used as oxidizing agents, obtaining

Table 3. R-(+)- limonene epoxidation using R,R-Jacobsen as catalyst. Effect of the oxidizing agent.

Entry	Oxidizing agent	Conversion (%)	selectivity (%)	Process of catalyst recovery
1	None	0	-	-
2	NaOCl [a]	65	75 [d]	Vacuum distillation
3	*m* - CPBA[b]	35	95 [d]	Vacuum distillation
	DMD [c]	100	93 [e]	Centrifugation and filtration

[a] R-(+)-limonene=1.0 mmol; NaOCl=2.0 mmol (0.05 M aqueous Na2HPO4 solution) ; 4-phenyl pyridine N -oxide = 0.4mmol; R, R-Jacobsen = 0.05mmol; dichloromethane = 4 mL Reaction time = 30 min. [b]R − (+)-limonene = 1.0mmol; *m*-CPBA = 2.0mmol; 4-methylmorpholine N-oxide = 5mmol; R,RJacobsen = 0.05 mmol ; dichloromethane = 4 mL ; Reaction time = 30 min.[c] R-(+)-limonene = 1.0 mmol ; KHSO5 = 2.0mmol (from Oxone ®); catalyst = 0.05mmol; acetone = 4.0 mL; water = 4.0 mL; Reaction time = 30 min. [d] Selectivity to 1,2-limonene oxide; [e] Selectivity to R-(+)-limonene diepoxides (sum of the four diepoxides of R-(+)-limonene).

about 95% with *m*-CPBA. In the case of DMD, R -(+)-limonene diepoxides, were the major products. These differences in selectivities can be associated to the easy segregation of the catalyst. Presumably, this phenomenon creates active sites in the precipitated solid that promotes double epoxidation. It has been reported that the catalyst is not separable by means of physical-mechanical methods, when NaOCl and *m*-CPBA were used as oxidizing agents (Abdi et al., 2004). In these cases, a distillation process under vacuum (160°C and 0.08 MPa) was required in order to separate the catalyst from reaction products, whereas with DMD the catalyst was isolated by physical-mechanical separation methods. On the other hand, figure 7 reveals that the catalyst exhibited the best stability to the oxidative degradation when in situ generated DMD was used as the oxidizing agent, since the main band associated to the Mn (III) salen complexes ($1530\ cm^{-1}$) is still present in the FTIR spectrum, where DMD was used as oxidizing agent. Other important bands associated to the $Mn-O(550\ cm^{-1})$ and $Mn-N(480\ cm^{-1})$ stretching vibrations are slightly displaced towards a lower wavelength.

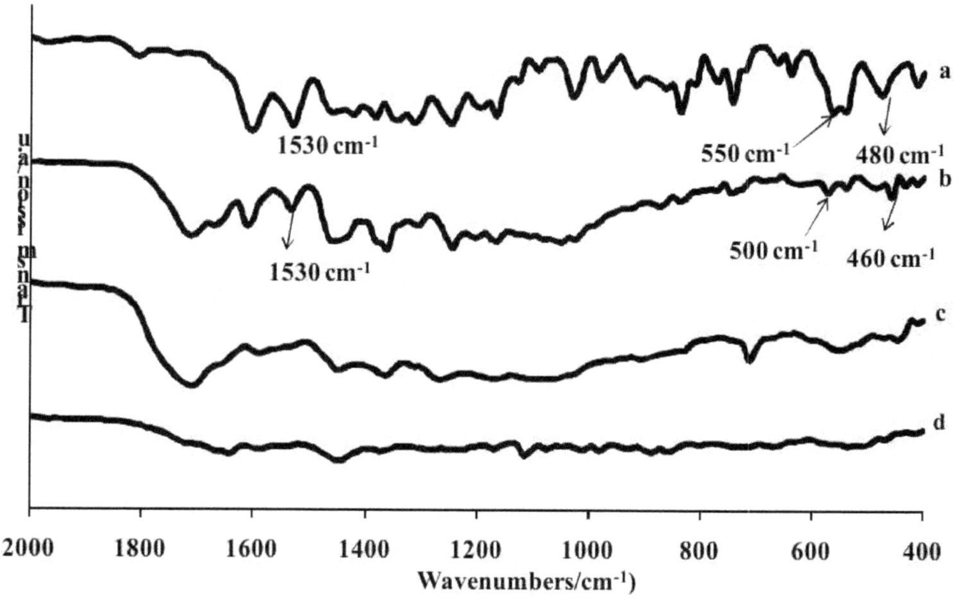

Fig. 7. FT-IR Spectra of the (R,R) Jacobsen's catalyst used with different oxidizing agents: fresh catalyst (a), DMD (b), NaOCl (c) and m-CPBA (d).

Finally, catalyst reuse was explored for both reactions. As can be observed in Tables 1 and 2, catalysts experienced a slight decrease in their initial catalytic activity through three consecutive runs in both reactions. The catalyst was segregated into a solid phase, while the reaction products remained in the liquid phase. This allowed the easy separation of catalyst and reaction products. On the other hand, Figure 8 shows the spectra of the catalyst used in three cycles. Clearly, it is observed that the main bands associated to the Mn(III) salen complexes are retained in the used catalysts. It indicates that the catalyst is very stable to the reaction conditions during three consecutive runs, whereas with oxidizing agents like NaOCl and *m*-CPBA the catalyst was deactivated in the first run (Figure 7). Therefore, the slight loss of catalytic activity with DMD is associated to physical loss of the catalyst during the isolation process rather than to oxidative degradation.

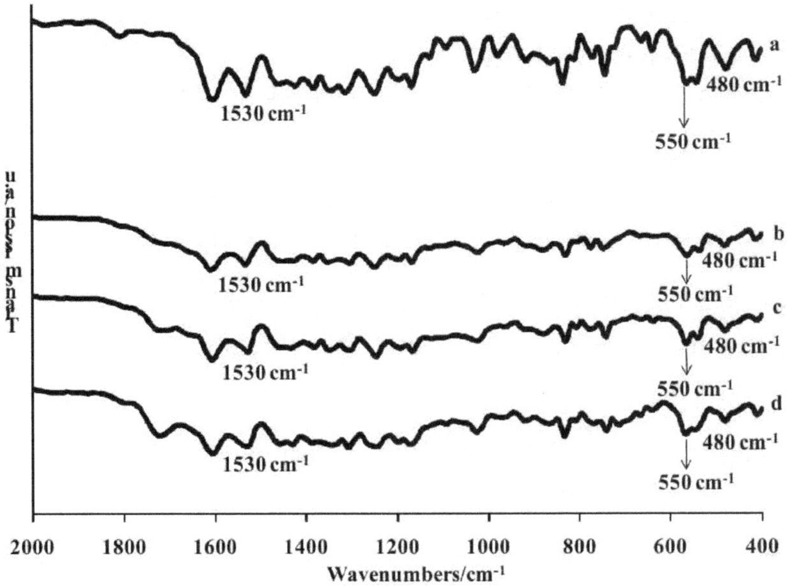

Fig. 8. FT-IR Spectra of the (R,R) Jacobsen's catalyst used in various run: fresh catalyst (a), second run (b) third run (c) fourth run.

4. Concise Summary

Diepoxides are the primary products resulting from the oxidation of R-(+)-limonene, while a monoepoxide is produced during the catalytic oxidation of cis-ethyl cinnamate. In the former scenario, a pure enantiomeric catalyst is necessary, whereas the Jacobsen racemic catalyst sufficed for the catalytic oxidation of R-(+)-limonene. Considering the comparable catalytic activity of the three catalysts in R-(+)-limonene epoxidation and the sole distinction being the stereogenic center at the C1-C2 bond of the 1,2-diamino cyclohexane component, I deduce that catalytic activity is independent of the stereogenic center. The catalyst could be reused three times without significant loss of its initial efficacy.

5. References

Abdi, S. H. R.; Kureshy, R. I.; Khan, N. H. & Jasra, R. V (2004). Asymmetric epoxidation of non-functionalized alkenes using transition metal complexes. Catalysis Surveys from Asia, Vol.8, No.3, (September 2004), pp. 187-197, ISSN 1574-9266.

Adam, W.; Mock-Knoblauch, C.; Saha-Mo11ler, C. R. & Herderich, M (2000). Are MnIV species involved in Mn (salen)-catalyzed Jacobsen-Katsuki epoxidations? a mechanistic elucidation of their formation and reaction modes by EPR spectroscopy, mass-spectral analysis, and product studies: chlorination versus oxygen transfer. J. Am. Chem. Soc, Vol.122, No.40, (September 2000), pp. 9685-9691, ISSN 0002-7863.

Bahramian, B.; Mirkhani, V.; Moghadam, M. & Amin, A. H (2006). Water-soluble manganese(III) salen complex as a mild and selective catalyst for oxidation of alcohols. Appl. Catal. A: Gen, Vol.315, (November 2006), pp. 52-57, ISSN 0926-960X.

C. Baleizaõ. & H. García (2006). Chiral salen complexes: an overview to recoverable and reusable homogeneous and heterogeneous catalysts. Chem. Rev, Vol.106, No.9, (September 2006), pp. 3987-4043, ISSN 0009-2665.

Canali, C. & Sherrington, D (1999). Utilisation of homogeneous and supported chiral metal(salen) complexes in asymmetric catalysis. Chem. Soc. Rev, Vol.28, No.2, (1999), pp. 85-93, ISSN 0306-0012.

Chaube, V. D.; Shylesh, S. & Singh, A. P (2005). Synthesis, characterization and catalytic activity of Mn (III)- and Co (II)-salen complexes immobilized mesoporous alumina. J. Mol. Catal. A: Chem, Vol.241, No.1-2, (November 2005). pp. 79-87, ISSN: 1381-1169.

Clapham, B.; Reger, T. S. & Janda K. D (2001). Polymer-supported Catalysis in Synthetic Organic Chemistry. Tetrahedron, Vol.57, No.22, (May 2001), pp. 4637-4662, ISSN 0040-4020.

Cubillos, J.; Montilla, I. & Montes de Correa, C (2009). Easy separation and reutilization of the Jacobsen's catalyst in olefin oxidation. Appl. Catal. A: Gen, Vol. 366, No.2, (September 2009). pp. 348-352, ISSN 0926-960X.

Cubillos, J.; Vargas, M.; Reyes, J.; Villa, A. & Montes de Correa, C (2010). Effect of the substrate and catalyst chirality on the diastereoselective epoxidation of limonene using Jacobsen-type catalysts. Chirality, Vol.22, No.4, (June 2009), pp. 403-410, ISSN 1520-636X.

Cubillos, J.; Vasquéz, S. & Montes de Correa, C (2010). Salen manganese (III) complexes as catalysts for R-(+)-limonene oxidation. Appl. Catal. A: Gen, Vol. 373, No. 1-2. (January 2010). pp. 57-65, ISSN 0926-960X.

Dasa, P.; Kuźniarska-Biernacka, I.; Silva, A. R.; Carvalho, A. P.; Pires, J. & Freire, C (2006). Encapsulation of chiral Mn (III) salen complexes into aluminium pillared clays: Application as heterogeneous catalysts in the epoxidation of styrene. J. Mol. Catal. A: Chem. Vol.248, No.1-2, (January 2006), pp. 135-143, ISSN 1381-1169.

Deng, L. & Jacobsen, E. N (1992). A practical, highly enantioselective synthesis of the Taxol side chain via asymmetric catalysis. J. Org. Chem, Vol. 57, No.15, (July 1992), pp. 4320-4323, ISSN 0022-3263.

Fan, Q. H.; Li, Y. M. & Chan A. S. C (2002). Recoverable catalysts for asymmetric organic synthesis. Chem. Rev., Vol.102, No.10, (September 2002). pp. 3385-3466, ISSN 00092665.

Gladiali, S. (2007). Guidelines and Methodologies in Asymmetric Synthesis and Catalysis. C. R. Chimie, Vol.10, No.3, (March 2007), pp. 220-231, ISSN 1631-0748.

Kureshy, R. I.; Ahmad, I.; Khan, N. H.; Abdi, S. H. R.; Pathak, K. & Jasra R. V (2006). Chiral Mn(III) salen complexes covalently bonded on modified MCM-41 and SBA-15 as efficient catalysts for enantioselective epoxidation of nonfunctionalized alkenes. J. Catal, Vol.238, No.1, (January 2006), pp. 134-141, ISSN 0021-9517.

Kureshy, R. I.; Khan, N-u. H.; Abdi, S. H. R.; Patel, S. T. & Jasra, R. V (2001). Dimeric chiral Mn (III) Schiff base complex-catalysed enantioselective epoxidation of nonfunctionalised alkenes. Tetrahedron Lett, Vol.42, No.15. (April 2001). pp. 2915-2918, ISSN 0040-4039.

Larrow, J. F. & Jacobsen, E. N (1994). A Practical process for the large-scale preparation of (R,R)-N,N'-bis(3,5-di-tert-butylsalicylidene)-1,2-cyclohexanediaminomanganese (III) chloride, a highly enantioselective epoxidation catalyst. J. Org. Chem, Vol.59, No.7, (April 1994), pp. 1939-1940, ISSN 0022-3263.

Linde, C.; Anderlund, F. M. & Åkermark, B (2005). The effect of phenolates in the (salen)Mncatalyzed epoxidation reactions. Tetrahedron Lett, Vol.46, No.33, (August 2005), pp. 5597-5600, ISSN 0040-4039.

Mahata, N.; Silva, A. R.; Pereira, M. F.; Freire, C.; de Castro, B. & Figueiredo, J. L (2007). Anchoring of a [Mn(salen)Cl] complex onto mesoporous carbon xerogels. J Colloid Interface Sci. Vol.311, No.1, (July 2007), pp. 152-158, ISSN 0021-9797.

Mastrorilli, P. & Nobile, C. F (2004). Supported catalysts from polymerizable transition metal complexes. Coord. Chem. Rev. Vol.248, No.3-4, (February 2004), pp. 377-395, ISSN 0010-8545.

Murzin, D. Y.; Mäki-Arvela, P.; Toukoniitty, E. & T. Salmi (2005). Asymmetric heterogeneous catalysis: science and engineering. Catal. Rev., Sci. Eng, Vol.47, No.2, (February 2007), pp. 175-256, ISSN 1520-5703.

Porter, M. J. & Skidmore, J (2000). Asymmetric epoxidation of electron-deficient olefins. Chem. Commun, No.14, pp. 1215-1225, ISSN 1359-7345.

Steiner, D.; Ivison, L.; Goralski, C. T.; Appell, R. B.; Gojkovic, J. R. & Singaram, B (2002). A facile and efficient method for the kinetic separation of commercially available cisand trans-limonene epoxide. Tetrahedron asymmetry, Vol.13, No.21, (October 2002), pp. 2359-2363, ISSN: 0957-4166.

Wang, D.; Wang, M.; Zhang, R.; Wang, X.; Gao, A.; Ma, J. & Sun, L (2006). Asymmetric epoxidation of styrene and chromenes catalysed by dimeric chiral (pyrrolidine salen)Mn(III) complexes. Appl. Catal. A. Gen, Vol.315, (November 2006), pp. 120127, ISSN 0926-960X.

Xia, Q. H.; Ge, H. Q. ; Ye, C. P.; Liu, Z. M. & Su, K. X (2005). Advances in homogeneous and heterogeneous catalytic asymmetric epoxidation. Chem. Rev, Vol.105, No.5, (2005), pp. 1603-1662, ISSN 0009-2665.

Chapter-8

Carbohydrate-derived Surfactants: Structural-activity Correlations

1. Introduction: An Overview

Carbohydrate-based surfactants (CBS) currently represent one of the most significant categories of amphiphilic chemicals (Dembintsky, 2004; Queneau et al., 2008; Ruiz, 2009). Their structure arises from the combination of saccharides and lipids, either biosynthesized within live cells or synthetically produced through sequential reactions involving carbohydrates and fatty materials, via one or several linkages. The increasing interest in these chemicals stems from several basic, practical, economic, and environmental factors. (Razafindralambo et al, 2009, 2011a; Hill, K. & LeHen-Ferrenbach, 2009; Kitamoto et al, 2009). First, they can be easily prepared from the most abundant renewable vegetable raw materials (cellulose, pectin, hemicellulose, starch, etc.) in a wide range of structure and geometry by modular synthesis thanks to the presence of numerous reactive hydroxyl groups. Second, such a structural diversity makes them, on the one hand, as excellent models for getting insight into the surfactant mechanisms in modifying interfacial properties, which control the formation and the stability of colloidal systems such as micelles, vesicles, foams, emulsions, and suspensions (Razafindralambo et al, 2011b). Conversely, a multitude of traits and functionalities can be anticipated from the vast array of compounds that may have unique uses across diverse industrial sectors. Third, their environmental compatibility, such as enhanced biodegradability and reduced toxicity, serves as a good requirement for their application as alternatives to petrochemical-derived surfactants. Due to the aforementioned reasons, a systematic examination of structure-activity relationships, which has been infrequently conducted previously, seems essential for enhancing understanding of the influence of each CBS structural component on their activities and functionalities, ultimately facilitating a rational approach in the selection and combination of appropriate compounds for future developments and applications. This chapter presents the findings on the dynamic and equilibrium surface properties of homologous and analogous series of uronic acid derivatives, thereby demonstrating the influence of various structural components on their intrinsic properties at the air-water interface.

2. Extent of the contribution

The objective of this contribution is: (1) to examine CBS through structural classification based on molecular size (mono-, oligo-, polymeric surfactants), geometry (standard, bipolar or bolaform, and gemini surfactants), and the characteristics of the polar headgroup (charged or uncharged, cyclic or acyclic), the apolar tail (number and length of alkyl chains), and the linker (amide, ester, etc.) and/or the spacer; (2) to systematically present findings on structure-activity relationships of uronic acid derivatives (UADs), a specific category of carbohydrate-based surfactants. This pertains to the influence of each structural component, including the polar headgroup (stereochemistry), apolar tail (chain length, quantity, and unsaturation), and linkage/spacer, on the efficacy of UADs in modifying surface characteristics and potentially in forming and stabilizing colloidal systems.

3. Classification of surfactants derived from carbohydrates

Carbohydrate-based surfactants, currently known as glycolipids (Chester, 1997; Hato et al., 1999), are constituted by a saccharide unit (mono-, di-, oligo-, or polysaccharide) linked to a hydrophobic part of one, two, or multi-hydrocarbon chains by a single, or several bonds. These may be an ester, thioester, ether, amine, or/and amide group (Stubenrauch, 2001). Generally, glycolipids may be classified according to their amphiphilic structure, which depends on the polar headgroup, the apolar tail, but also the linkage between these main entities. Based on these three structural parameters, glycolipids are grouped in different categories, as presented in Table 1.

Table 1. Main classes of glycolipids, based on the amphiphilic structure and geometry

Glycolipids	Hydrophilic part	Hydrophobic part	Linkers
Monocatenary Bicatenary Multicatenary	One headgroup	One tail Two tails Multi tails	
Glycoglycerolipids		Acylglycerol	
Glycosphingolipids		Sphingoides	Ester, thioester, Ether, amine, amide
Bolaforms	Two headgroups	One spacer	
Geminis		One spacer/two tails	
Alkylpolyglucosides	Multi-headgroups	One or multi-tails	

Besides the nature ionic, nonionic, and amphoteric of the polar headgroup, as for all surfactants, glycolipids can also be classified in small, medium, and large compounds, in relation to their molecular mass and size. According to these criteria, three main classes belonging to glycolipid-based monomeric, oligomeric, and polymeric surfactants may be distinguished. In addition to structural criteria, glycolipids may have two origins, those from microbial fermentation, crops, and animals, which are natural products, and those synthetized by chemical, enzymatic, and chemo-enzymatic routes, globally considered as synthetic compounds. In the following section, we try to illustrate by a few representative examples the most important classes of glycolipids with their chemical structure on the basis of their source, molecular mass, and geometry while considering the other classification aspects.

3.1 Natural glycolipids

3.1.1 Microbial biosurfactants

Bioglycolipids	Structure	Specificities (headgroup, tail, linkage)
	Low molecular mass	
• Rhamnolipids	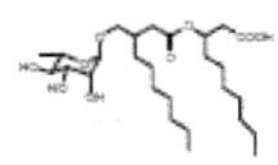	nonionic, mono/di, ester, (Dembitsky, 2005)
• Sophorolipids	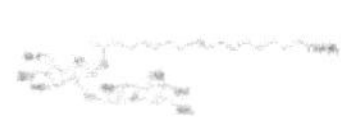	nonionic, mono, ester, (Dembitsky, 2005)
• Trehalolipids		nonionic, di, ester, (Dembitsky, 2005)
• Mannosylerythriol lipid	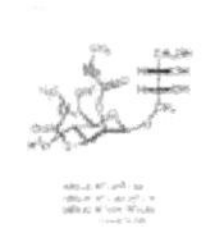	nonionic, mono, ester, (Kitamoto et al. 2009)
• Monogalactosyldiacylglycerol	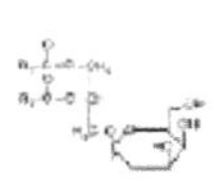	nonionic, di, ester, (Infante et al. 1997)

	High molecular mass	
• Cyclo-oligosaccharides	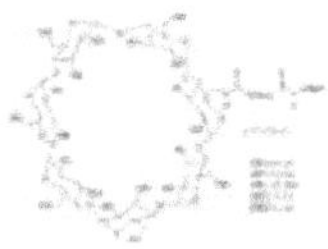	nonionic, mono, ester-amide, (Dembitsky, 2005)
• Emulsan (Polymeric biosurfactant)		nonionic, multi, ester/amide, (Desai et Banat, 1997)

3.1.2 Plant biosurfactants

Bioglycolipids	Structure	Specificities (headgroup, tail, linkage)
• Butanolide glycosides		nonionic, multi, ester, (Dembitsky, 2005)
• Sucrose ester	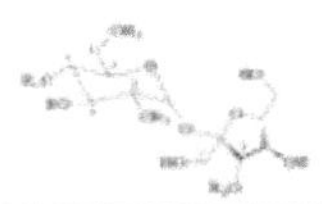	nonionic, multi, ester, (Dembitsky, 2005)

3.1.3 Animal biosurfactants

Bioglycolipids	Structure	Specificities (headgroup, tail, linkage)
• Cerebrosides	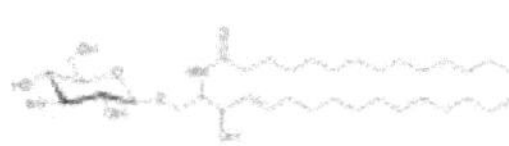	nonionic, di, amide, (Dembitsky, 2005)
• Gangliosides		cationic, di, amine, (Dembitsky, 2005)

3.2 Synthetic glycolipids

Glycolipids	Structure	Specificities (headgroup, tail, linkage)
	Monomeric	
• Sorbitan monoester	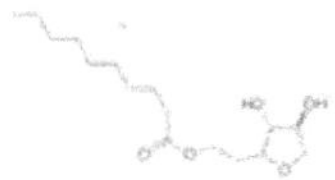	nonionic, mono, ester, (Hill &LeHen- Ferrenbach, 2009)
• Isosorbide derivative		nonionic, mono, ester (Hill &LeHen- Ferrenbach, 2009)
• Uronic acid derivatives: a. *Monocatenary*	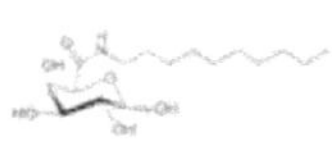	nonionic, mono, amide (Laurent et al. 2011)
b. *Bicatenary*	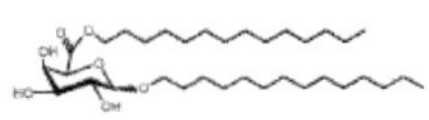	nonionic, di, ester-ether (Richel et al. 2010)
• Alkylglucoside		nonionic, mono, ether (Vulfson, 1990)

Glycolipids	Structure	Specificities (headgroup, tail, linkage)
Bolaform		
• α,ω-diamino-(oxa) [Spacer: -$(CH_2)_n$-]	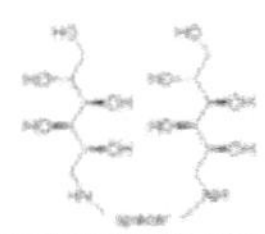	nonionic, bi, amide (Wagenaar&Engberts, 2007)
Gemini		
• Glu(n)-2-Glu(n)	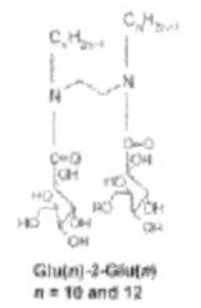	nonionic, bi, amides (Sakai, 2008)
Oligomeric/polymeric		
• Alkylpolyglucoside		nonionic, mono, ether (Queneau et al. 2008)
• Inulin carbamate	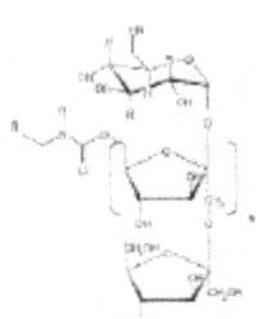	nonionic, mono, ester (Tadros, 2004)

4. Uronic acid derivatives

Uronic Acid-Derivative Surfactants (UADs) are a particular class of monomeric glycolipids (Fig.1). Their polar headgroup represents a considerable part of carbohydrate components, widely distributed in natural plant polysaccharides (Langguth & Benet, 1992). In addition, uronic acids are the result of the primary alcohol oxydation into a carboxylic group within a monosaccharide like glucose and galactose. Thus, they contain, in their structure, both hydroxyl and carboxylic groups that are highly reactive, explaining their potential as a basic unit for generating closely related surfactant compounds. The polar headgroup configurations according to the stereochemistry of OH groups and geometry (cyclic or not, bipolarity), the hydrophobic tail (number and length of alkyl chain), and the type of linker (ester, ether and amide, etc) are among the main variables in their structural entities. Therefore, they represent a set of ideal compounds for investigating structure-surface activity relationships.

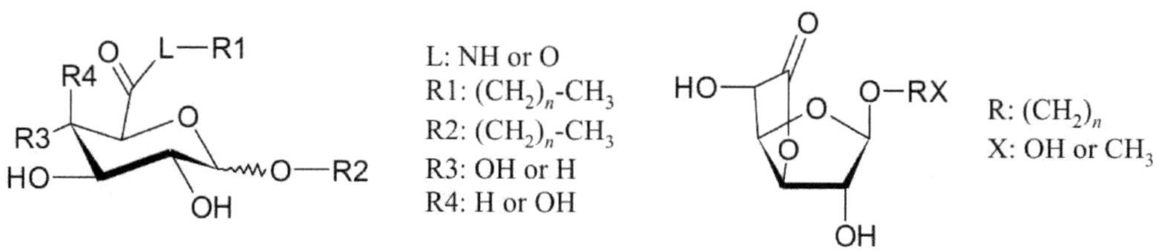

Fig. 1. General structure of uronic acid (left) and uronolactone (right) derivatives

Among several activities that CBS are able to develop (Ruiz, 2009), those related to the modification of surface and interface fundamental properties are undeniably the most important to evaluate and understand. Through several research works during last decades, it is established that fundamental properties such as dynamic and equilibrium surface tensions play a key role in the formation and stability of colloidal systems like foams, emulsions, and suspensions (Razafindralambo et al. 2011b). In this chapter, we are only focused on the relationships between chemical structure of UADs and their surface-activities at the air-water interface. UADs are a set of glucuronic and galacturonic acid-based surfactants varying in the polar headgroup configuration, including a cyclic (lactone bond) or non-cyclic structure and a α - or β - anomeric form, the hydrophobic tail chain length (C8 to C14) and number (mono- and bicatenary), the presence of a double bond, as well as an OH group at the terminal carbon, and in the type of linkage, ester, amide in C6, or ether in C1. These compounds have been synthesized by chemical or enzymatic routes, purified, and their chemical structure has been confirmed by means of various spectroscopic techniques (Nuclear Magnetic Resonance, Infrared, and Mass Spectrometry).

5. Surface-active properties

Surface-active properties of any surfactants can be measured through their capacity, either to reduce the surface tension of aqueous solution (γ_o) to any value (γ), or to increase the surface pressure ($\pi = \gamma_o - \gamma$) of the air-water interface. Measurements can be performed in both dynamic and static/quasi-static modes. Based on the solubility of the coumpound to characterize, one of the two following approaches must be used. The Gibbs'approach, measuring γ by adsorption, i.e. by migration of surfactant to the surface, is more convenient for short chain derivative surfactants (high solubility in water) whereas the Langmuir approach, determining π after spreading and compressing surfactant molecules at the surface, is rather appropriate for long chain derivative surfactants (low solubility in water). Surface activities of uronic acid derivatives may be evaluated in both dynamic and static/quasi-satic modes using a series of complementary techniques at room temperature comprised between 20 – 25°C (Razafindralambo et al., 1995; 2009; 2011).

5.1 Dynamic surface properties

In dynamic mode, surface tensions (γ_{dyn}) are measured using a drop volume tensiometer (TVT1, Lauda, Germany) by determining the critical volume (maximum pending size) of a series of drop created at different growing rates from 3 to 120 s (Razafindralambo et al., 2004).

5.2 Equilibrium surface properties

5.2.1 Gibbs approach

In static and quasi-static modes, the equilibrium surface tensions (γ_e) are measured by means of Wilhelmy plate (Tensimat N3, Prolabo, France), drop volume (TVT1, Lauda, Germany), or pending drop (Tracker, IT-Concept, France)-based tensiometers as detailed in previous papers (Razafindralamboet al., 1995; 2009). These methods consist of measuring continuously γ_e as a function of time, for surfactant solutions at different concentrations. The equilibrium surface tension (γ_e) can be determined as the fitting value of γ from representative experimental data, using appropriate mathematical models. All physical parameters such as critical micelle concentration (CMC), surface tension at the CMC (γ_{cmc}), surface excess (Γ_{max}), and minimum area per molecule(A_{min}), can be calculated using Gibbs equations for nonionic surfactants:

$$\Gamma_{max} = -\frac{1}{RT}\left(\frac{d\gamma}{d\ln\ C_o}\right) \tag{1}$$

and

$$A_{cmc} = \left(\frac{1}{\Gamma_{\max} N}\right) \tag{2}$$

where C_o is the bulk solution concentration; $\mathrm{R} = 8.314\ \mathrm{J.K^{-1}.m^{-2}}$, the gas constant; T , the temperature in Kelvin, and N, the Avogadro's number.

5.2.2 Langmuir approach

By Langmuir approach, a small volume of surfactant organic solution is spread drop wise onto the clean milli-Q water surface by means of Hamilton micro-syringe using a reproducibility adapter. After a given laps time, allowing the solvent evaporation, the surface pressure (π) of the monolayer is determined by measuring the horizontal force per unit of length, for a delimited area, during the compression. With an automatic Langmuir film Waage LFW2 3 "5 (Lauda, Königshofen, Germany), a total trough area of 927 cm^2 is available. This is progressively reduced by moving a single barrier made in Teflon, and the force per length unit detected at a float gives the surface pressure of the monolayer at a constant temperature. For characterizing uronic acid derivatives, mainly bicatenary surfactants, the research of optimized experimental conditions is required (Razafindralambo et al. 2011).

6. Structure-activity relationships

6.1 Polar headgroup effect

6.1.1 Stereochemistry

Anomeric forms

The effect of the octyl glucuronolactone anomeric form is observed for the equilibrium surface properties (Table 2). The anomer a CMC is about the half of that of the anomer β, all other related parameters being quite similar. γ_{cmc} and $\mathrm{A_{min}}$ are quite similar, indicating that the anomeric form does not affect the molecular arrangement at the air-water interface, but changes that in the bulk water. In other words, the anomer α self-aggregates readily in water while adopting the same close-packed configuration at the air-water interface than the anomer β. Therefore, the anomer a appears more hydrophobic than the anomer β, which is attributed to the difference in the steric effect of their polar headgroup. This interacts stronger with water molecules for the anomer β, delaying its self-aggregation process. Concerning their dynamic surface properties, the anomer β appears slightly more efficient than the anomer a (data not shown), the difference being less important than that observed for equilibrium surface properties.

Table 2. Physical parameters of octyl glucuronolactone with α and β anomers

	Υ_{cmc} [mN/m]	CMC [mM]	A_{min} [Å^2/mol]
$\alpha - C_8 -$ Glucuronolactone	32.8	1.3	43
$\beta - C_8 -$ Glucuronolactone	35.2	2.4	40

4-hydroxyl group (4-OH) axial or equatorial position effect

The second parameter related to the polar headgroup stereochemistry is the axial or equatorial position of the hydroxyl group of the fourth carbon within bicatenary derivatives of galacturonic and glucuronic acids. This stereochemistry effect impacts their configuration and behaviour at the air-water interface, as observed with the (π-A) isotherms (Fig.2). The glucuronic acid derivative $(GlcA - C_{14/14})$ is more expanded when it is spread at the air-water around 20°C, and shows a transition phase under compression, which does not occur for the galacturonic acid derivative (GalA- $C_{14/14}$). This difference is in agreement with film morphologies and thicknesses, and is also supported by molecular models (Razafindralambo et al., 2011). In contrast, the 4-OH axial or equatorial position has no effect when the molecules are vertically oriented, i.e. within a film at a condensed state. Such results have been attributed to the configuration of two alkyl chains, which is in "open" or "close" structure according to the film is in expanded or in condensed state.

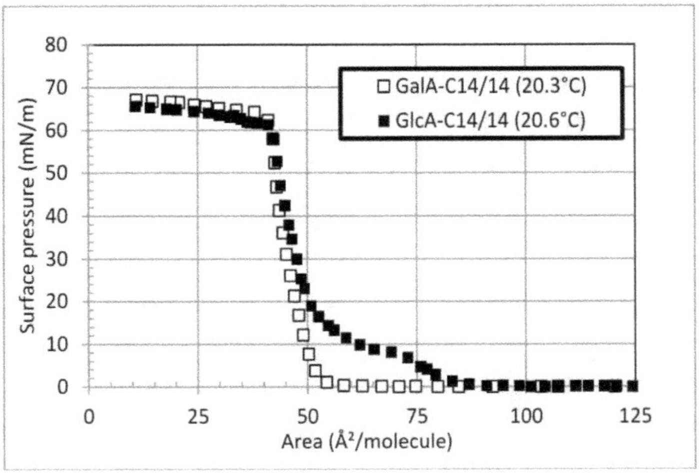

Fig. 2. (π-A) isotherms of GalA- $C_{14/14}$ and GlcA- $C_{14/14}$ spread at the air-water interface 20°C

6.1.2 Impact of the bipolarity: (OH group ending the alkyl chain)

Adding an OH group at the end of the alkyl chain changes the geometry of uronic acid derivatives, which becomes a bipolar asymmetric. Such a modification impacts some dynamic (Fig.3) and equilibrium surface properties (Table 3) of glucuronic acid derivatives.

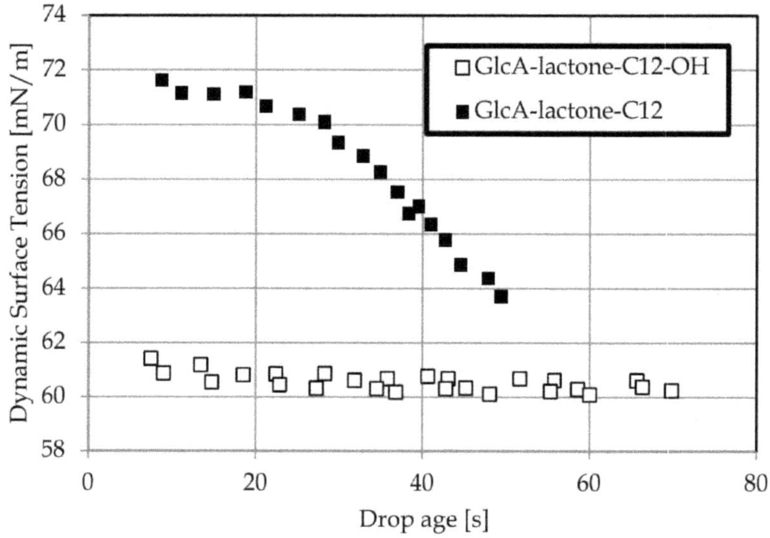

Fig. 3. Dynamic surface tension vs. time of dodecyl glucuronolactones (0.2 g/L)**, with and without a hydroxyl group at the end, measured by TVT1 at** 20°C

Table 3. Physical parameters of dodecyl glucuronolactones ending with, and without hydroxyl group

	Υ_{cmc} [mN/m]	CMC [mM]	A_{min} [Å^2/mol]
OH-C_{12}-Glucuronolactone	43.9	0.7	65
C_{12}-Glucuronolactone	47.2	0.6	34

It increases the area occupied per molecule (cross-sectional area), and the performance in the adsorption time by reducing faster the dynamic surface tension. This difference is easy to understand, and may be explained by the fact that the two polar headgroups are directed into the aqueous phase, leaving the hydrophobic spacer in the gaseous phase, and forming, as a consequence, a "convex" configuration. In contrast, it has no significant effect on the equilibrium (long time) surface properties, based on their CMC and γ_{cmc} values.

6.2 Hydrophobic effect

6.2.1 Alkyl chain length

The impact of the alkyl chain length on all of interfacial properties can be evaluated with a homologous series of monocatenary and bicatenary uronic acid derivatives. As for all monomeric surfactants, whatever their nature, the general trend is respected with uronic acid derivatives, that is, the longer the alkyl chain length, the lower the CMC of the uronate and uronamide derivatives (Blecker et al., 2002; Laurent et al., 2011). About dynamic surface properties, an optimum chain length of eleven carbon atom gives the best performance, regarding the adsorption time and the maximum reduction of the surface tension. Concerning bicatenary derivatives, the comparison of GalA- $C_{10/10}$ and GalA-$C_{14/14}$ monolayer properties shows that the shorter the two alkyl chains ($C_{10/10}$), the more expanded and compressible the spread molecules, and the less stable the molecular film (Fig.4). Such behaviour of the shorter chain derivatives is attributed to stronger interactions with the subphase, which is comparable with the case of the dimyristoyl-phosphatidylcholine (DMPC), an ionic phospholipid having the same alkyl chains than that of GalA-C $C_{14/14}$ (data not shown).

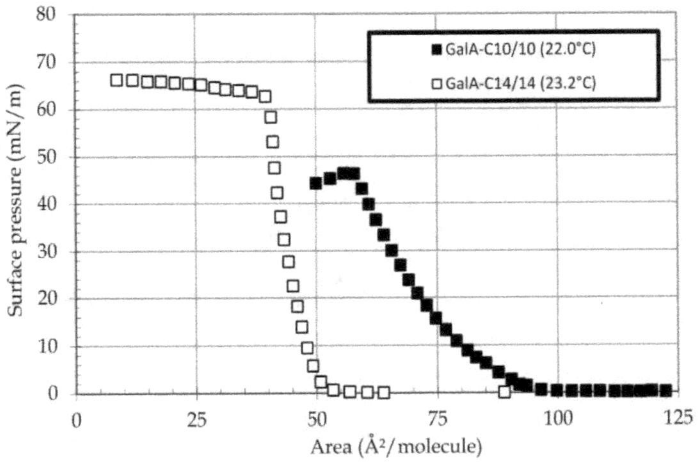

Fig. 4. (π-A) isotherms of GalA- $C_{10/10}$ and GlcA- $C_{14/14}$ spread at the air-water interface

6.2.2 Alkyl chain number (Monocatenary vs. bicatenary)

The alkyl chain number of uronic acid derivatives also appears crucial on their surface properties characterized by the Langmuir approach. When a second alkyl chain is added to the polar headgroup, that is, the compound becomes bicatenary, the mechanical properties of the film at the air-water interface are improved, regarding its collapse pressure (Fig.5). It is to be noted, from this figure, that the inclusion of a cyclic ester bond into the polar headgroup affects slightly the behaviour of the monolayer.

6.2.3 Unsaturation

By including a double bond at the end of the alkyl chain, no significant effect has been observed on the equilibrium surface properties, the γ_{cmc}, CMC , and A_{min} values being similar and equal to $34.2 \pm 1.1 mN/m, 0.12 \pm 0.01$ g/L, and $39.0 \pm 2.0 Å^2/mol.$, respectively. In contrast, it impacts extensively the dynamic surface properties of the undecanoyl glucuronate (Fig.6). The derivative with an unsaturated chain migrates faster at the airwater interface, and reduces readily the dynamic surface tension, compared to the saturated one. It is attributed to the "shortening effect" of the double bond inclusion (Milkereit et al. 2005), reducing the hydrophobicity of the surfactant, and therefore, its adsorption time.

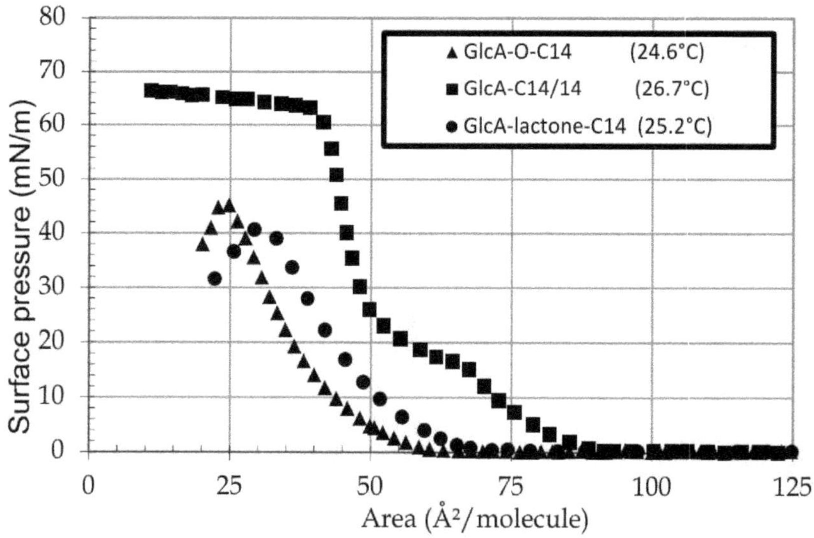

Fig. 5. (π-A) isotherms of GlcA-O- C_{14}, GlcA-lactone- C_{14}, and GlcA- $C_{14/14}$ spread at the airwater interface 25°C.

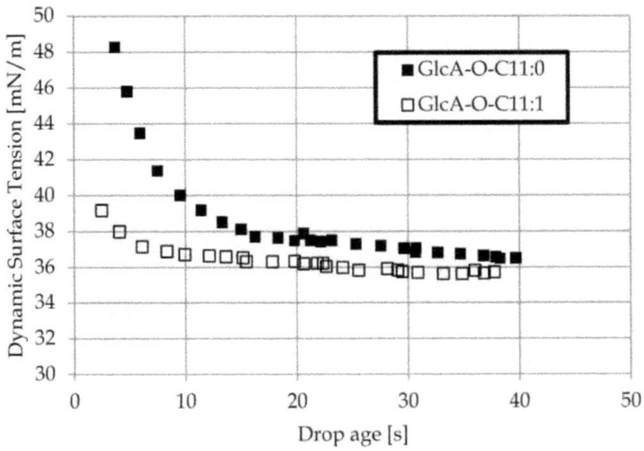

Fig. 6. Dynamic surface tension vs. time of undecanyl (C11:0) and undecenyl (C11:1)glucuronolactones (0.1 g/L) measured by TVT1 at 20°C.

6.3 Linkage effect

6.3.1 Ester bond direction

One of the spectacular effects of the linkage between the polar headgroup and the hydrophobic tail is the change of the ester bond orientation within the octyl sugar derivatives. By comparing the dynamic and equilibrium surface properties of glucose octanoate (C8-CO-O-Glc) and octyl glucuronate (C8-O-CO-GlcA), significant differences are observed (Table 4). When the carbonyl group is closed to the polar headgroup constituted by glucuronic acid (C8-O-CO-GlcA), the surfactant is more hydrophobic, and the polar head group is larger. The consequence is the increase of dynamic properties, and the decrease of the CMC (10.7 mM vs. 19.1 mM). When the carbonyl group is close to the alkyl chain (C8-CO-O-Glc), there is an "interruption" of the hydrophobic tail, which reduces its hydrophobicity (Razafindralambo et al. 2009). The molecular area becomes smaller and the CMC is higher. On the other hand, the adsorption time rate characterized by t^* and v_{max} are increased, reducing its performance on dynamic surface properties.

Table 4. Physical parameters of glucose octanoate and octyl glucuronate; (1) Compared at 1.63mM(25°C).

Parameters	Glucose octanoate [Glc-O-C $_8$]	Octylglucuronate [GlcA-O-C $_8$]
Dynamics (1) :		
$t^*(10^{-3}$ s)	36.1	3.4
n	0.14	0.56
v_{max} (mN/m/s)	0.024	0.0081
Equilibrium :		

Υ_{cmc} (mN/m)	25.1	28.0
CMC (mM)	19.1	10.7
A_{min} (Å^2/mol)	37	44

6.3.2 Ester vs. amide bond

By comparing the surface properties of three derivatives with an octyl chain, the impact of the type of the linkage bond can be evidenced, despite the difference in the ratio alpha/beta of anomeric forms (Table 5). It appears that the uronamide is the most effective in terms of equilibrium surface tensions. This result may be attributed to the difference in the molecular area occupied by the two octyl derivatives. The amide bond has a smaller bulk size than the ester ones. In comparison with a unique form of octyl glucoside, the glucuronamide is more effective. Even though the linkage is not at the same position, on the C1 for glucoside, and on the C6 for glucuronamide, the impact may be significant, considering the similarity in the molecular mass, and the number of the OH group available.

Table 5. Physical parameters of glucuronic acid and glucose derivatives containing different linkages at the air-water interface. (1) **By modeling; (2) by linear fit**

Physical parameters	C8-glucuronate [C6-Ester linkage]	C8-glucuronamide [C6-Amide linkage]	C8-glucoside **(25°C)** [C1-Ether linkage] $\alpha;\beta$
	$\alpha/\beta = 1.5$	$\alpha/\beta = 0.6$	[Boyd et al, 2000]
Υ_{cmc} (mN/m)	28.0	24.0	35.2 ; 31.3
CMC (mM)	10.7	3.3	12.0; 18.2
A_{min} (Å^2 /mol)	44 ; 45 (1)	25 ; 35 (1)	42 ; 42 (2)

7. Concise Summary

This chapter enhances the comprehension of the structure-surface activity connections of uronic acid derivatives, an auspicious category of carbohydrate-based surfactants. Each structural element influences either its dynamic performance, assessed over a brief duration, or its equilibrium activities, evaluated over an extended period. In addition to the scientific significance of this fundamental information, the approach facilitates the identification of appropriate structures for effectively forming and stabilizing colloidal systems such as foams, emulsions, and suspensions, which are prevalent in nearly all soft manufactured products.

8. References

Blecker, C.; Piccicuto, S.; Lognay, G.; Deroanne, C.; Marlier, M. & Paquot, M. (2002). Enzymatically prepared *n*-alkyl esters of glucuronic acid: the effect of hydrophobic chain length on surface properties.Journal of colloid and interface science, 247, 424-428.

Boyd, B. J.; Drummond, C.; Krodkiewska, I. & Grieser, F. (2000). How Chain Length, Headgroup Polymerization, and Anomeric Configuration Govern the Thermotropic and Lyotropic Liquid Crystalline Phase Behavior and the Air-Water Interfacial Adsorption of Glucose-Based Surfactants. Langmuir, 16, 7359-7367

Chester, A. (1997). Nomenclature of glycolipides. Pure and Applied Chemistry, 69, 2475-2487.

Dembitsky, V. (2005). Astonishing diversity of natural surfactants: 3. Carotenoid glycosides and isoprenoid glycolipids. Lipids, 40, 535-557.

Dembitsky, V. (2004). Astonishing diversity of natural surfactants: 1. Glycosides of fatty acids and alcohols. Lipids, 39, 933-953.

Dembitsky, V. (2005). Astonishing diversity of natural surfactants: 4. Fatty acid amide glycosides, their analogs and derivatives. Lipids, 40, 641-660.

Desai, J.& Banat, I. (1997). Microbial production of surfactants and their commercial potential. Microbiology and Molecular Biology Reviews, 61, 47-64.

Hato, M.; Minamikawa, H.; Tamada, K. ; Baba, T. & Tanabe, Y. (1999). Self-assembly of synthetic glycolipid/water systems. Advances in Colloid and Interface Science, 80, 233270.

Hill, K. & LeHen-Ferrenbach, C. (2009). Sugar-Based Surfactants for Consumer Products and Technical Applications, In: Sugar-based surfactants fundamentals and Applications, C.C. Ruiz (Ed.), 1-20, CRC Press, ISBN 978-1-4200-5166-7.

Infante, M. ; Pinazo, A.& Seguer, J. (1997). Non-conventional surfactants from amino acids and glycolipids: Structure, preparation and properties. Colloids and Surfaces A: Physicochemical and Engineering Aspects, 123-124, 49-70.

Kitamoto, D.; Morita, T. ; Fukuoka, T. ; Konishi, M. & Imura, T. (2009). Self-assembling properties of glycolipid biosurfactants and their potential applications. Current Opinion in Colloid & Interface Science14, 315-328.

Laurent, P.; Razafindralambo, H. ; Wathelet, B. ; Blecker, C.; Wathelet, J. & Paquot, M. (2011). Synthesis and Surface-Active Properties of Uronic Amide Derivatives, Surfactants from Renewable Organic Raw Materials.Journal of Surfactants and Detergents, 14, 51-63.

Milkereit, G.; Garamus,V.; Veermans, K.; Willumeit, R. &Vill,V. (2005). Structures of micelles formed by synthetic alkyl glycosideswith unsaturated alkyl chains. Journal of Colloid and Interface Science, 284, 704-713

Queneau, Y.; Chambert, S. ; Besset, C. & Cheaib, R. (2008). Recent progress in the synthesis of carbohydrate-based amphiphilic materials: the examples of sucrose and isomaltulose. Carbohydrate Research, 343, 1999-2009.

Razafindralambo, H.; Richel, A.; Wathelet, B.; Blecker, C.; Wathelet, J.; Brasseur, R.; Lins, L.; Miñones, J. & Paquot, M. (2011a). Monolayer properties of uronic acid bicatenary derivatives at the air-water interface: effect of hydroxyl group stereochemistry evidenced by experimental and computational approaches. Physical chemistry chemical physics: PCCP, 13, 15291-15298.

Razafindralambo,H.; Blecker, C.& Paquot, M.(2011b). Screening of Basic Properties of Amphiphilic Molecular Structures for Colloidal System Formation and Stability: The Case of Carbohydrate-Based Surfactants. In Amphiphiles: Molecular Assembly and Applications, Nagarajan, R., Ed. ACS: Washignton; Vol. 1070, p53-66.

Razafindralambo, H.; Blecker, C.; Mezdour, S.; Deroanne, C.; Crowet, J.; Brasseur, R.; Lins, L.; Paquot, M. (2009). Impacts of the Carbonyl Group Location of Ester Bond on Interfacial Properties of Sugar-Based Surfactants: Experimental and Computational Evidences. The journal of physical chemistry. B, 113, 8872-8877.

Razafindralambo, H.; Thonart, P. & Paquot, M. (2004). Dynamic and equilibrium surface tensions of surfactin aqueous solutions.Journal of Surfactants and Detergents, 7 (1), 41-46.

Razafindralambo, H.; Blecker, C.; Delhaye, S.& Paquot, M. (1995). Application of the QuasiStatic Mode of the Drop Volume Technique to the Determination of Fundamental Surfactant Properties. Journal of colloid and interface science, 174, 373-377.

Richel, A.; Laurent, P.; Wathelet, B.; Wathelet, J. & Paquot, M. (2010). Microwave-assisted synthesis of D-glucuronic acid derivatives using cost-effective solid acid catalysts. Tetrahedron letters, 51, 1356-1360.

Ruiz, C. (2009). Sugar-Based Surfactants Fundamentals and Applications. CRC Press. ISBN978-1-4200-5166-7.

Sakai, K.; Tamura, M.; Umezawa, S.; Takamatsu, Y.; Torigoe, K.; Yoshimura, T.; Esumi, K.; Sakai, H. & Abe, M. (2008). Adsorption characteristics of sugar-based monomeric and gemini surfactants at the silica/aqueous solution interface. Colloids and Surfaces A: Physicochemical and Engineering Aspects, 328, 100-106.

Stubenrauch, C. (2001). Sugar surfactants - Aggregation, interfacial, and adsorption phenomena. Current Opinion in Colloid and Interface Science, 6, 160-170.

Tadros, T.; Vandamme, A.; Levecke, B.; Booten, K. & Stevens, C. (2004). Stabilization of emulsions using polymeric surfactants based on inulin. Advances in Colloid and Interface Science, 108-109, 207-226.

Vulfson, E.; Patel, R. & Law, B. (1990). Alkyl- α-glucoside synthesis in a water-organic twophase system. Biotechnology Letters, 12, 397-402.

Wagenaar, A.& Engberts, J. (2007). Synthesis of nonionic reduced-sugar based bola amphiphiles and gemini surfactants with an α, ω-diamino-(oxa) alkyl spacer. Tetrahedron, 63, 10622-10629.

Chapter-9

Microalgal CO_2 Mitigation and Biofuels: Photobioreactor Advances

1. Introduction: An Overview

Heightened apprehensions regarding global warming and greenhouse gas emissions, together the depletion of readily available fossil fuel resources, necessitate the development of effective carbon dioxide (CO_2) mitigation methods and the adoption of clean, renewable energy sources. Carbon dioxide is a principal gas contributing to global warming. Carbon dioxide is 68% of the projected total greenhouse gas emissions (Harrington & Foster, 1999).

Various strategies have been suggested for regulating atmospheric CO_2 emissions, including ocean sequestration methods like deep ocean injection and enhancing the dissolution of CO_2 in the ocean. Another suggested method of sequestration is the incorporation of CO_2 into terrestrial ecosystems (Halmann, 1993). Direct injection sequestration into geological or marine sinks is acknowledged as a strategy to diminish CO_2 levels; nevertheless, it fails to tackle sustainability concerns (Stewart & Hessami, 2005). Carbon sequestration can also be achieved using chemical methods; however, these methods must provide environmental safety, long-term stability, and cost competitiveness with alternative sequestration solutions. Alternative approaches, including chemical absorption, membrane separation, cryogenic fractionation, and adsorption with molecular sieves, have been evaluated; however, these exhibit even worse energy efficiency, rendering them economically unfeasible (Stewart & Hessami, 2005).

A notably underexplored approach for CO_2 reduction is the utilization of biological processes, specifically through microalgae, for the direct conversion of CO_2 to biomass from point source emissions in engineered systems like photobioreactors. The biofixation of carbon dioxide (CO_2) by microalgae in photobioreactors has lately garnered renewed interest as a viable technique for CO_2 reduction. The employment of photobioreactors for microalgal CO_2 sequestration has significant benefits, including enhanced microalgal productivity due to regulated environmental conditions and optimized spatial and volumetric efficiency, hence facilitating a more effective use of expensive land resources.

The scaled-up photosynthetic approach would offer a significantly superior and sustainable alternative from both environmental and economic perspectives.

Fig. 1 shows the importance of the microalgae photobioreactor and its general applications, microalgae used to capture waste CO_2 utilizing the nutrients in wastewater and natural solar light. The microalgae biomass produced can be used for biofuel production (such as biodiesel and methane) and other by products (such as animal feeds and polymers).

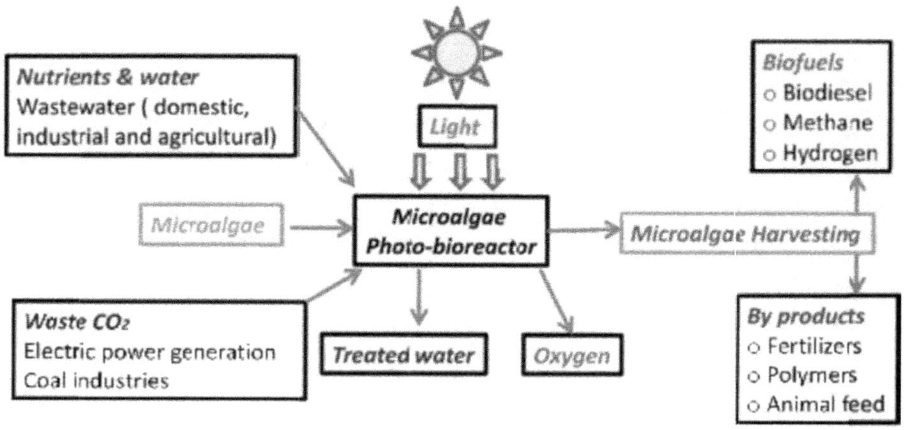

Fig. 1. Microalgae photobioreactors applications

Broadly, photo-bioreactors can be classified as open (pond) systems or closed systems. Considering all the limitations and shortcomings of the pond systems, most researchers, had oriented their research works towards the development of an unconventional way for microalgae culture, which should be fully closed and compact with high surface-to-volume ratio and all the growth factors be optimized.

With these desired characteristics as the main goals, research on tubular and airlift photobioreactors were the right orientation and some forms of designs had in certain aspects succeeded when used in the lab scale. However, few of these forms could be really applicable in the pilot production scale, due to serious obstacles of operational problems and growth limitations. Amongst them, were primarily the oxygen build-up in the growth medium, photoinhibition, light saturation effect and the overheating inside the tube walls by the intensive solar radiation when operating in summer seasons especially in the midday light hours. Besides, the poor circulation of the growth medium causes the algal staining on the inner walls of the tubes, gave eventually an uneconomic results. Over the years, several solutions have been proposed to overcome these fundamental limitations to productivity. However, these systems are complicated to scale up and may be suitable for small-scale cultivation. Moreover, there is little knowledge about the feasibility of photobioreactors scale-up and developments.

This chapter will examine the current state of photobioreactor technology for CO_2 mitigation and biofuel production, addressing critical issues including the effective provision of light, the supply of carbon dioxide with minimal losses, the removal of photosynthetically generated oxygen that could hinder metabolism or harm the culture if accumulated, the practical scalability of the technology, and the harvesting of microalgae biomass for biofuel production. This chapter will also summarize the theoretical underpinning of microalgae cultivation. This chapter will give potential future insights for the commercial exploitation of microalgae in CO_2 biomitigation and biofuel production.

2. Mechanism of the photosynthesis and biophotolysis

Photoautotrophic microorganisms like eukaryotic green microalgae, possess chlorophyll and other pigments to capture sunlight energy and use photosynthetic systems (PSII and PSI) to carry out plant-like oxygenic photosynthesis (Kruse et al. 2005). The pigments in PSII (P680) absorb the photons with a wavelength shorter than 680 nm , generating a strong oxidant capable of splitting water into protons (H^+), electrons (e^-)and O_2 as shown in Fig. 2.

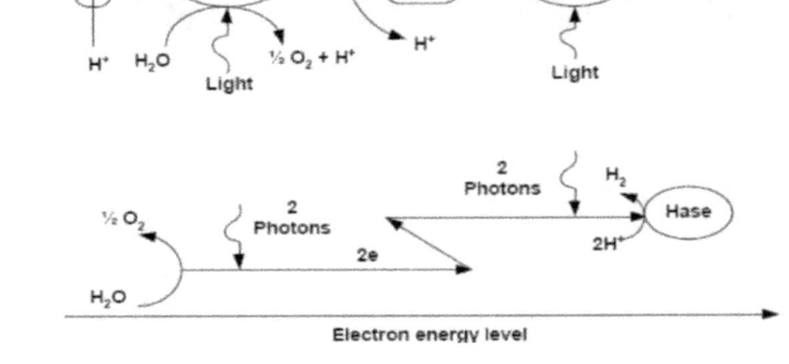

Fig. 2. Schematic mechanisms of photosynthesis and biophotolysis of photoautotrophic microbes (Amos, 2004).

The electrons or reducing equivalents are transferred through a series of electron carriers and cytochrome complex to PSI. The pigments in PSI (P700) absorb the photons with a wavelength under 700 nm , which further raises the energy level of the electrons to reduce the oxidized ferredoxin (Fd) and/or nicotinamide adenine dinucleotide phosphate (NADP+) into their reduced forms. The proton gradient formed across the cellular (or thylakoid) membrane drives adenosine triphosphate (ATP) production via ATP synthase. CO_2 is reduced with ATP and NADPH via a reductive pentose phosphate pathway or Calvin cycle for cell growth.

The excess reduced carbon is stored inside the cells as carbohydrates (CH_2O) and/or lipids. The type of carbohydrate product produced depends on the type of strain being used. The reducing power (Fd) could also be directed to hydrogenase (Hase) for hydrogen evolution.

3. Microalgae and microalgae cultivation

Algae are a large and diverse group of simple, typically autotrophic organisms, ranging from unicellular to multicultural forms. The largest and most complex marine forms are called seaweeds. They are photosynthetic, like plants, and simple because they lack the many distinct organs found in land plants. Some unicellular species rely entirely on external energy sources and have limited or no photosynthetic apparatus. All algae have photosynthetic machinery ultimately derived from the cyanobacteria, and so produce oxygen as a byproduct of photosynthesis.

Table 1. Chemical composition of algae expressed on a dry matter basis (%) (Becker,1994)

Strain	Protein	Carbohydrates	Lipids	Nucleic acid
Scenedesmus obliquus	50-56	10-17	12-14	3-6
Scenedesmus quadricauda	47	-	1.9	-
Scenedesmus dimorphus	8-18	21-52	16-40	-
Chlamydomonas rheinhardii	48	17	21	-
Chlorella vulgaris	51-58	12-17	14-22	4-5
Chlorella pyrenoidosa	57	26	2	-
Spirogyra sp.	6-20	33-64	11-21	-
Dunaliella bioculata	49	4	8	-
Dunaliella salina	57	32	6	-
Euglena gracilis	39-61	14-18	14-20	-
Prymnesium parvum	28-45	25-33	22-38	1-2

Tetraselmis maculata	52	15	3	-
Porphyridium cruentum	28-39	40-57	9-14	-
Spirulina platensis	46-63	8-14	4--9	2-5
Spirulina maxima	60-71	13-16	6-7	3-4.5
Synechoccus sp.	63	15	11	5
Anabaena cylindrica	43-56	25-30	4-7	-

Algae can be classified into two types based on their sizes, microalgae and macroalgae. Microalgae are microscopic photosynthetic organisms (less than 2 mm in diameter). However, macroalgae, these organisms that are found in both marine and freshwater environments. Biologists have categorized microalgae in a variety of classes, mainly distinguished by their pigmentation, life cycle and basic cellular structure (Amos, 2004). The most frequently cited microalgae as carrying one or more of the desirable features for efficient and economical combination of CO_2 biofixation, wastewater treatment and lipid synthesis toward biofuel production are:

1. The diatoms (Bacillariophyceae). These algae dominate the phytoplankton of the oceans, but are also found in fresh and brackish water. Approximately 100,000 species are known to exist. Diatoms contain polymerized silica (Si) in their cell walls. All cells store carbon in a variety of forms. Diatoms store carbon in the form of natural oils or as a polymer of carbohydrates known as chyrsolaminarin.

2. The green algae (Chlorophyceae). These are also quite abundant, especially in freshwater. They can occur as single cells or as colonies. Green algae are the evolutionary progenitors of modern plants. The main storage compound for green algae is starch, though oils can be produced under certain conditions.

3. The blue-green algae (Cyanophyceae). Much closer to bacteria in structure and organization, these algae play an important role in fixing nitrogen from the atmosphere. There are approximately 2,000 known species found in a variety of habitats.

4. The golden algae (Chrysophyceae). This group of algae is similar to the diatoms in pigmentation and biochemical composition. They have more complex pigment systems, and can appear yellow, brown or orange in color. Approximately 1,000 species are known to exist, primarily in freshwater systems. The golden algae produce natural oils and carbohydrates as storage compounds.

All algae primary comprise of the following, in varying proportions (Table 1): proteins, carbohydrates, fats and nucleic acids. While the percentages vary with the type of algae, there are algae types that are comprised of up to 40% of their overall mass by fatty acids that could be extracted and converted into biodiesel.

4. Photo-bioreactor for carbon dioxide sequestration

Photobioreactors for microalgae cultivation can be classified as open systems or closed systems. Open systems are ponds, constructed on the large open areas, in rows with growth medium exposed to environment and sunlight. Closed systems are those where growth medium enclosed from the environment. Open systems have many disadvantages over closed system, for instance they are hard to control and, contamination from external environment is high and could cause the microalgae mutate (Camacho Rubio et al., 1999). Closed systems are easy to monitor, less chances of contamination, better mass transfer (varies based on the type of bioreactor), occupy less space for the same algal growth. Closed systems can be classified as tubular photobioreactors; stirred photobioreactors; flat plate photobioreactors; hollow fiber membrane photobioreactors; airlift and sparged bubble column photobioreactors. Unfortunately, none of the these bioreactor configurations is able to control effectively all process parameters that are required for maximum CO_2 biofixation, microalgal growth and metabolic rates, particularly at large scale production. Below is a brief description for the most widely used photobioreactors for CO_2 biofixation and biofuel production:

4.1 Tubular photo-bioreactors

Tubular photo-bioreactors consist of long thin tubes arranged in different geometrical patterns (helical, straight tubes) to optimize irradiance from a point light source (sun). Generally liquid growth medium is circulated in these tubes by air bubbling and by injection of air into one end of the system and degassed at the other end. Construction, light regime, mass transfer and scale up issues of these photo-bioreactors have been discussed (Molina Grima, 2000). Experiments have showed that a large-scale tubular photo-bioreactor has failed and the main reason attributed for its failure was the large dissolved oxygen in the system (Molina Grima, 2000). Hence a system should not at any stage be over saturated with oxygen as this would cause algae to shutdown photosynthesis and growth (Camacho Rubio et al., 1999). It was also reported that tubular photo-bioreactors are difficult to build and maintain, and have limited scalability.

4.2 Mechanically stirred photo-bioreactors

Mechanically stirred photo-bioreactors use baffles to stir the growth medium to attain a mass transfer of air/ CO_2 into liquid. A drawback of the stirred medium is if stirred vigorously the algae cell wall would be damaged by the high fluid shear forces (Molina Grima et al., 1996). If it is stirred slowly, eddy currents will not be established that move the algae toward the light source thereby decreasing the efficiency of light available for the photosynthetic process and also reducing mass transfer of nutrients from the air/CO2 to the liquid in the systems.

4.3 Airlift photo-bioreactors

Airlift photo-bioreactors are basically a column divided into two parts, air/CO2 is bubble through only one side of the partition which causes a liquid current pattern to develop with the air bubble side called the riser and other part called the downcomer (Sánchez Mirón et al., 2000). These bioreactors are extensively investigated for fermentation process and wastewater treatment (Znad et al. 2004, Znad et al. 2006) but have not been looked at as a replacement for the popular tubular photo-bioreactors until recent times (Sánchez Mirón et al., 2000). The airlift photo-bioreactor characterized by; high mass transfer, good mixing with low shear stress, low energy consumption, high potentials for scalability, easy to sterilize, readily tempered, good for immobilization of algae, reduced photo-inhibition. However, the main limitations are; the small illumination surface area and decrease of illumination surface area upon scale-up. It has become clear that biological carbon sequestration and hydrogen production technologies have been poorly studied in the airlift photo-bioreactors and are in their infancy of development.

4.4. Bubble column photo-bioreactors

Bubble column bioreactors are vertical columns either cylindrical or rectangular filled with growth medium and air is bubble through a sparged system installed at the bottom. These systems have the highest gas hold ups rates which means they have the best mass transfer compared to other systems (Miron et al. 2000; Kommareddy & Anderson , 2003). A modified version of these bubble column bioreactors is porous membrane reactors, which have efficient aeration, give smaller bubbles, and pressure drop across the membrane is low compared with other rigid sparged bubble column reactors. These characteristics are achievable at high gas flow rates, with low energy costs (Poulsen & Iversen, 1997).

5. Determinants influencing photobioreactor efficacy

The principal factors influencing photobioreactor performance, specifically the growth of microalgae, include the effective provision of light, carbon dioxide concentration, photosynthetically produced oxygen, gas transfer, mixing rates, temperature, pH, and nutrient requirements.

5.1 Light provision

Light is the basic energy source for phototrophic microorganisms. The intensity and utilization efficiency of the light supplied are thus of crucial importance in microalgal bioreactors. Light intensity decreases deeper within the culture medium, especially in highdensity cultures; hence, the issue of optical depth, which measures the proportion of radiation absorbed or scattered along a path through a partially transparent medium, should be considered in microalgal bioreactor design (Kumar et al., 2010).

Both sunlight and artificial light have been used via outer surface exposure as well as inner volume exposure, through the placement of lighting devices (e.g. LEDs or optical fibers) inside the bioreactor itself (Suh & Lee, 2003). The photosynthetically active radiance is normally assumed to be $43-45\%$ in the wavelength range of $400-700$ nm (Laws et al. 1987). The light intensity available to microalgae in high-density cultures is significantly attenuated by mutual shading; to maximize light absorbance and minimize light attenuation, bioreactors should be designed with a high surface area-to-volume ratio, coupled with a short light path (Richmond et al. 2003).

Good microalgal growth rates have been reported (Hu et al. 1998) under a light intensity of 4000μ mol m^{-2} s^{-1}; this intensity is twice the solar flux in a medium latitude spot at midday during summer. However, a strong species-dependence exists that should be taken into account. By contrast, light above a saturation point causes light inhibition, which can be counterbalanced by exposing microalgal cells to very short cyclic periods of light and darkness (Pulz, 2001).

The ratio of light to dark (or low-intensity light) periods in a cycle is crucial for microalgal productivity (Munoz & Guieysse, 2006). Similar overall numbers of moles of photons do not necessarily produce equal growth rates of (or CO_2 assimilation by) microalgae. When the light/dark cycle period approaches the photosynthetic unit turnover time (equal to the dark reaction time, estimated to lie within $1-15$ ms), maximum photosynthetic efficiencies can be achieved (Richmond et al. 2003). Moreover, compared with periodic darkness, periods of low light intensity significantly increase growth, CO_2 assimilation and lipid productivity in microalgae for a given whole light level (Cuello et al. 2008). This type of lighting design can be achieved via artificial light, such as hybrid lighting systems

(Muhs, 2000). Different lamps generate distinct spectra, and different microalgal species possess dissimilar absorption optima; therefore, each individual case should be studied before deciding on the set point of this important operational parameter. Variation of the exponential growth rates of Phorphyridium cruentum have been recorded (Suh & Lee, 2003) with variable radiation energies and light spectra, concluding that blue light ($400-500$ nm) increases cell growth and polysaccharide production.

In terms of artificially illuminated bioreactors, the need for small reactor diameters to increase the illuminated surface area per unit volume of culture can be circumvented through provision of internal illumination. High biomass yields are more crucial in the case of artificially illuminated reactors, because the light provided adds to the overall operational cost of the underlying process. Such costs can be kept below acceptable thresholds via in situ growth-monitoring and associated online control of the intensity of light supplied.

5.2 Carbon uptake

Biological CO_2 fixation can be carried out by higher plants and microalgae, yet the latter possess a greater ability to fix CO_2 (Li et al. 2008; Chisti, 2007; Tredici 2010). Usual sources of CO_2 for microalgae include atmospheric CO_2; CO_2 from industrial exhaust gases (e.g. flue gas and flaring gas); and CO_2 chemically fixed in the form of soluble carbonates (e.g. $NaHCO_3$ and Na_2CO_3) (Kumar et al., 2010). The tolerance of various microalgal species to the concentration of CO_2 is variable; however, the CO_2 concentration in the gaseous phase does not necessarily reflect the CO_2 concentration to which the microalga is exposed during dynamic liquid suspension, which depends on the pH and the CO_2 concentration gradient created by the resistance to mass transfer. Under heterotrophic or mixotrophic conditions, some microalgal species can metabolize a variety of organic compounds, including sugars, molasses and acetic acid, as well as compounds present in wastewater and petroleum (Becker, 1994). Atmospheric CO_2 levels (0.0387% (v/v)) are not sufficient to support the high microalgal growth rates and productivities needed for full-scale biofuel production.

Waste gases from combustion processes, however, typically contain $> 15\%(v/v)CO_2$; this percentage indicates, in principle, that combustion processes will provide sufficient amounts of CO_2 for large-scale production of microalgae (Doucha et al. 2005). Owing to the cost of upstream separation of CO_2 gas, direct utilization of power plant flue gas has been considered in microalgal biofuel production systems (Lackne, 2003). Flue gases that contain CO_2 at concentrations ranging from 5 to $15\%(v/v)$ have indeed been introduced directly into ponds and bioreactors of various configurations that contain several microalgal species (Kumar et al., 2010).

5.3 Oxygen generated

Another specific issue of microalgal bioreactors is the accumulation of photosynthetically generated oxygen that may inhibit metabolism or otherwise damage the culture if allowed to accumulate, especially when the rate of photosynthesis, which often correlates with the rate of CO_2 transfer, is high (as typical in horizontal tubular reactors) (Kumar et al., 2010). Most solutions to this problem rely on the use of a degasser (or gas exchange unit), where dissolved oxygen can be released (Morita et al. 2000). However, to attain effective separation between the gas and liquid phases, the path through the degasser should be such that the smallest bubbles have sufficient time to disengage from the liquid.

In tubular bioreactors, connections between tubes can incorporate a tube specifically for oxygen degassing, or a layer of parallel tubes connected by two manifolds: the lower manifold is used to inject air into the culture, and the higher one acts as the degasser (Kumar et al., 2010). Nevertheless, microalgal productivities were lower than expected in these tubular systems, possibly because of build-up of dissolved oxygen during high light intensity periods and along the bioreactor path between manifolds. In systems with exhaust gas recirculation, dissolved oxygen accumulation can be avoided by bubbling exhaust gas through a sodium sulfite solution before its return to the bioreactor (Cien-Fernandez et al. 2005).

Unfortunately, the efficiencies of most techniques used to date for dissolved oxygen removal from microalgal cultures are still not satisfactory. As a result, the classical bubbling mode of operation has been employed to avoid the costlier need for degassing devices. The use of several small bioreactors instead of one large unit also alleviates this problem. Scale-up is indeed easier for facilities that use many small reactors in parallel, even though investment costs might be higher than with fewer large equipment units. Continued research is needed to accurately match the amount of CO_2 supplied to the actual uptake requirement of the metabolizing microalgae, as well as the amount of O_2 removed to the actual amount of O_2 produced.

5.4 Gas transfer

Gases introduced into bioreactors serve a number of purposes in microalgal cultivation, including: supply of CO_2 as sources of carbon for biomass primary and secondary metabolism; provision of internal mixing, which avoids nutrient concentration gradients; promotion of exposure of all cells to light (especially in high density cultures), while minimizing self-shading and phototoxicity; control of pH by assuring dissolution of CO_2 and avoiding gradients thereof; and stripping of accumulated dissolved oxygen, hence reducing its toxicity to microalgae ((Kumar et al., 2010, Pulz, 2001).

Among the various alternatives, bubbling CO_2-enriched air into the bottom of the bioreactor with bubble diffusers has been the most frequently used approach. Moderate overall transfer efficiencies (13 – 20%) can be achieved by this mode of gas delivery (Carvalho et al. 2006); however, associated drawbacks are loss of CO_2 to the atmosphere, biofouling of diffusers, and poor mass transfer rates owing to a relatively low interfacial specific surface area. Better overall efficiencies are expected for hollow-fiber membrane bioreactors in which the slightly lower mass transfer coefficients that arise from a less turbulent local hydrodynamic pattern are compensated by the much larger area per unit volume available for mass transfer. In addition, the area of mass transfer is well defined, and the pressure on the gas side can be controlled so as to supply only the required amount of CO_2, hence permitting more accurate control of the transfer rate and a dramatic reduction in the amount of CO_2 lost to the atmosphere (Carvalho & Malcata, 2001).

5.5 Mixing rates

Mixing is a key parameter for acceptable performance of microalgal bioreactors. Low mixing rates hamper gaseous mass transfer and might even permit biomass settling. In either case, poor mixing leads to emergence of stagnant zones, where light and nutrients are insufficiently available and anoxic/anaerobic conditions will thus prevail, which results in a decrease of productivity (Kumar et al., 2010). Culture viability might also be compromised by production and accumulation of toxic compounds in stagnant zones (Becker, 1994). Conversely, high mixing rates can cause shear damage to cells (Carlsson et al. (2007), besides requiring a large energy input.

The most common methods of mixing in microalgal bioreactors are pumping, mechanical stirring and gas injection. Pumping offers fair mixing efficiency, but low gas transfer rates; the associated hydrodynamic stress increases with the rotation speed of the pumps, or the number of passes of the microalgal suspension through the pump units (Jaouen et al. 1999).

Mechanical stirring has been reported to provide good mixing efficiency and gas transfer; however, it is likely to produce significant hydrodynamic stress (Tredici, 2003), which can be managed via adequate use of baffles to create a controlled turbulence pattern. Gas injection (bubbling) produces lower hydrodynamic stress, while providing good gas transfer and reasonable mixing efficiency (Richmond & Cheng-Wu, 2001); however, cell damage in sparged cultures increases as the biomass concentration increases, because exponentially higher degrees of stirring are needed to maintain a high-density culture at a predefined level of mixing (Pulz, 2001). One approach to minimize this problem is to maintain a low gas input per nozzle, so as to reduce shear stress and consequent cell damage (Barbosa et al. 2003).

5.6 Temperature effects

Temperature is one of the major factors that regulate cellular, morphological and physiological responses of microalgae: higher temperatures generally accelerate the metabolic rates of microalgae, whereas low temperatures lead to inhibition of microalgal growth (Munoz & Guieysse, 2006). The optimal temperature varies among microalgal species (Ono & Cuello, 2003); however, optimal temperatures are also influenced by other environmental parameters, such as light intensity. Optimal growth temperatures of $15 - 26\ ^{\circ}C$ have been reported for some species, with maximum cell densities obtained at $23\ ^{\circ}C$. Only daytime higher temperatures were observed to have clearly favorable effects on microalgal growth rates due to photosynthesis, except when the night temperature was as low as $7\ ^{\circ}C$ (Tamiya, 1957).

5.7 pH effects

Most microalgal species are favored by neutral pH , whereas some species are tolerant to higher pH (e.g. Spirulina platensis at pH 9) (Hu et al. 1998) or lower pH (e.g. Chlorococcum littorale at pH 4) (Kodama et al. 1993). There is a complex relationship between CO_2 concentration and pH in microalgal bioreactor systems, owing to the underlying chemical equilibria among such chemical species as CO_2, H_2CO_3, $HCO_3^{\ -}$and $CO_3^{\ 2-}$. Increasing CO_2 concentrations can lead to higher biomass productivity, but will also decrease pH , which can have an adverse effect upon microalgal physiology. By contrast, microalgae have been shown to cause a rise in pH to $10 - 11$ in open ponds because of CO_2 uptake (Oswald, 1988). This increase in pH can be beneficial for inactivation of pathogens in microalgal wastewater treatment, but can also inhibit microalgal growth. Similarly, the speciation of NH_3 and $NH_4^{\ +}$ in microalgal bioreactors is strongly dependent on pH – NH_3 uncouples electron transport in the microalgal photosystem and competes with water molecules in oxidation reactions, thus leading to release of O_2 (Hu et al. 1998).

5.8 Nutrient requirements

In addition to the carbon, nitrogen is the most important element that is required for microalgal nutrition (Becker, 1994) and, as a constituent of both nucleic acids and proteins, nitrogen is directly associated with the primary metabolism of microalgae. Fast-growing microalgal species prefer ammonium rather than nitrate as a primary nitrogen source (Green & Durnford, 1996); intermittent nitrate feeding, however, will enhance microalgal growth if a medium that lacks nitrate is used (Jin et al. 2006). Under partial nitrogen deprivation, microalgae grow at lower rates, but produce significantly more lipids, which are reserve compounds synthesized under stress conditions, even at the expense of lower productivities (Lardon et al. 2009).

Phosphorus is the third most important nutrient for microalgal growth, and should be supplied to significant excess as phosphates because not all phosphorus compounds are bioavailable (e.g. those combined with metal ions) (Kumar et al. 2009). In the case of marine microalgae, seawater supplemented with commercial nitrate and phosphate fertilizers is commonly used for production of microalgae (Green & Durnford, 1996). Nevertheless, trace species, such as metals (Mg, Ca, Mn, Zn, Cu and Mb) and vitamins, are typically added for effective cultivation (Becker, 1994).

6. Microalgae harvesting and conversion to fuels

6.1 Microalgae biomass harvesting

Harvesting of the microalgae biomass, i.e., concentrating microscopic algal cells from the dilute solutions of the algal mass culture, is an essential step to secure high-quality effluents and to prevent cell washout (Richmond et al. 2003, Munoz & Guieysse, 2006). The main difficulties encountered in harvesting microalgae arise from the relatively low biomass concentration in conventional bioreactors, coupled with the small size of its constituent microalgal cells. Harvesting typically contributes to 20 – 30% of the total cost of microalgal biomass production (Carlsson et al. 2007). The major techniques presently applied in the harvesting of microalgae include coagulation, flocculation, sedimentation, centrifugation, foam fractionation, ultrasonic separation, flotation, membrane filtration, and electrophoresis techniques (Carlsson et al. 2007; Kumar et al., 2010; Uduman et al. 2010).

Selection of the harvesting method mainly depends on the properties of microalgae, such as density, size, the value of the desired products. Microalgae harvesting can generally be divided into a two-step process, bulk harvesting, to separate microalgal biomass from the bulk suspension, in this method, the total solid mater can reach 2 – 7% using flocculation, flotation, or gravity sedimentation; and the second step is thickening, to concentrate the slurry, using filtration and centrifugation. This step needs more energy than bulk harvesting (Brennan & Owende, 2010).

Microalgal cell immobilization has been proposed to circumvent the harvesting issue, but large-scale applications are limited. Further investigation is clearly needed to optimize operating conditions and design new processes (Mallick, 2002).

Following biomass harvest by centrifugation or filtration, microalgal paste traditionally consists of 90%(w/w) water, which meets the requirements for anaerobic digestion. However, it is necessary to reduce this value to a maximum of 50%(w/w) water for efficient oil extraction (Kumar et al., 2010). Despite its energy-intensive nature, drying has often been the dewatering process of choice.

Almost 90% of the energy required for biodiesel production is indeed accounted for by harvesting and dewatering of biomass, besides lipid extraction itself (Lardon et al. 2009). In addition to lipid extraction for biodiesel production, a novel process that gasifies biomass to methane and concentrated CO_2 has recently been proposed (Stucki et al. 2009) for improved overall energy efficiency.

Most microalgae exhibit the phenomenon of bioflocculation, which is the spontaneous aggregation of algal cells into large flocs. These flocs will then settle rather rapidly. The process yet not fully understood and need more investigation. It depends on the elaboration of polymers by the algal cells that makes the cells stick together (Benemann & Pedroni, 2008). Sufficient experience exists to suggest that bioflocculation, possibly in combination with centrifugation, could achieve the cost goals for efficient CO_2 biofixation and biofuel production. Further study and development of this process remains a central problem, next to productivity and controlled cultivation of specific algal species in the designed photobioreactor (Benemann & Pedroni, 2008).

6.2 Microalgae biomass conversion to fuels

For economic and environmental reasons the demand for liquid energy from renewable resources will have an ascending trend in the coming year. The advantages of biomass include that it is biodegradable, sustainable and also causes less pollution when compared with fuels being used. Microalgae with high lipid content produces higher biodiesel than commercially used oilseed crops (rapeseed, soybean oil) utilizing less amount of water (Sheehan et al. 1998).

Converting the harvested biomass to a biofuel considered the least difficulty step. The high water content of the harvested biomass makes drying or any thermochemical conversion process (e.g. combustion, gasification, pyrolysis) impractical, and an even more critical problem is the high nitrogen content of algal biomass. Any thermochemical processing would result in unacceptable NO x generation and loss of this valuable nutrient and resource. Thus, microalgae biomass fuel conversion processes are dependent on fermentations to produce methane or ethanol, or the metabolism of the algae themselves, to produce oils and hydrocarbons, useable for conversion to biodiesel, or to evolve hydrogen. Methane production from microalgae biomass is technically and economically feasible, but still requires some research and development to improve yields and overall efficiency.

Compared to anaerobic digestion, very little work has been done on ethanol fermentations of algal biomass. The reason is that ethanol fermentations, typically carried out by yeast, are restricted to sugars, starches and similar easily degraded carbohydrates. Microalgae typically contain only about 20% or less of such carbohydrates, present as starch in green algae and glycogen in cyanobacteria. For practical ethanol production, an algal biomass with very high fermentable carbohydrate content, preferably over 60% on a dry weight basis, is required. Such high starch or glycogen accumulation is only observed under conditions of nitrogen limitation, where cell growth is reduced and much or most of the photosynthetically-fixed CO_2 is diverted to storage reserves. Thus the issue is whether it is possible to optimize for both high carbohydrate content and high productivity (e.g. CO_2 fixation) using nitrogen limitation (Benemann & Pedroni, 2008).

7. Concise Summary

Microalgae have garnered significant interest for CO_2 fixation and biofuel generation due to their ability to convert CO_2 into biomass through photosynthesis at substantially faster rates than traditional biofuel crops.

This chapter identifies some issues associated with microalgal-based CO2 sequestration and biofuel production. The majority of investigations completed thus far have been executed at the bench scale and under rigorously controlled conditions. Consequently, less information exists regarding the viability of scaling up the photobioreactor. Factors include the provision of sufficient CO_2, nutrients, and light to microalgal cells must be examined and optimized on a wide scale.

The co-digestion of microalgae with wastewater sludge for biogas production warrants consideration, as this approach might be integrated into the current wastewater infrastructure.

The sustainability of microalga-based CO_2 fixation and biofuel production can be enhanced by integrating microalgal biomass production with current power generation and wastewater treatment systems. Microalgae can employ low-quality water, like agricultural runoff or municipal, industrial, and agricultural wastewaters, as both a growth medium and a source of nitrogen, phosphate, and trace nutrients (Becker, 1994). Consequently, there is an added economic and environmental motivation due to the reduced expenses associated with water and chemicals necessary for the growth medium formulation, while also offering a solution for wastewater treatment (Kumar et al., 2010; Mallick, 2002; Demirbas, 2004). This coupling technology requires additional investigation. Several critical research gaps persist that must be addressed to attain full-scale operation, including enhanced algal growth and nutrient uptake rates; the integration of biosystems with waste gas, wastewater, and water reclamation systems; improved gas transfer and mixing; and optimized algal harvesting and dewatering.

The processes of harvesting, dewatering, and lipid extraction from microalgal biomass remain problematic due to their high energy consumption, mostly attributed to the diminutive cell size and very low biomass concentrations of microalgal cultures. Consequently, the development of more efficient and cost-effective harvesting technology is essential to improve the commercial viability of the microalgae biofuels business (Kumar et al., 2010).

A primary problem in microalgal biodiesel production is the utilization of microalgal species capable of sustaining both elevated growth rates and high metabolic rates, resulting in substantial lipid outputs. This significant difficulty can be effectively tackled through comprehensive bio-prospecting or targeted genetic engineering, which are emerging as promising strategies (Kumar et al., 2010).

Ultimately, there appears to be an absence of essential information required to effectively enhance the performance of current bioreactors. Innovative bioreactor topologies and designs are required to enhance microalgal growth and CO_2 biofixation, exhibiting volumetric productivities at least an order of magnitude greater than those of traditional open pond systems.

8. References

Acién Fernández, FG.; Fernández Sevilla, JM.; Egorova-Zachernyuk, TA.; Molina Grima, E. (2005). Cost-effective production of 13C, 15 N stable isotope-labelled biomass from phototrophic microalgae for various biotechnological applications. Biomol. Eng. 22, 193-200.

Amos, W.A. (2004). Updated cost analysis of photobiological hydrogen production from chlamydomonas reinhardtii green algae, NREL/MP-560-35593, National Renewable Energy Laborotary, Golden, Colorado, USA.

Barbosa, MJ.; Albrecht, M.; Wijffels, RH. (2003). Hydrodynamic stress and lethal events in sparged microalgae culture. Biotechnol. Bioeng. 83, 112-120.

Becker, E.W. (1994). In "Microalgae: biotechnology and microbiology" Ed. Baddiley, J. et al., 178 (1994) Cambridge Univ. Press, Cambridge, New York

Benemann, J.R. & Pedroni, P.M. (2008). Biofixation of Fossil CO2 by Microalgae for Greenhouse Gas Abatement. Treccani Encylopedia of Hydrocarbons, Volume III, p 837 -861.

Brennan, L. & Owende, P. 2010. Biofuels from microalgae - a review of technologies for production, processing, and extractions of biofuels and co-products. Renew. Sust. Energ. Rev. 14, 557-577.

Camacho Rubio, F.; Acién Fernández, F.G.; Sánchez Pérez, J.A.; Garcia Camacho, F. & Molina Grima, E. (1999). Prediction of dissolved oxygen and carbon dioxide concentration profiles in tubular photobioreactors for microalgal culture. Biotechnol. Bioeng 62, 71-86.

Carlsson, A. S.; Beilen, J. V.; Möller, R.; Clayton, D. (2007). Micro and Micro-algae: Utility for Industrial Applications, University of York.

Carvalho AP.; Meireles, LA.; Malcata FX. (2006). Microalgal reactors: a review of enclosed systems design and performances. Biotechnol. Prog. 22, 1490-1506.

Carvalho, A.P. & Malcata, F.X. (2001). Transfer of carbon dioxide within cultures of microalgae: plain bubbling versus hollow-fiber modules. Biotechnol. Prog. 17, 265272

Chisti, Y. (2007). Biodiesel from microalgae. Biotechnol. Adv. 25, 294- 306.

Cuello, J.L., et al. (2008). Hybrid solar and electric lighting (HYSEL) for space light support. Presentation at Carbon Recycling Forum 2008, Arizona

Demirbas, A. (2004). Current technologies for the thermo-conversion of biomass into fuels and chemicals. Energy Source 26, 715-730

Doucha, J.; Straka, F.; Lívanský, K. (2005). Utilization of flue gas for cultivation of microalgae (Chlorella sp.) in an outdoor open thin-layer photobioreactor. J. Appl. Phycol. 17, 403-412

Green, B.R. & Durnford, D.G. (1996). The chlorophyll-carotenoid proteins of oxygenic photosynthesis. Ann. Rev. Plant Physiol. Plant Mol. Biol. 47, 685-714.

Halmann, M. (1993). Chemical fixation of carbon dioxide: methods for recycling CO2 into useful products. CRC Press, Ann Arbor, MI.

Harrington, L. Foster, R. (1999). Australian residential building sector greenhouse gas emissions 1990-2010. Final Report, Energy Efficient Strategies. Australian Greenhouse Office

Hu, Q., Zarmi, Y. and Richmond, A. (1998). Combined effects of light intensity, light path and culture density on output rate of Spirulina platensis (Cyanobacteria). Eur. J. Phycol. 33, 165-171.

Jaouen, P.; Vandanjon, L.; Quéméneur, F. (1999). The shear stress of microalgal cell suspensions (Tetraselmis suecica) in tangential flow filtration systems: the role of pumps. Biores. Technol. 68, 149-154.

Jin, HF.; Lim, BR.; Lee K. (2006). Influence of nitrate feeding on carbon dioxide fixation by microalgae. J. Environ. Sci. Health A Tox. Hazard Subst. Environ. Eng. 41, 2813-2824.

Kodama, M.; Ikemoto, H.; Miyachi, S. (1993). A new species of highly CO_2-tolerant fastgrowing marine microalga suitable for high-density culture. J. Marine Biotechnol. 1, 21-25.

Kommareddy, A. & Anderson G. (2003). Mixing and Eddie Currents in a Modified Bubble Column Reactor . Paper Number: RRV03-0043 An ASAE Meeting Presentation. Written for presentation at the 2003 CSAE/ASAE Annual Intersectional Meeting Sponsored by the Red River Section of ASAE Quality Inn & Suites 301 3rd Avenue North Fargo, North Dakota, USA October 3-4, 2003

Kruse, O.; Rupprecht, J.; Mussgnug, J.R.; Dismukes, G.C. & Hankamer, B. (2005). Photochemical & Photobiological Sciences, 4, 957.

Kumar, A.; Ergas, S.; Yuan, X.; Sahu, A.; Zhang, Q.; Dewulf, J.; Xavier Malcata, F. & Herman van Langenhove. (2010). Enhanced CO_2 fixation and biofuel production via microalgae: recent developments and future directions. Trends in Biotechnology 28, 371-380

Kumar, A.; Yuan, X.; Sahu, A. K.; Dewulf, J.; Ergas, S. J.; Van Langenhove, H. (2010). Hollow fiber membrane photo-bioreactor for CO_2 sequestration from combustion gas coupled with wastewater treatment: a process engineering approach. J. Chem. Technol. Biotechnol. 85, 387-394

Lackne, K.S. (2003), Climate Change: A guide to CO2 sequestration. Science 300, 1677-1678.

Lardon, L.; Hélias, A.; Sialve, B.; Steyer, JP.; Bernard, O. (2009). Life-cycle assessment of biodiesel production from microalgae. Environ. Sci. Technol. 43, 6475-6481.

Laws, E.A.; Taguchi, S.; Hirata, J.; Pang, L. (1986). Continued studies of high algal productivities in a shallow flume. Biomass 11, 39-50.

Li Y.; Horsman M.; Wu N.; Lan CQ.; Dubois-Calero N. (2008). Biofuels from microalgae. Biotech. Prog. 24, 815-820.

Mallick, K. (2002). Biotechnological potential of immobilized algae for wastewater N, P and metal removal: a review. Biometals 15, 377-390.

Miron, A. S.; Camacho, F. G.; Gomez, A. C.; Grima, E. M. & Chisti, Y. (2000). Bubble-Column and Airlift Photobioreactors for Algal Culture. American Institute of Chemical Engineers Journal. 46(9): 1872-1887.

Molina Grima, E.; Acién Fernández, F.G.; Garcıa Camacho, F.; Camacho Rubio, F. & Chisti, Y. (2000). Scale-up of tubular photobioreactors, J Appl Phycol 12:355-68

Molina Grima, E.; Fernandez Sevilla, J. M.; Sanchez Perez, J. A. & Garcıa Camacho, F. (1996). A study on simultaneous photolimitation and photoinhibition in dense microalgal cultures taking into account incident and averaged irradiances. J. Biotechnol. 45, 5969

Morita, M.; Watanabe Y.; Saiki H. (2000). Investigation of photobioreactor design for enhancing the photosynthetic productivity of microalgae. Biotechnol. Bioeng. 69, 693-698.

Muhs, J. (2000). Design and analysis of hybrid solar lighting and fullspectrum solar energy systems. Proceedings of the International Solar Energy Conference, Solar Engineering, pp. 229-237.

Munoz, R. & Guieysse, B. (2006). Algal-bacterial processes for the treatment of hazardous contaminants: a review. Water Res. 40, 2799-2815

Ono, E. & Cuello, J. L. (2003). Selection of optimal microalgae species for CO_2 sequestration. Proceedings of Second Annual Conference on Carbon Sequestration. Alexandria, VA

Oswald, W.J. (1988). Large-scale systems (engineering aspects). In Micro-algal Biotechnology (Borowitzka, M.A. and Borowitzka, L.J., eds), pp. 357-394, Cambridge University Press.

Poulsen, B. R. & J. J. L. Iversen. (1997). Characterization of Gas Transfer and Mixing in a Bubble Column Equipped with a Rubber Membrane Diffuser. Biotechnology and Bioengineering. 58(6): 633-641.

Pulz, O. (2001). Photobioreactors: production systems for phototrophic microorganisms. Appl. Microbiol. Biotechnol. 57, 287-293.

Richmond, A. & Cheng-Wu, Z. (2001). Optimization of a flat plate glass reactor for mass production of Nannochloropsis sp. outdoors. J. Biotechnol. 85, 259-269

Richmond, A. et al. (2003). Efficient use of strong light for high photosynthetic productivity: interrelationships between the optical path, the population density and the cell growth inhibition. Biomol. Eng. 20, 229-236.

Sánchez Mirón, A.; Garćia Camacho, F.; Contreras Gómez, A.; Molina Grima, E. & Chisti, Y.(2000). Bubble column and airlift photobioreactors for algal culture. AIChE J 46:1872-87

Sheehan, J.; Dunahay, T.; Benemann, J.; Roessler, P. (1998). A look back at the US Department of Energy's aquatic species program - Biodiesel from algae, Golden (CO), National Renewable Energy Laboratory, July, NERL/TP 580-24190.

Stewart, C. & Hessami, M. A. (2005). Energy Conversion and Management, 46, 403-420.

Stucki, S.; Vogel, F.; Ludwig, C.; Haiduc, A. G.; Brandenberger, M. (2009). Catalytic gasification of algae in supercritical water for biofuel production and carbon capture. Energy Environ. Sci. 2, 535-541.

Suh, I.S. & Lee, S.B. (2003). A light distribution model for an internally radiating photobioreactor. Biotechnol. Bioeng. 82, 180-189.

Tamiya, H. (1957). Mass culture of algae. Ann. Rev. Plant Physiol. 8, 309-334.

Tredici, M. R. (2003). Closed photobioreactors: basic and applied aspects. In Proceedings of Marine Biotechnology: Basics and Applications, p. 1, Matalascan as, Spain

Tredici, M.R. (2010). Photobiology of microalgae mass cultures: understanding the tools for the next green revolution. Biofuels 1, 143-162.

Uduman, N.; Qi, Y.; Danquah, M.K.; Forde, G.M. & Hoadley, A. 2010. Dewatering of microalgal cultures: a major bottleneck to algae-based fuels. J. Renew. Sust. Energ. 2, 012701.

Znad, H.; Báleš, V. & Kawase, K. (2004). Modeling and scale up of airlift bioreactor. Computers & Chemical Engineering, 28, 2765-2777.

Znad, H.; Kasahara, N. & Kawase, Y. (2006). Biological decomposition of herbicides (EPTC) by activated sludge in a slurry bioreactor. Process Biochemistry 41, 1124-1128.

Chapter-10

Formation of Pollutants in Combustion Processes

1. Introduction: An Overview

Every combustion process generates a range of emissions. Combustion produces not just carbon dioxide and water but also numerous additional products of full and incomplete combustion. Understanding the causes and formation paths enables the application of key approaches to mitigate emissions, ultimately decreasing atmospheric discharge.

2. Composition of various fuels

Combustion, from a chemical perspective, is the oxidation reaction of organic molecules. Organic compounds are a diverse array of hydrocarbons that contain carbon (C) and hydrogen (H) atoms in their molecular structure. The oxidation of hydrocarbons releases significant heat, rendering the reaction exothermic. Hydrocarbons and their derivatives, which include atoms such as sulfur (S), nitrogen (N), oxygen (O), and chlorine (Cl), constitute a combustible compound that may be classified as fuel only when it satisfies specific qualitative criteria. Fuels can be categorized using many approaches. One classification is based on the state of matter, categorizing fuels into solids, liquids, and gases. Another classification method is based on origin: natural fuels (e.g., fossil fuels) and synthetic fuels derived from the processing of natural fuels.

Natural solid fuels mostly comprise coal, lignite, peat, and wood. Natural liquid fuels primarily consist of oil, from which various synthetic fuels, including gasoline, kerosene, diesel, and fuel oil, are derived. Among natural gas fuels, natural gas is essential, whereas synthetic gas fuels encompass blast furnace gas, generator gas, water gas, city gas, and extracted gas fractions such as methane, propane, or butane, along with their combinations. Another type of fuel are wastes and produced from them the so-called alternative fuel which is known as refuse derived fuel (RDF) or secondary recovered fuel (SRF). Generally, fuel, in addition to the information about its form (physical state), is first of all characterized by elemental composition i.e. the contents of basic elements such as carbon (C), hydrogen (H), sulfur (S), nitrogen (N), oxygen (O), chlorine (Cl), etc. Apart form the contents of these elements, the content of water (W) and non-flammable substance (A), also known as ash, in the fuel are extremely important. Typical elemental composition of selected solid, liquid and gas fuels is shown in Table 1.

Table 1. Typical elemental composition of selected solid, liquid and gas fuels [in %]

Fuel	C	H	*O*	*N*	S	Cl	W	A
hydrogen	-	100	-	-	-	-	-	-
acetylene	92,3	7,7	-	-	-	-	-	-
methane	75	25	-	-	-	-	-	-
propane	81,8	18,2	-	-	-	-	-	-
butane	83	17	-	-	-	-	-	-
natural gas	69,4	22,9	0,7	7,0	0,05	-	-	-
blast furnace gas	41,6	19,4	9,0	2,2	-	-	-	-
coke oven gas	15,0	50,0	23,0	12,0	-	-	8,5	-
petrol	85	15	-	-	0,05	-	-	-
diesel	87	13	-	-	0,05	-	-	-
heating oil	86	13	-	-	0,2	-	-	-
coal	72-83	3,4-5,3	1,8-12,5	1,0-1,2	0,1-6,5	0,1-1	3-20	3-30
lignite	25-77	2-6,5	1,5-20	0,1-0,5	0,2-6,5	0,1-1	10-60	3-30
peat	40	5	25	2	1	0,1	20	7
coke	80-90	0,3-1,5	1,0-2,0	1,2-2,2	0,5-1,0	0,5	1,5-8,5	2-17
wood	35-50	5-7	35-45	0,3-07	0,01-0,1	0,01	5-65	0,3-7

3. Lower and upper heating value

The elemental composition determines another very important parameter which characterizes fuel i.e. heat of combustion so called upper heat value. Heat of combustion (H_0) - is the amount of heat that is generated during complete and perfect burning of unit weight or unit volume of the analyzed substance in constant volume, wherein:

- temperature and pressure are constant,
- non-volume work is equal to zero,
- products of combustion are present at the initial temperature,
- water contained in exhaust gas is in liquid state,
- nitrogen in the products is in the form of molecules (as N_2),

- sulfur contained in the fuel after combustion is in the form of SO_2

The term complete combustion means the process in which all the fuel (organic matter) will be oxidized (burned) and perfect combustion occurs when all the combustion products are non-flammable. Data on typical values of combustion heat of the selected fuels are shown in Table 2.

Heat of combustion (H_0) can be easily estimated on the basis of the elemental composition. The approximate value for the heat of combustion of solid and liquid fuels (upper heat value) can be calculated from the formula (Recknagel et al., 1994):

$$H_o = 34{,}8 \cdot C + 93{,}9 \cdot H + 10{,}5 \cdot S + 6{,}3 \cdot N - 10{,}8 \cdot O \quad (1)$$

where:

H_o - combustion heat [MJ/kg]

C - carbon content in the fuel [kg/kg],

H - hydrogen content in the fuel [kg/kg],

S - sulfur content in the fuel [kg/kg],

N - nitrogen content in the fuel [kg/kg],

O - oxygen content in the fuel [kg/kg].

Table 2. Combustion heat and calorific value of fuel (in MJ/kg)

Fuel	Heat of combustion (upper heal value)	Heat value (lower heat value)
hydrogen	147,8	119,9
acetylene	49,9	48,2
methane	55,5	50,1
propane	50,3	46,4
butane	49,5	45,7
natural gas	41,1	37,5
blast furnace gas	4,1	4,0
coke oven gas	19,7	17,5
petrol	46,1	42,7

diesel	44,7	41,8
heating oil	44,8	42,7
coal		19-31
lignite		6-14
peat		12-16
coke		28-32
wood		10-18

Heat of combustion can be determined experimentally in an adiabatic calorimeter. However, fuel usually contains also water (moisture) whose presence and evaporation significantly reduces the amount of heat generated during the combustion of the fuel unit. In order to determine the amount of heat that can be effectively generated in the combustion process the concept of heat (calorific) value (H_u), called the lower heat value, was introduced.

The calorific (heat) value is the amount of heat emitted during combustion of mass unit or volume unit of fuel at its complete and perfect combustion, assuming that the steam contained in the exhaust gas does not condense, even though the temperature of the exhaust gases reach the initial temperature of the fuel. Knowing the heat of combustion the calorific value can be calculated from the formula (Recknagel et al., 1994):

$$H_u = H_o - (9 \cdot H + W) \cdot r_w \tag{2}$$

where:

H_u - calorific value [MJ/kg]

W - water content in the fuel [kg/kg],

r_w - water evaporation heat [MJ/kg].

4. Oxygen excess

Every combustion process is conducted with an excess of oxygen relative to stoichiometric amount resulting from the oxidation reaction of individual components of the fuel. The measure of this excess is so called coefficient of excess air (λ) defined as a ratio of the actual amount of air supplied to the combustion process to the theoretical amount of air required to carry out the complete and perfect combustion - the amount resulting from the stoichiometry of combustion (oxidation) reaction. If in the combustion process coefficient λ is equal to 1 then exactly the same amount of air that is required by the stoichiometry of the oxidation reaction is supplied to the process. If the coefficient λ amounts to less than 1 the combustion process is incomplete. As mentioned above, each actual combustion process is conducted with an excess of air, hence the coefficient λ is usually more than 1 . Practically, its value depends on the type of fuel burned (gaseous and liquid fuels require less excess air than solid fuels) and design of the equipment in which combustion is carried out.

If the thermal process takes place without the presence of oxygen it is the process of pyrolysis. Pyrolysis - is the thermal decomposition of flammable substance without oxygen. This is an endothermic process that requires supplying heat from the outside. Pyrolysis process usually takes place at temperatures of $450-900°C$ in which the organic part of the fuel is converted into: the gas phase (pyrolysis gas), the liquid phase (pyrolysis oils) and solid phase (pyrolysis coke). Pyrolysis is sometimes also called degassing or carbonization. Depending on the temperature we distinguish: smouldering (below $600°C$) and coking (above $600°C$).

The composition and quantity of pyrolysis products depend on the type of fuel, their physicochemical properties and the temperature of the process. During the process of pyrolysis the mass of fuel is converted into (Thome-Kozmiensky, 1994):

- pyrolysis gas containing mainly hydrogen, methane, ethane and their homologues, carbon monoxide and carbon dioxide as well as other compounds such as: hydrogen sulfide, ammonia, hydrogen chloride and hydrogen fluoride
- pyrolytic coke - a solid phase containing carbon and metals as well as other inert substances
- liquid phase containing a mixture of oils, tars and water as well as dissolved in it simple aldehydes, alcohols and organic acids.

The process of pyrolysis of hydrocarbons can be described by the following simplified chemical equation:

$$C_xH_yO_z \underset{heat}{\rightarrow} CO_2 + H_2O + CO + CH_4 + C_2H_6 + CH_2O + \cdots + C_mH_n + C - q \qquad (3)$$

However, if the thermal process takes place with limited access of air (oxygen) - when the coefficient λ is less than 1 , often fuel gasification process occurs. Gasification is a total conversion of solid fuels into gas fuels under the influence of gasifying medium e.g. air or steam or both. During the flow through the glowing layers of solid fuel intended for the gasification a reaction of oxygen or water with carbon takes place in result of which carbon monoxide (CO) and hydrogen (H_2) are formed as well as small amounts of methane (CH_4), carbon dioxide (CO_2), nitrogen (N_2) and water steam (H_2O). The process of gasification of solid hydrocarbons can be described by the following simplified chemical equations (Thome-Kozmiensky, 1994):

$$C + \frac{1}{2}O_2 \longrightarrow CO + q \tag{4}$$

$$C + O_2 \longrightarrow CO_2 + q \tag{5}$$

$$C + 2H_2O \longrightarrow CO_2 + 2H_2 - q \tag{6}$$

$$C + H_2O \longrightarrow CO + H_2 - q \tag{7}$$

$$C + 2H_2 \longrightarrow CH_4 + q \tag{8}$$

$$CO + H_2O \longrightarrow CO_2 + H_2 + q \tag{9}$$

$$CO + 3H_2 \longrightarrow CH_4 + H_2O + q \tag{10}$$

$$C + CO_2 \longrightarrow 2CO - q \tag{11}$$

which are accompanied at the same time by thermal decomposition according to the simplified reaction equation:

$$C_xH_yO_z \underset{\text{heat}}{\rightarrow} CO_2 + H_2O + CO + CH_4 + C_2H_6 + CH_2O + \cdots + C_mH_n + C - q \tag{12}$$

Some of the mentioned above chemical reactions are exothermic ($+q$ - proceeding with heat emission) and some endothermic (- q - requiring supplying heat from the outside).

5. Volume of flue gases from combustion processes

The stream of flue gases from the combustion process consists of a stream resulting from the oxidation reaction of flue gases and a stream of excess air supplied to the combustion process (with excess air coefficient λ greater than 1). Knowing the elemental composition of the fuel or only its calorific value it is possible to estimate both stoichiometric amount of air needed for combustion process and the amount of flue gases generated in the process.

For liquid and solid fuels, the amount of air needed for combustion reaction (stoichiometric) can be calculated from the formula (Recknagel et al., 1994):

$$V_T = 8{,}88 \cdot C + 26{,}44 \cdot H + 3{,}32 \cdot S - 3{,}33 \cdot O \tag{13}$$

where:

V_T - minimal (stoichiometric) amount of air for combustion converted to standard conditions (T = 273 K, P = 1013hPa) [m³/kg]

The volume of flue gases can be calculated on the basis of known elemental composition of the fuel from the formula (Recknagel et al., 1994):

$$V_S = 1{,}85 \cdot C + 11{,}11 \cdot H + 0{,}68 \cdot S + 0{,}8 \cdot N + 1{,}24 \cdot W \tag{14}$$

where:

V_S - volume of flue gas from the combustion process converted to standard conditions [m^3/kg]

N - nitrogen content in the fuel [kg/kg],

In case the elemental composition of fuel is not known the minimum amount of combustion air and exhaust gas volume can be calculated from approximate formulas taking into account only the calorific value of fuel (Recknagel et al., 1994):

- for solid fuel:

$$V_T = 0{,}241 \cdot H_u + 0{,}5 \tag{15}$$
$$V_S = 0{,}212 \cdot H_u + 1{,}65 \tag{16}$$

- for liquid fuel:

$$V_T = 0{,}203 \cdot H_u + 2{,}0 \tag{17}$$
$$V_S = 0{,}265 \cdot H_u \tag{18}$$

The total gas flow (after taking into account the excess air coefficient λ) under normal conditions can be calculated from the formula:

$$V_C = (\lambda - 1) \cdot V_T + V_S \tag{19}$$

Multiplying the value of the total flue gas stream V_C by the amount of fuel burned per unit of time we can calculate the volumetric flow of flue gases from combustion process converted to standard conditions:

$$V_N = B \cdot V_C \tag{20}$$

where:

V_N - volumetric flow of flue gases from the combustion process converted to standard conditions [m^3/h]

B - fuel consumption [kg/h],

Taking into account the current temperature (T) and flue gases pressure (P) it is possible to estimate flue gas stream in the real conditions from the formula:

$$V = V_N \cdot \frac{T + 273}{273} \cdot \frac{1013}{P} \tag{21}$$

where:
T - flue gas temperature [°C]

P - flue gas pressure [hPa]

6. Combustion as an autothermal process

Heat emitted in a chemical reaction of oxidation in the combustion process is used for heating of flue gases from the process (combustion products and excess air supplied to the combustion process), solid secondary products of the combustion process (slag and ashes) and the equipment in which the combustion process is conducted (plus the obvious loss to the environment) up to the process temperature. This is illustrated in Figure 1.

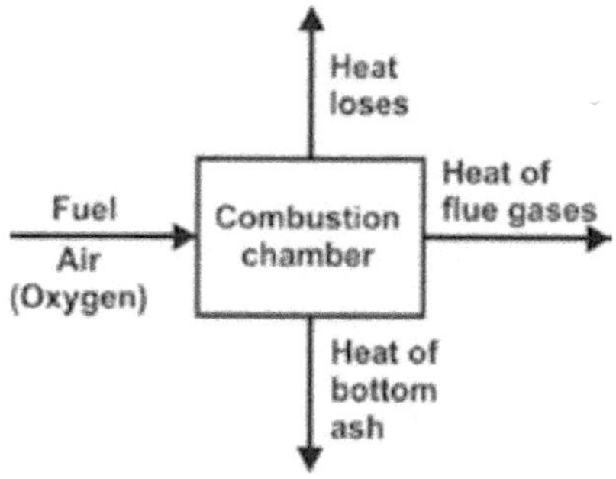

Fig. 1. Simplified heat balance i the combustion chamber

It is obvious that, if the generated amount of heat is too small, the combustion process does not run properly, its temperature is too low and the flue gases do not reach the required temperature. If, however, fuel of high calorific value is burned the amount of heat is sufficient to ensure proper combustion temperature and temperature of flue gases. This is particularly important in the case of waste incineration, where the law (Directive 2000/76/EC on the incineration of waste) requires that the exhaust gases have a minimum temperature of 850°C and remain in this temperature for at least 2 seconds when the waste fuel contains less than 1% chlorine. When the chlorine content in waste exceeds 1% it is required to achieve a minimum temperature of 1100°C. If the required temperature of flue gas (and the combustion process) is reached in the device, in which the combustion process is run, it is called autothermal combustion process, but when due to too low calorific value of fuel the required temperatures cannot be reached, combustion is not an autothermal process and requires additional fuel of higher calorific value.

7. Combustion as a set of chemical reactions

As mentioned at the beginning from the chemical point of view combustion is the reaction of oxidation of hydrocarbons. Combustion of gaseous fuel is the simplest combustion, in which simple reaction of hydrocarbons oxidation takes place according to the simplified reaction scheme:

$$C_xH_y + \left(x + \frac{y}{4}\right) O_2 \longrightarrow xCO_2 + \frac{y}{2} H_2O \tag{22}$$

$$C_xH_y + \left(\frac{x}{2} + \frac{y}{4}\right) O_2 \longrightarrow xCO + \frac{y}{2} H_2O \tag{23}$$

and the basic products of oxidation are carbon dioxide and water. In case of insufficient oxygenation of the combustion zone the additional reaction product is carbon monoxide. Flammable gas usually mixes well with air, causing a kinetic character of the combustion i.e. controlled by the rate of chemical oxidation reaction. In rare cases, when the flammable gas is mixed with air, the process can be controlled by diffusion of oxygen to the reaction zone. With inadequate oxygenation of the reaction zone the product of the combustion process is carbon monoxide, and elemental carbon (soot), because the reaction rate of water formation is higher than the rate of carbon oxidation.

Combustion of liquid fuels is a more complicated process, because to ensure the oxidation (combustion) it is necessary to evaporate the liquid, because its vapor is burned. Evaporation is promoted by the liquid spray, however, it is very difficult to obtain a homogeneous mixture of liquid vapor and air, hence the combustion of liquid fuels has rarely kinetic character, usually this is a diffusion combustion. Organic liquids generally have a more complex chemical structure than gases, and hence more often we are dealing with the formation of products of incomplete combustion, some of them being other organic compounds:

$$C_xH_y + \frac{3}{2} O_2 \longrightarrow C_{x-1}H_{y-2} + CO_2 + H_2O \tag{24}$$

of course simpler than the compound that is subject to burning. In extreme cases there can be a situation when due to shortage of air in result of the combustion process the decomposition of organic compounds contained in the fuel will be incomplete and not all carbon will be oxidized. Consequently, in addition to gaseous products from the combustion process a solid product will be also formed - elemental carbon (C) or soot:

$$C_xH_y + \left(x - \frac{1}{2} + \frac{y}{4}\right) O_2 \longrightarrow C + (x-1)CO_2 + \frac{y}{2} H_2O \tag{25}$$

The process of solid fuel combustion runs in even more complicated way. It is a multidirectional and multi-stage process, a combination of combustion (oxidation), gasification, thermal decomposition, including pyrolysis (lack of oxygen process). One should be aware that direct heterogeneous reaction between gaseous oxygen and solid hydrocarbon is problematic and because of the hetero phases it must be slow. Thus, direct combustion of solid phase fuel runs in a minimal degree. The real combustion process is much more complicated. In the initial period, at sufficiently high temperatures degassing processes (pyrolysis) begin and flammable gases are emitted from the solid fuel. They are relatively easily oxidized (burned) in the gas phase generating heat flux heating the solid phase and accelerating the release of flammable gases. These processes run as long as fuel in a given element does not run out of hydrogen. The solid phase in result of degassing is enriched in the process of carbonization in carbon hence degassing process (the release of flammable gases) is accompanied by solid phase carbonization. At this point a change in the mechanism of the process takes place and the process of gasification of solid fuel begins (using air and/or steam), which results in carbonization phase being converted into flammable gases (carbon monoxide and hydrogen), which then in the gas phase are oxidized (burning). In the gas phase the oxidation reaction dominates, however, at the same time a number of other reactions take place - such as synthesis, cyclization, etc.

Apart from that we should not forget about other reactions running in parallel such as Boudouard reaction in which carbon dioxide, previously formed during the combustion, at the temperature in the range 400 – 950°C can be reduced with the participation of carbon to less oxidized form - a flammable carbon monoxide according to the equation:

$$CO_2 + C \longrightarrow 2CO \tag{26}$$

which can participate in further reactions e.g. oxidation. It is assumed that depending on the temperature in the combustion process of solid fuels several important steps can be defined (Thome-Kozmiensky, 1994):

- 100–200°C – thermal drying process, separation of water (physical process),
- 250°C – deoxidation reductive processes, reduction, decomposition of sulfuric acid esters, separation of bound moisture and carbon dioxide, depolymerization, the beginning of the release of hydrogen sulfide,
- 340°C – formation of aliphatic compounds, including unsaturated, the beginning of release of methane and other aliphatic compounds,
- 380°C – beginning of carbonization in the smouldering process,
- 400°C – beginning of the formation of the compounds of carbon with oxygen and nitrogen,
- 400–600°C – transformation of bituminuous substances into smouldering oil or tar,
- 600°C – cracking of bituminous substances in the direction of thermally persistent substances (gases, hydrocarbons of short chain structure), formation of new compounds (benzene derivatives in result of cyclization of unsaturated aliphatic compounds),
- > 600°C – further course of the hypothetical reaction of dimerization of butylene formation, dehydration to butadiene, the formation of cyclohexane C_6H_{12}, thermal aromatization to benzene and other higher aromatic compounds.

8. Complete and incomplete combustion

However, we should be aware that the actual combustion process is far from ideal - that is complete and perfect combustion. The combustion process (thermal decomposition and oxidation) of many organic compounds (in particular those contained in the waste) is not a perfect process that produces only carbon dioxide, carbon monoxide and water. In this process usually a large number of intermediate products of decomposition and oxidation are formed which then are not further decomposed. It would seem that in the drastic combustion conditions, at about 1000°C any organic material must be burned. Unfortunately this is not true. Many chemical compounds, often flammable, are not subject to complete destruction during process of combustion. Some of the organic compounds are generated in result of many secondary reactions running in the combustion zone and outside it. Simplified diagram of the formation of pollutants in combustion processes is shown in Figure 2.

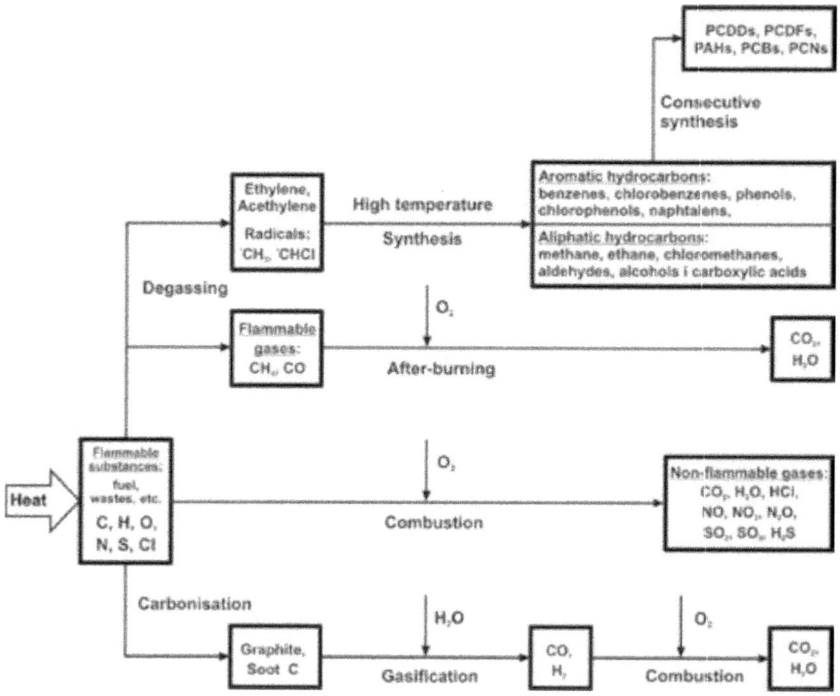

Fig. 2. Simplified diagram of the formation of pollutants in the combustion process of solid flammable substance (fuel, waste, etc.)

In the process of fuel combustion there are four main pathways. The first pathway - direct oxidation reactions. If we treat fuel as a collection of various chemicals consisting of carbon, hydrogen, oxygen, nitrogen, sulfur, chlorine, etc. so in result of direct oxidation typical flammable gases will appear such as carbon dioxide (CO_2), water (H_2O) in accordance with reaction equation (22) and also sulfur dioxide (SO_2), nitrogen oxides (NO, NO_2, N_2O) or hydrogen chloride (HCl). At the same time the combustion process will be conducted according to other two pathways. In the process of thermal decomposition flammable gases will be emitted that contain for example carbon monoxide (CO), methane (CH_4) and other simple hydrocarbons, which are relatively easily oxidized to carbon dioxide and water in the gas phase. Thermal decomposition process impoverishes the solid phase of fuel in hydrogen and other volatile substances, causing its carbonization.

On the third pathway carbonized solid phase is gasified with water and/or air to carbon monoxide and hydrogen which in the gas phase are easily oxidized to carbon dioxide and water. The last fourth pathway of the course of combustion processes is a secondary synthesis path. In result of decomposition and gasification processes in the gas phase simple hydrocarbons and hydrocarbon radicals are formed. Since virtually all fuels contain trace admixtures of chlorine (carbon even sometimes even about 1% !) in the gas phase there are also simple chlorinated hydrocarbons (including unsaturated - for example, acetylene C_2H_6) and chlorinated hydrocarbon radicals. They are extremely reactive and in the gas phase appear a number of synthesis reactions, including cyclization. In this way chlorinated and non-chlorinated aliphatic hydrocarbons are formed (methane, ethane, chloromethanes, chloroethanes etc., aldehydes: e.g. formaldehyde and acetaldehyde, alcohols: e.g. methanol, simple carboxylic acids: e.g. formic acid and acetic acid) and aromatic hydrocarbons (benzenes, chlorobenzenes, phenols and chlorophenols, naphthalenes and many others). The latter take part in the synthesis of polychlorinated dibenzo-p-dioxins (PCDDs), polychlorinated dibenzofurans (PCDFs), polychlorinated biphenyls (PCBs), polychlorinated naphthalenes (PCNs) and polycyclic aromatic hydrocarbons (PAHs) (Wielgosiński, 2009).

9. Formation of sulphur oxides

Practically all fuels solid, liquid and gas contain some amounts of sulphur (Table 3).

Table 3. Sulphur content in different fuels

Kind of fuel	Sulphur content
Gas fuels: LNG, CNG, LPG	below 0,01%
Diesel oil	below 0,10%
Petrol	below 0,05%
Heating oil	0,2-0,3%
Wood	0,02-0,04%
Coal (Poland)	0,5-2,5%
Coal (World)	0,1-4,5%
Lignite (Poland)	0,4-1,2%
Lignite (World)	0,4-6,5%
Coke	0,2-1,0%
Municipal waste	0,2-0,6%

Sulphur in the combustion process is relatively easily oxidized to sulphur dioxide (SO_2) in accordance with the simplified reaction scheme.

$$S + O_2 \underset{\text{combustion}}{\rightarrow} SO_2 \tag{27}$$

The rate of this reaction at temperatures, under which the combustion process is run, is very high. On the basis of many research results it can be assumed that approximately $90 - 95\%$ of the sulfur contained in fuel in the combustion process will be oxidized to SO_2. The presence in the burned material of the non-combustible fraction containing some metals (e.g. vanadium - V) will result in partial catalytic oxidation of the formed sulfur dioxide to sulfur trioxide (SO_3) in accordance with reaction equation:

$$SO_2 + \frac{1}{2}O_2 \underset{\text{catalyst-e.g.vanadium}}{\rightarrow} SO_3 \tag{28}$$

Coal as a chemical element, like hydrogen, has reductive properties. In high temperature accompanying the combustion process, in the conditions of limited access of oxygen, in the presence of hydrogen in the gasification reaction, sulfur contained in the fuel can be reduced to hydrogen sulphide according to reaction equation:

$$S + H_2 \underset{\text{carbon}}{\rightarrow} H_2\,S \tag{29}$$

Hydrogen sulphide formed during the reaction (30) is a flammable gas and at a later stage may be oxidized to sulphur dioxide and water according to reaction equation:

$$H_2\,S + \frac{3}{2}O_2 \longrightarrow SO_2 + H_2O \tag{30}$$

It should be clearly stated that in the flue gases from combustion process both SO_3 and $H_2\,S$ may be present in trace amounts (well below 1% of the total emissions of sulfur compounds) and sulphur dioxide - SO_2 will be the main pollutant that is emitted. It is the so called raw material pollutant whose quantity in the exhaust almost in 100% corresponds to the amount of sulphur introduced to the combustion process. This allows relatively accurate calculation of sulphur emissions from combustion processes.

10. Formation of nitrogen oxides

The studies of the combustion processes lead to the conclusion that the formation of nitrogen oxides (NOx) is observed for many fuels. This applies primarily to nitric oxide (NO), nitrogen dioxide (NO_2) and nitrous oxide (N_2O). Of course there are known many other chemical compounds which are a combination of oxygen and nitrogen, however, in the flue gas from combustion they are in fact absent. The basic nitrogen oxide formed in the combustion process is nitrogen monoxide - NO. The extensive literature on the subject gives the information about three mechanisms of the formation of this compound in thermal processes. They are:

- thermal mechanism of Zeldovich,
- fuel mechanism,

- Fenimore prompt mechanism (radical).

Thermal mechanism was discovered and described for the first time by Zeldovich in the forties of the last century (Zeldovich, 1946). It includes the direct synthesis of nitric oxide from oxygen and nitrogen contained in air at high temperature. In the first stage there is decomposition of molecular oxygen to two molecules of active atomic oxygen in result of contact with high-energy inert molecule M, acting as a catalyst (it may be e.g. a molecule of hot metallic wall of the combustion chamber, burner, etc.):

$$O_2 + M \longrightarrow O + O + M \tag{31}$$

At high temperature molecular oxygen is decomposed into highly active form of atomic oxygen. Atomic oxygen molecules attack the nitrogen molecules and nitric oxide is formed, while formed in this reaction active nitrogen atoms, by attacking oxygen molecules, also form nitric oxide and active atomic oxygen. So it is a classic example of chain reaction, for which the first reaction (30) is the stage of initiation while the next two reactions (31) and (32) are the propagation stage:

$$O + N_2 \longrightarrow NO + N \tag{32}$$
$$N + O_2 \longrightarrow NO + O \tag{33}$$

in fuel-rich mixtures where the concentration of hydroxyl radicals is significant, greater than the concentration of hydrogen and oxygen atoms ($OH \gg H > O$) the following reaction can be considered as the last stage of termination:

$$N + OH \longrightarrow NO + H \tag{34}$$

An additional source of nitric oxide formation may be the following reactions in accordance with the mechanism described by Bozzelli (Bozzelli et at., 1994):

$$H + N_2 \leftrightarrow N_2H \tag{35}$$
$$N_2H + O \leftrightarrow NO + NH \tag{36}$$

The rates of formation of nitrogen oxides in the thermal mechanism are relatively high but only at high temperatures. This mechanism becomes negligible at temperatures above 1400°C.

Fuel mechanism is directly related to nitrogen content in fuels. As shown in Table 1 most solid and liquid fuels contain nitrogen. So the source of nitrogen in this mechanism is fuel while the source of oxygen is air introduced to the combustion process. The formation of nitric oxide in this mechanism is quite long and it goes through a number of succeedingparallel reactions (Bowman et al., 1982, Miller & Bowman, 1989) shown in Figure 3.

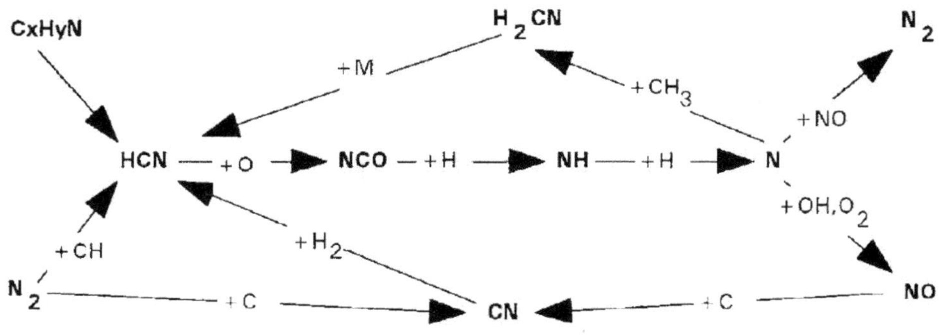

Fig. 3. Diagram of nitric acid formation according to fuel mechanism

In the first stage from organic matter, subject to combustion process (hydrocarbons containing nitrogen), hydrogen cyanide is released which by further oxidation and hydrogenation gives first radical NCO, next NH and then free atomic nitrogen (N). At this point there are three possible further courses of reactions - oxidation reaction in which nitric oxide is formed, the reaction with previously formed nitric oxide in result of which neutral molecular nitrogen is formed and the reaction with hydrocarbon radical returning the reaction to the beginning. Thus, only one course leads to the formation of nitric oxide while the other two effectively reduce its quantity - hence you can see that there are technical possibilities of influencing on the amount of produced nitrogen oxides - through the organization of the combustion process.

Prompt mechanism was first described by Fenimore in 1971 (Fenimore, 1971). He noted that in the early stage of flame, in the presence of numerous hydrocarbon radicals, occurs a synthesis of nitric oxide from oxygen and nitrogen introduced into the combustion process. Since NO formation reactions occur in the early stage of flame this mechanism was called the prompt mechanism. The prompt mechanism is complementary to the fuel mechanism and some chemical reactions follow the same path. Generally it can be assumed that the following reactions proceed in this mechanism:

$$\mathrm{CH} + \mathrm{N_2} \longrightarrow \mathrm{HCN} + \mathrm{N} \tag{37}$$

$$\mathrm{CH_2} + \mathrm{N_2} \longrightarrow \mathrm{HCN} + \mathrm{NH} \tag{38}$$

$$\mathrm{HCN} + \mathrm{O} \longrightarrow \mathrm{NCO} + \mathrm{H} \tag{39}$$

$$\mathrm{NCO} + \mathrm{O} \longrightarrow \mathrm{NO} + \mathrm{CO} \tag{40}$$

The rate of formation of nitric oxide in this mechanism is very high, but the amount of formed NO according to this mechanism is relatively small.

Generally, during combustion of fuels, especially solid fuels, the greatest amount of nitrogen oxide is produced by the fuel mechanism (sometimes even $80-90\%$). Thermal mechanism begins to play a role only after exceeding the temperature of $1400^{\circ}C$, so the contribution of this mechanism usually does not exceed $10-20\%$. The contribution of the prompt mechanism is usually at the level of $1-5\%$. Of course the contribution of thermal mechanism increases with the increase of temperature.

In recent years, it was noted that during low-temperature combustion of fuel-poor mixtures with low excess air there appears an additional mechanism of nitric oxide formation. It is connected with the described by Malte and co-workers (Steele et al., 1995) synthesis of nitrous oxide in the combustion conditions. Most likely, this reaction occurs with the participation of active atoms of hydrogen, oxygen or carbon monoxide - a product of incomplete combustion according to the following equations:

$$N_2O + H \longrightarrow NO + NH \tag{41}$$
$$N_2O + O \longrightarrow 2NO \tag{42}$$

$$N_2O + CO \longrightarrow NO + NCO \tag{43}$$

Reactions of nitric oxide formation from nitrous oxide proceed relatively quickly at temperatures below $1200^{\circ}C$ in the area of good oxygenation of the combustion zone. The nitrous oxide synthesis itself proceeds according to several possible paths, one of them requires, just as in the thermal mechanism, the contact with the high-energy inert molecule M :

$$O + N_2 + M \longrightarrow N_2O + M \tag{44}$$
$$NCO + NO \longrightarrow N_2O + CO \tag{45}$$
$$NH + NO \longrightarrow N_2O + H \tag{46}$$
$$2NO \longleftrightarrow N_2O + O \tag{47}$$

The reaction (45), (46) and (47) shows that the synthesis of N_2O can take place outside the combustion zone with the participation of nitric oxide previously formed in the presence of numerous radicals.

The studies of combustion products show above all the presence of nitric oxide (NO) and small quantities of nitrous oxide (N_2O). Nitrogen dioxide (NO_2), from the oxidation of NO formed during combustion, is a component of flue gas as well. The formation of NO_2

has been described by Miller and Bowman in 1989 (Miller & Bowman, 1989). As a result of diffusion of hydrogen radicals beyond flame zone to the area of lower temperatures (below 750°C) in the presence of excess oxygen there are formed, involving high-energy inert particles of M, HO_2 radicals which also in the same zone can react with nitric oxide according to the reaction:

$$NO + HO_2 \longrightarrow NO_2 + OH \tag{48}$$

In parallel to the synthesis reaction, nitrogen dioxide decomposition reactions take place:

$$NO_2 + O \longrightarrow NO + O_2 \tag{49}$$
$$NO_2 + H \longrightarrow NO + OH \tag{50}$$

Moreover, in the further zone of the installation, after heat recovery systems, at temperatures below 200°C occurs direct oxidation of NO to NO_2 by oxygen present in the flue gas:

$$NO + \frac{1}{2}O_2 \longrightarrow NO_2 \tag{51}$$

Under proper combustion conditions, the participation of NO_2 in the whole stream of emitted nitrogen oxides does not exceed 5-10%.

Nitrogen oxides (NOx) understood as the sum of NO, NO_2 and N_2O are pollutants whose rate of emission is minimally dependent on nitrogen content in the fuel. Their emissions is a direct result of plant construction and organization of the combustion process.

11. Particulate matter emission

The problem of particulate matter emissions practically applies only to solid fuel combustion process. It occurs in a minimal degree during the combustion of liquid fuels and practically does not occur during the combustion of gaseous fuels. The presence in the burned material of inorganic solid substance - non-flammable, most often called ash makes that in the combustion process solid residue in the form of slag and ash is formed. The flow of air through the combustion zone results in entrainment of fine solid particles and thus dust emission from combustion processes is formed. This emission is the higher, the better is the oxygenation of the combustion zone and the higher are air velocities observed in this zone. Solid particles, lifted in the flue gas stream from the combustion zone, have usually very small diameter and their main component is silicon dioxide (SiO_2 - silica). At the same time these particles are carriers of metals, elemental carbon (soot) as well as the adsorbed products of incomplete combustion and products of secondary synthesis outside the flame zone. In modern solid fuel combustion installations the release of solid particles is usually not more than 60% for the pulverized fuel and fluidal installations, and about 25 – 40% for the grate installations. The emission of solid particles, as dust from the combustion process is proportional to the content of non-flammable substances in the fuel and the factor of proportionality, the so called release factor, which is different for each installation and depends on its structure. So it depends partly on raw materials characteristics.

12. Metals emission

The emission of metals from the combustion process is very closely linked with the emission of solid particles (particulate matter) and therefore this problem does not exist during the combustion of gaseous fuels; it occurs to a small extent during combustion of liquid fuels and it applies fully to the combustion of solid fuels containing significant admixtures of non-flammable substance. Metals (mainly heavy metals), due to the speciation in the products of combustion, can be divided roughly into three groups:

- metals easily evaporated in the combustion zone and present mainly in emission to air,
- metals with high melting points and boiling points cumulative in solid combustion products (slag and grate ash)
- metals of intermediate properties for which there is division into gas and solid combustion products.

The first group includes above all mercury (Hg) and cadmium (Cd). Similar properties have selenium (Se). Most metals belong to the second group - they are mainly iron (Fe), magnesium (Mg), cobalt (Co), chromium (Cr), copper (Cu), manganese (Mn), molybdenum (Mo) and nickel (Ni). Similar properties also have aluminum (Al) and titanium (Ti). The third intermediate group, includes mainly arsenic (As), lead (Pb), tin (Sn) and zinc (Zn) as well as antimony (Sb). The detailed data on the speciation of the selected metals in the combustion process, based on the example of waste incineration plants (Belevi & Moench, 2000, Belevi & Langmaister 2000, Sukrut et al., 2002), are given in Table 4.

Table 4. Speciation of the selected metals in the combustion process

Metal	Solid phase [%]	Gas phase [%]
Antimony - Sb	25	75
Arsenic - As	68,6	31,4
Chromium - Cr	92	8
Tin - Sn	50,8	49,2
Zinc - Zn	52	48
Aluminium - Al	87,8	12,2
Cadmium - Cd	9,5	90,5
Cobalt - Co	90,1	9,9
Magnesium - Mg	91,9	8,1
Manganese - Mn	94	6
Copper - Cu	97,3	2,7
Molybdenum - Mo	91,7	8,3
Nickel - Ni	98,2	1,8
Lead - Pb	59	41
Mercury - Hg	0,7	99,3
Titanium - Ti	85,3	14,7
Iron - Fe	99,1	0,9

The situation is substantially changed, if in the burned area appear halides - especially chlorine and bromine. Melting point and boiling point of heavy metal salts (chlorides) is significantly lower than the melting point and boiling point of pure metal or its oxide.

According to Barton (Barton et al., 1991) and Niessen (Niessen, 2002) it is particularly evident in the case of nickel (Ni), thallium (Tl) and lead (Pb). This is illustrated by the data included in Table 5.

Table 5. Volatility temperature (reaching resilience 0,1 Pa) for the selected metals in the absence and in 10% presence of chlorine.

Metal	Volatility temperature [°C]	
	0% Cl	10% Cl
Antimony - Sb	660	660
Arsenic - As	32	32
Chromium - Cr	1613	1610
Cadmium - Cd	214	214
Nickel - Ni	1210	693

Lead - Pb	627	-15
Mercury - Hg	14	14
Selenium - Se	318	318
Thallium - Tl	721	138

So the emission of metals from combustion processes largely depends on the type of metal, temperature of the combustion process, but also on the presence of halides - chlorine and bromine which significantly increase the presence of some metals in flue gas.

13. Hydrogen chloride emission

Chlorine (Cl) is an element that is widespread in the environment. Its small, sometimes even trace amounts are found in each fuel. Of course, the least amount of chlorine (almost immeasurable amount) is in fuel gas, slightly more in liquid fuels, while in solid fuels its content may be as high as 2%. In the process of combustion, chlorine - like metals is subject to speciation. The largest part of chlorine (50 – 60%) is bound in the form of chlorides in the fly ash, a part is also bound in a similar way in slag and grate ashes. The remaining part of chlorine is released into the environment as emissions of hydrogen chloride (HCl), while only a small amount is released in form of molecular chlorine (Cl_2). With the increase of combustion temperature, the amount of chlorine released in the form of emissions increases and at the same time the amount of chlorine present in the slag and ashes decreases. Data on the speciation of chlorine in the combustion process at different temperatures according to the work of Liu (Liu et al., 2000) are shown in Table 6.

Table 6. Speciation of chlorine in the combustion process depending on the temperature

Presence of chlorine	Temperature (°C)					
	600	650	700	750	800	900
HCl	7,6	10,7	10,1	19,9	27,4	43,7
Cl_2	0,5	0,2	0,4	0,6	0,8	1,3
Fly ash	50,9	47,5	56,3	53,2	62,4	51,9
Slag	40,5	41,3	32,8	25,6	8,5	1,2

The presence of chlorine in fly ash is of great importance in the formation of chloroorganic compounds outside the combustion zone. The presence of free chlorine which can participate in the chlorination and oxychlorination reactions outside the combustion zone is of similar importance. In the flue gas, outside the combustion zone, hydrogen chloride may undergo a catalytic decomposition (Deacon reaction) in accordance with the reaction equation (Griffin, 1986):

$$4HCl + O_2 \underset{\text{catalyst- Cu,Fe,Al}}{\rightarrow} 2Cl_2 + 2H_2O \tag{52}$$

In the case of sulfur dioxide (SO_2) present in flue gas, at the same time proceeds hydrogen chloride reproduction reaction (Lindbauer et al., 1994):

$$SO_2 + Cl_2 + H_2O \longrightarrow SO_3 + 2HCl \tag{53}$$

The balance in a chlorine - hydrogen chloride system is also affected by hydroxyl radicals and water present in the flue gas:

$$HCl + OH \leftrightarrow Cl + H_2O \tag{54}$$

Thus the concentration of hydrogen chloride and free chlorine in the flue gas depends on many factors, including temperature and that is why their emission rates cannot be predicted on the basis of chlorine content in the fuel.

14. Products of incomplete combustion

The term products of incomplete combustion (PICs) means organic compounds introduced to combustion and formed during combustion that were not oxidized in the combustion zone as well as products of secondary synthesis outside the combustion zone. As already mentioned when discussing the combustion process paths shown in Figure No. 2, in the combustion zone and directly next to it, in the gas phase there are many hydrocarbon and chlorinated hydrocarbons radicals as well as simple aliphatic hydrocarbons, often unsaturated, which may participate in secondary, high-temperature synthesis reactions. Among these radicals there is also acetylene, which plays a key role in the later stages of synthesis.

The key here is a reaction first described by Aubrey and van Wazer (Aubrey & van Wazer, 1964), in which in high temperature aromatic compounds are formed from aliphatic compounds (acetylene):

$$9C_2Cl_6 \leftrightarrow 12CCl_4 + C_6Cl_6 \tag{55}$$

A special role is played by acetylene. It is always present in the flue gas from the combustion process and is a precursor of the formation of many chloro-aromatic compounds (Lenoir et al., 2001). It is subject in the first stage to chlorination to dichloroacetylene in result of ligand exchange reaction and then is subject either to cyclization to hexachlorobenzene or condensation to hexachlorobutadiene (Lenoir et al., 1998). Subsequent studies have shown that, for example from acetylene at high temperatures chlorobenzenes, chlorophenols and chloronaphthalenes can be formed.

In addition, the presence in the waste of organic substances of unsaturated character, that is containing in the molecule double or triple carbon-carbon bonds ($C = C$ or $C \equiv C$) causes that at a temperature of about $500 - 800°C$, occurs the synthesis reaction of polycyclic aromatic hydrocarbons (Liow et al., 1997). Polycyclic aromatic hydrocarbons (PAHs) are among the most dangerous substances in the environment. Many of them, such as benzo(a)pyrene, benzo(a)anthracene, benzo(k)fluoranthene, dibenzo(a,h)anthracene, indeno(1,2,3-c,d)pyrene are classified by the International Agency for Research on Cancer (IARC) as substances with proven carcinogenic effect. They are emitted to the atmosphere practically from all combustion processes, not only from the waste incineration plants (Mastral et al., 2000).

The result of these reactions, which proceed outside the combustion zone, is the presence of pollutants in the flue gas, among others such as (Eduljee, 1994):

- hydrocarbons C_1 and C_2,
- acrylonitrile, acetonitrile,
- benzene, toluene, ethylbenzene, xylene,
- 1,2-dichlorobenzene, 1,4-dichlorobenzene, 1,2,4-trichlorobenzene, hexachlorobenzene,
- phenol, 2,4-dinitrophenol, 2,4-dichlorophenol, 2,4,5-trichlorophenol, pentachlorophenol,
- chloromethane, chloroform, methylene chloride, carbon tetrachloride,
- 1,1-dichloroethane, 1,2-dichloroethane, 1,1,1-trichloroethane, 1,1,2-trichloroethane, 1,1,2,2-tetrachloroethane,
- 1,1-dichloroethylene, trichloroethylene, tetrachloroethylene,
- formaldehyde, acetaldehyde, acetone, methyl-ethyl-ketone,

- vinyl chloride, diethyl phthalate,
- formic acid, acetic acid and many others.

Subsequent studies conducted by other researchers (Wienecke et al., 1995, Jay & Stieglitz, 1995, Mascolo et al., 1997, Trenholm 1998) showed the presence of more than 350 different types of chemical compounds (organic) in the flue gases from the waste incineration plants in concentrations above $5\mu\ g/m^3$. Similar results were obtained for burning of wood and other biomass. Unfortunately, there are no such test results for the combustion of coal, although it is expected that the situation is similar.

15. Formation of PCDD/Fs, PCBs, PCNs and PAHs

Polychlorinated dibenzo-p-dioxins (PCDDs), polychlorinated dibenzofurans (PCDFs), polychlorinated biphenyls (PCBs), polychlorinated naphthalenes (PCNS) and polycyclic aromatic hydrocarbons (PAHs) are also products of incomplete combustion formed outside the combustion zone. Their common feature is the negative biological impact - some of them have proven carcinogenic properties (PAHs), others are xantoestrogens (endocrine disrupters) that disrupt the hormonal balance of living organisms. In recent years many myths referring to them have appeared, however, most of them have nothing to do with reality.

According to the diagram in Figure 2 showing the combustion process of solid fuels, the crucial stage for the synthesis of PCDD/Fs, PCBs, PCNs and PAHs is the formation of hydrocarbon radicals, often including halogenated (mainly chlorinated) ones as well simple unsaturated hydrocarbons e.g. ethylene and acetylene. These reactions proceed at high temperatures and in the next stage their products undergo further reactions such as cyclization - the formation of aromatic often chlorinated compounds - including benzene, naphthalene, chlorobenzenes and chlorophenols. All these chemical compounds are formed in virtually any combustion process of solid and liquid fuels and also, although less frequently, gas fuels. If in the zone of respective temperatures appear chlorine, oxygen and organic matter chlorinated compounds are formed - such as PCDD / Fs, PCBs and PCNs. This phenomenon is observed not only for combustion process but also for most thermal processes running at temperatures of $200-700^{\circ}$C. The overall reaction is as follows:

$$\mathrm{R-Cl+O_2} \underset{\text{combustion}}{\rightarrow} PCDD/Fs, PCBs, PCNs \quad (56)$$

On the basis of the critical analysis of numerous literature data, we can assume the following paths of appearance of dioxins (as well as PCBs and PCNs) in the flue gas stream from combustion:

- dioxins introduced to the system which did not participate in any chemical reactions or were subject to the reactions of partial chlorination and/or dechlorination of the molecule (Tosine et al., 1985),
- dioxins from the aromatic rings formation and rearrangement of substituents with the participation of chlorinated precursors such as PVC, PCBs, chlorinated benzenes, chlorinated phenols directly behind the combustion zone - this is a reaction that occurs in the gas phase at temperatures of 500 – 700°C (Ballschmiter et al., 1985),
- dioxins formed outside the combustion zone in result of the reaction of non-chlorinated organic compounds such as PAHs, lignin, lignite, etc. and then chlorination reaction as well as from various types of chlorinated organic precursors such as chlorophenols, chlorobenzenes, polychlorinated diphenyl ethers, polychlorinated biphenyls and other products, etc. as well as other products of incomplete combustion which may react with each other and on the surface of metals in the fly ash particles (Dickson & Karasek 1987), where they are subject to the reactions of chlorination and dechlorination of the molecule,
- dioxins formed outside the combustion zone of molecular carbon (carbonized residues of the combustion, unburned carbon in the particles of dust, soot and organic or inorganic chlorine in the presence of fly ash at a temperature of 200 – 400°C usually called de novo synthesis (Vogg & Stieglitz 1986).

The main difference in these mechanisms is the source of carbon and the temperature range in which the synthesis takes place. In the light of the research results it seems that the most important mechanism owing to which most dioxins are formed is the third mechanism (de novo), then the second and least dioxins are produced by the first mechanism (Huang & Buekens, 1995).

The analysis of the above mentioned paths of appearance of dioxins in the flue gas stream from thermal processes, including combustion of solid fuels shows that the first case is relatively unlikely. PCDD/Fs and PCBs are not chemicals of high thermal stability. In the combustion conditions (presence of oxygen, mixing, flow) practically most organic compounds, including dioxins, is decomposed at the temperature about 850°C. The temperature of 99.9% of the destruction of PCDD/Fs is around 700°C while the temperature of the destruction of other compounds that are precursors of the synthesis of dioxins can reach up to 950°C. However, in the case of lack of oxygen the limit of the decomposition of organic compounds may be increased to around 1000°C. It therefore seems unlikely that the dioxins contained in the material subject to combustion at temperatures reaching and exceeding 1000°C could not be subject to destruction, although in the case of poor construction of the combustion installation and the formation of cold zones in the combustion chamber this is not excluded.

The second path of the appearance of PCDD/Fs, PCBs, PCNs in flue gas is a high temperature synthesis which is a condensation of precursor molecules - benzene, naphthalene, phenanthrene, acetone, trichloroethanes, benzaldehyde, dibenzofuran, benzofuran, phenol, mono-, di-, trichlorophenoles, chlorobenzenes, tetrachlorethylene or polychlorinated diphenyl ethers. These precursors may react with each other both at 500700°C in the gas phase and on the surface of fly ash at a temperature of 200 – 500°C. The key to the formation of dioxins in result of synthesis in the gas phase are the conditions for conducting the combustion process. Numerous studies have shown a very close relationship between the formation of chlorinated and non-chlorinated aromatic compounds in the combustion and afterburning chamber (at a temperature 650 – 900°C) and the parameters of combustion. It was also noted that by-products of gas fuel combustion - such as natural gas, acetylene, ethane and ethylene may be a sufficient source of carbon for the synthesis of dioxins.

Outside the combustion zone at temperatures below 500°C dioxins are formed as a result of a series of catalytic reactions occurring on the surface of dust containing metals. Usually it is assumed that dioxins are formed there by two mechanisms - the catalytic synthesis of precursors such as chlorobenzenes, chlorophenols, polychlorinated naphthalenes and polychlorinated biphenyls (similarly as in the gas phase) and catalytic synthesis of elemental carbon (soot) contained in dust particles or polycyclic aromatic hydrocarbons as well as gaseous chlorine and oxygen.

The first of these two mechanisms was first described in 1987 by Dickson (Dickson & Karasek 1987). Numerous studies conducted in subsequent years have shown that this reaction proceeds in principle as a condensation reaction of chlorobenzenes or chlorophenols or as a reaction of selective oxychlorination of polychlorinated naphthalenes and polychlorinated biphenyls. This reaction can be catalyzed by many metals - e.g. copper, titanium, manganese, cobalt, zinc but definitely copper has by far the largest catalytic ability. The characteristic of the condensation mechanism (high temperature in gas phase or on the surface of dust particles) is the fact that in this way are formed low chlorinated dioxins, containing from one to maximum 3 – 4 atoms of chlorine.

So the problem of the origins of high chlorinated dioxins, containing from 4 to 8 atoms of chlorine per molecule, in flue gas remained unresolved. The third mechanism of dioxin formation offers the solution to this problem - the de novo synthesis.

The second mechanism of the formation of dioxins in the zone after the combustion chamber, in the area of flue gas cooling, was discovered in the late eighties of the last century by Stieglitz and co-workers, the so called mechanism of de novo synthesis (Vogg & Stieglitz 1986). The de novo synthesis is a slow heterogenic catalytic reaction in which dioxins are formed without the formation of gaseous intermediate products. It occurs within the temperature range 200 – 500°C from carbon particles contained in fly ash, through the formation and closure of benzene rings, which then react with oxygen and chlorine on the catalyst surface (Stieglitz & Vogg, 1987). The source of carbon in de novo synthesis is usually so called elemental carbon contained in fly ash particles, as a carbonized residue of thermal decomposition and oxidation in the combustion process. In practice it is impossible to achieve complete burnout of carbon contained in the fuel and small amounts of unburned carbon are present in flue gas in form of soot which is made up of elemental carbon and the polycyclic aromatic hydrocarbons - PAHs. The emission of dioxins from the combustion of pure wood suggests that the carbon source can also be large, complex molecules of organic compounds such as lignin and lignite. The chlorinating agent in de novo synthesis can be free, molecular chlorine, chlorinated organic compounds as well as volatile inorganic chlorine salts - such as NaCl or $FeCl_3$. However, the fundamental role in the formation of dioxins plays the concentration of chlorine free radicals in the zone of combustion reaction and cooling. Despite the obvious role of chlorine in de novo synthesis the data from over 1900 research works on an industrial scale (Rigo & Handler, 1998) shows that there is no clear relationship between the amount of chlorine introduced in the fuel to the combustion process and the amount of dioxins emitted. There is, however, relationship between the amount of dioxins emitted and the concentration of chlorine or hydrogen chloride at the outlet of the combustion chamber.

The emission of dioxins depends primarily on the construction of the combustion unit and burning conditions. The concentration of chlorine in the flue gases significantly affects the profile of formed congeners PCDD and PCDF. The higher the concentration the greater the share of congeners with higher number of chlorine atoms. This suggests that de novo synthesis is the reaction of subsequent chlorination of individual congeners of dioxins and furans. Analyzing the above mechanism of de novo synthesis it can be stated that on the fly ash containing $CuCl_2$ there can also run the follow-chlorination of lower substituted dioxins generated in the gas phase as well as dechlorination of higher chlorinated dioxins.

The optimal temperature assumed for de novo synthesis is about 350°C. Oxygen concentration also influences the course of the de novo synthesis. This is obvious if we consider that oxygen is present in the molecule of both dibenzo-p-dioxin and dibenzofuran. Studies have confirmed that the increase in oxygen concentration outside the combustion zone significantly increases the amount of generated dioxins, the increase in the concentration of hydrogen chloride and gives the same effect.

The catalyst in the de novo synthesis is primarily copper chloride. Copper has a very strong effect in supporting the formation and closing the rings such as chlorophenols. Subsequent studies have shown that transition metals catalyze both ring closing and the process of chlorination and de novo synthesis may be treated as an electrophilic substitution reaction that runs in two stages - the first stage includes the chlorination of carbon surface while the second stage is oxidation and decomposition in which dioxins and furans are formed. Similar to the catalytic properties of copper, in the de novo synthesis similar properties also have aluminum, iron, magnesium, manganese, cobalt as well as sodium, potassium and zinc. Other metals - such as nickel does not exhibit catalytic activity or they decrease the amount of generated dioxins e.g. chromium, wolfram or vanadium. But there is no doubt that the best catalyst for de novo synthesis is copper. In reality, however, on the surface of fly ash there are many metals, whose influence on the synthesis of dioxins is different - some of them catalyze the de novo synthesis or the chlorination process of low chlorinated dioxins formed earlier (in the synthesis in gas phase) - for example, copper, zinc, iron or aluminum, while others catalyze the decomposition of generated dioxins or their dechlorination - e.g. chromium, vanadium, tungsten or nickel.

After analyzing the emissions of dioxins from combustion processes and all the described paths of the presence of dioxins in the flue gas, especially the paths of their synthesis it should be noted that probably most dioxins are formed outside the combustion zone. According to Goldfarb, who analyzed the problem of dioxin emissions from Canadian incineration plants, this is most probably the vast majority (Goldfarb, 1986).

Summing up the problem of paths of dioxins formation in combustion processes we should first mention the decisive influence of the combustion process conditions on the quantity of produced dioxins and furans. Combustion process conditions include first of all temperature, residence time of gases from the combustion in the zone of high temperatures, oxygen concentration in the combustion zone and directly connected with it concentrations of carbon monoxide in the flue gas. Poor combustion conditions i.e. too low temperature and high concentration of carbon monoxide contribute to the formation of incomplete combustion products such as chlorobenzenes and chlorophenols which are precursors of dioxins and can undergo condensation reactions in both high and low temperature zone in the presence of metals catalyzing condensation reactions. Poor combustion conditions also favor the formation of polycyclic aromatic hydrocarbons (PAHs) and unburned coal residues in the dust entrained in the air which is an excellent raw material for de novo synthesis and this in turn means that you can prove the thesis that there is a direct impact of the combustion conditions on the amount of dioxins generated in all three known mechanisms described earlier.

When conducting combustion processes we are always dealing with two opposing reactions (reaction systems) - formation of dioxins as a result of various homo- and heterogenic reactions and decomposition of dioxins as a result of high temperature and catalytic reactions. Graphic illustration of this problem is shown in Figure 4. Most of the dioxins generated in waste incineration plants is formed in result of de novo synthesis in the temperature range 200 – 400°C with maximum temperatures around 300°C. This maximum is the result of synthesis and decomposition processes running in parallel.

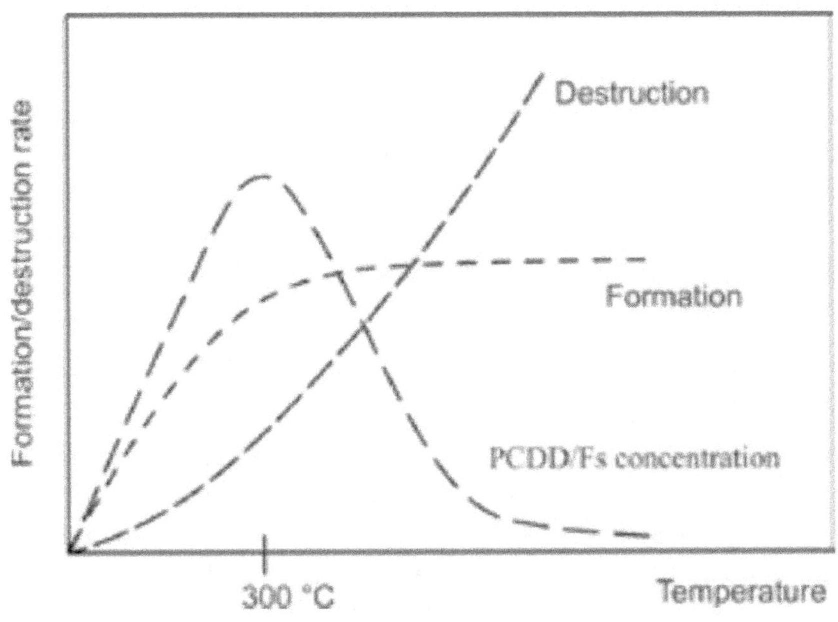

Fig. 4. Formation and decomposition of dioxins in a catalytic process on dust particles.

The process of formation of PCDD/Fs, PCBs, PCNS and PAHs during combustion depends on many factors and that is why it is very difficult to predict how much of these compounds will be formed in a specific process with fixed parameters. Momentary dioxin emissions results not only from the current combustion conditions. Due to the fact that probably most of the dioxins is produced by de novo synthesis, which is a very slow reaction, the formation of dioxins and their release continues long after the optimization of the parameters of the combustion process and minimizing the amount of generated dioxins. This is so called memory effect occurring in waste incineration plants. In the case of unstable operation of incinerators high concentrations of dioxins in the flue gas can be observed over long periods of time (Hunsinger et al., 2007).

16. Primary methods of emission reduction

Understanding the mechanisms of pollutants formation in the combustion process enables the development of such technologies and methods of combustion which ensure the smallest amount of generated pollutants. There are well known so called low emission combustion processes relating to carbon monoxide and nitrogen oxides - the use of the combustion zone, in stages, flue gas recirculation and the so-called reburning (introducing fuel to the periphery of flame zone) allows to reduce NO_x emissions by about 50 – 60% compared to the original process (Jarosiński, 1996, Hill & Smoot, 2000).

Such efforts to reduce emissions, which are an interference in the combustion process are called primary methods of reducing emissions. They become more and more important in recent years, because from the economic point of view, they are more cost effective (cheaper) than the secondary methods, called "end of pipe technologies". Their effectiveness is generally lower than that of flue gas cleaning methods, however, they significantly help to improve the work of the latter, allowing to obtain the total degree of purification of gases at a level above 99%.

In recent years the determination of the effect of parameters of the combustion process on the emission of metals, PAHs, dioxins and other organic substances from the combustion process became an important issue. Numerous literature reports clearly show that keeping good combustion process parameters significantly affects the reduction of emissions of organic substances (including PAHs and dioxins) to the atmosphere. A parameter which characterizes "good combustion conditions" very well is the concentration of carbon monoxide in the flue gas. Other parameters defining the "good combustion conditions" are above all so called "3T"- temperature, turbulence and time of flue gas residence in proper temperature. Detailed conditions are specified among others in the Directive on the incineration of waste (2000/76/EC) - they are as follows: flue gas residence time at 850°C should be not less than 2 seconds at high gas turbulence (for waste containing less than 1% of chlorine compounds, in the case of waste containing more than 1% of chlorine compounds the temperature should be higher than 1100°C).

It is essential, however, to ensure combustion conditions close to complete and perfect combustion, with minimized amount of formed carbon monoxide. Such conditions prevail in properly oxygenated combustion zone at an optimal concentration of oxygen (for coal combustion $6-8\%$, excess air coefficient about $1.4-1.8$). In the case of waste incineration better oxygenation of the combustion zone and providing excess air coefficient of the order of 2,02,4 (oxygen concentration $10-13\%$) is required. The effect of oxygen concentration and combustion temperature on the concentration of carbon monoxide in the flue gas is illustrated in Figure 5 (Seeker, 2001).

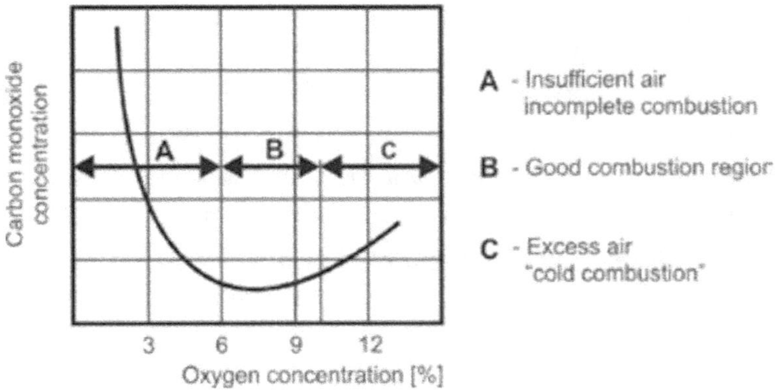

Fig. 5. The effect of oxygen concentration in the combustion zone on the carbon monoxide emission.

According to the research of Seeker (Seeker, 2001) the concentration of carbon monoxide, that is combustion conditions, has a great impact on the amount of dioxins produced in the combustion process and their emission. This is illustrated in Table 7.

Table 7. The effect of combustion conditions on the rate of PCDD/Fs emission to atmosphere and the concentration in secondary waste from the combustion process (per 1 Mg of incinerated waste).

Combustion parameters	Slag, bottom ash	Fly ash	Dust from electrostatic precipitator	Emission to atmosphere
T = 830°CCO = 25ppm	0,33μg /Mg	0,14μg /Mg	21,2mg/Mg	11,2μg/Mg
T = 850°CCO = 1ppm	0,35μg /Mg	0,22μg /Mg	21,2mg/Mg	1,6μg/Mg

Schmöckel's research (Schmöckel & Streit, 1994) on the wood combustion process in plants with capacities ranging from 11 kW to 13.9 MW demonstrated a strong correlation between the concentrations of certain pollutants in the flue gas and carbon monoxide levels, indicating the influence of combustion conditions on the emission of organic pollutants. The findings of these research are illustrated in Figure 6.

Combustion characteristics significantly influence metal emissions. This results from the varying volatility of metals and their speciation based on their forms of occurrence. The presence of chlorine and hydrogen chloride in the combustion zone results in varying concentrations of metals in slag, fly ash, ash from de-dusting equipment, or in gases released into the environment (Chiang et al. 1997). The combustion temperature and oxygen content in various combustion zones significantly influence the metal balance in individual secondary products of the combustion process (Wunsh et al., 1995; Wey et al., 1998). For instance, regulating the primary air quantity in the distinct zones of the grate and the temperature can lead to nearly zero emissions of some metals, with nearly 100% of their concentration retained in the slag (Modigell & Liebig, 1999). The implementation of primary emission reduction measures facilitates the secondary purification of flue gas to comply with legal standards (e.g. Directive 2000/76/EC). Table 8 illustrates the beneficial impact of primary emission reduction measures on the concentration of contaminants in flue gas.

Table 8. The effect of the primary methods for reducing emissions of pollutants in waste incineration plants (Büttenberger & Hansen 1997).

Pollution	Unit	Installation without the use of primary methods	Installation using the primary methods	Allowable concentration according to Directive 2000/76/EC
Dust	mg/m^3	2000-10000	1000-1500	10
Carbon monoxide	mg/m^3	50-80	10-15	50
Sum of organic compounds	mg/m^3	10-100	0,5-1	10
PCDD/Fs	$ng\ TEQ/m^3$	5-12	0,6-1,2	0,1

Acetaldehyde concentration [mg/m^3]

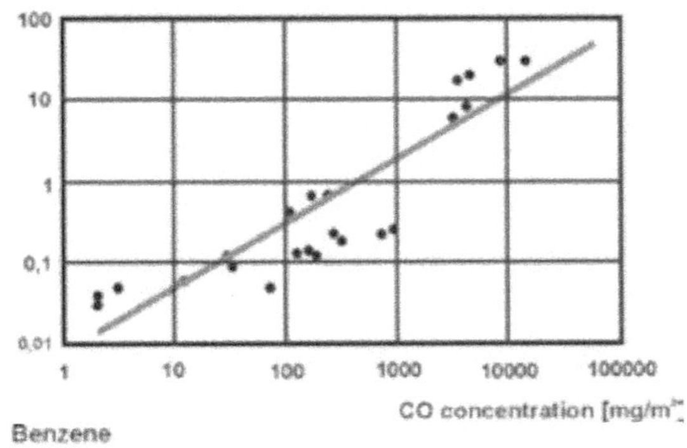

Benzene concentration [mg/m^2]

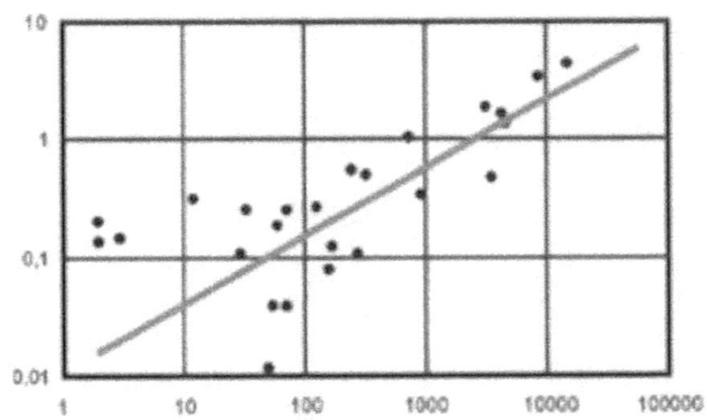

CO concentration [mg/m ^{2}.

PAHs concentration

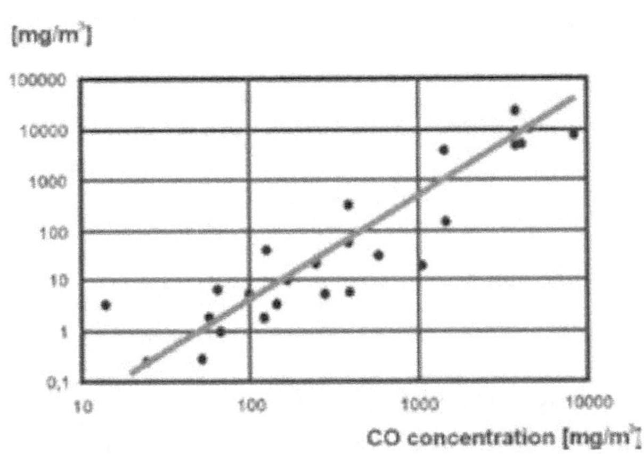

Formaldehyde concentration [mg/m^3]

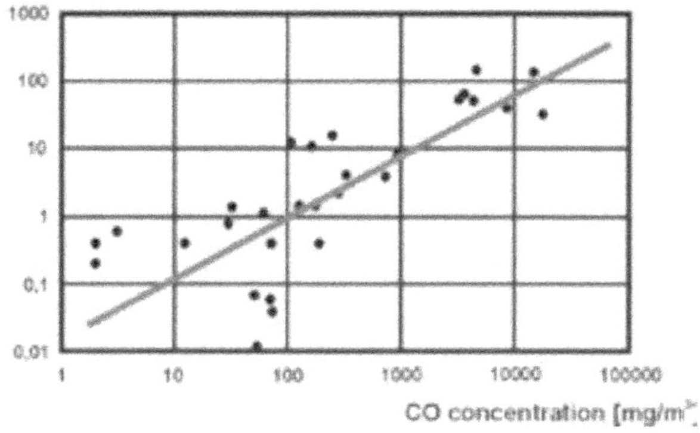

Total Phenols concentration [mg/m^3]

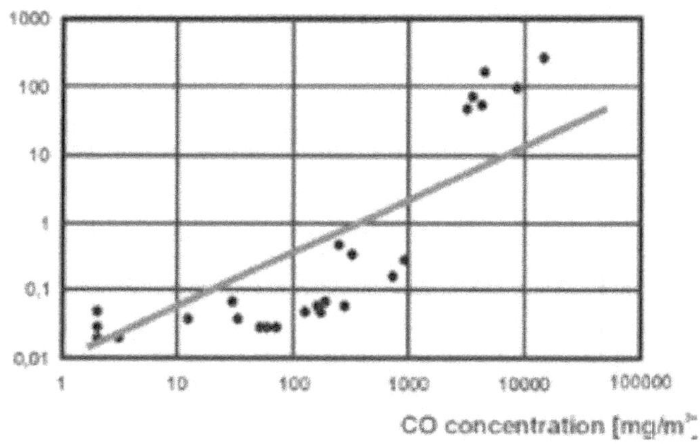

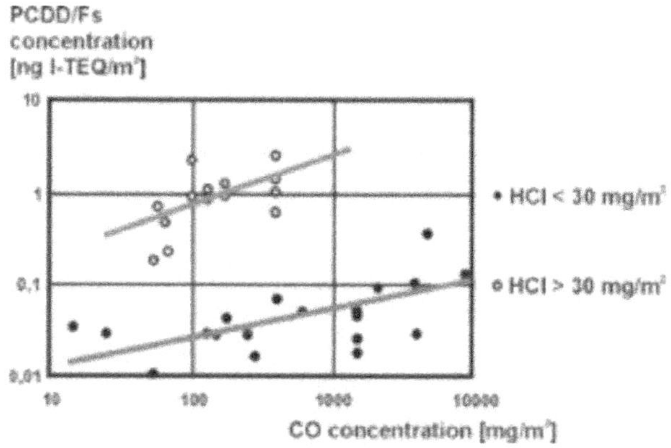

Fig. 6. The effect of combustion conditions (concentration of CO in flue gas) on the emission of some organic pollutants from combustion process (Schmöckel & Streit, 1994).

17. References

Aubrey, N. E., van Wazer, J. J. (1964) Equilibrium Rearrangement of Perchlorocarbon Compounds. Journal of American Chemical Society, Vol.86, pp.4380-4383

Ballschmiter, K., Zoller, W., Buchert, H., Clas Th. (1985) Correlation between substitution pattern and reaction pathway in the formation of polychlorodibenzofurans. Fresenius Zeitschrift für Analistische Chemie, Vol. 322, pp.587-594

Barton, R. G., Seeker, W. R., Bostian, H. E. (1991) The behavior of metals in municipal sludge incinerators. Trans. Inst. Chem. Eng. Vol.69, pp.29-36

Belevi, H., Langmeier, M. (2000) Factors determining the element behavior in municipal solid waste incinerators. Laboratory experiments. Environ. Sci. Technol. Vol.34, pp.2507-2512

Belevi, H., Moench, H. (2000) Factors determining the element behavior in municipal solid waste incinerators. Field studies. Environ. Sci. Technol. Vol.34, pp.2501-2506

Bowman, C.T., Hanson, R.K., Louge, M. Garman, A. (1982) Gas Phase Reaction Kinetics of NOx Formation from Fuel Nitrogen in Fossil Fuel Combustion. - in: Proceedings of 1982 Joint Symposium on Stationary Symposium NOx Control, Dallas, USA

Bozzelli, J. W., Chang, A. Y., Dean, A. M. (1994) Analysis of the reactions $H + N_2O$ and $NH + NO$: Pathways and rate constants over a wide range of temperature and pressure. in: Proceedings of International Symposium on Combustion, Vol.25, No.1, pp.965-974

Büttenberger, B., Hansen, H. (1997) Das EVT-Müllverbrennungssystem im Hinblick auf die neuesten gesetzlichen Anforderungen. BWK/TÜV-Umwelt-Special, October 1997, pp.26-38

Chiang, K.Y., Wang, K.S. Lin, F.L., Chu, W.T (1997) Chloride effects on the speciation and partitioning of heavy metals during the municipal solid waste incineration process. The Science of the Total Environment, Vol. 203, pp.129-140

Dickson, L. C., Karasek, F. W. (1997) Model studies of polychlorinated dibenzo-p-dioxin formation during municipal refuse incineration. - Science, Vol.237, pp.754-756

Eduljee, G. H. (1994) Organic micropollutant emissions from waste incineration. - in: Hester R. E., Harrison R. M. (ed.) - „Waste incineration and the environment", Royal Society of Chemistry, London

Fenimore, C. P. (1971) Formation of nitric oxide in premixed hydrocarbon flames. - in: Proceedings of XIII th International Symposium on Combustion, The Combustion Institute, Pittsburg, pp.373-384

Goldfarb, T. D. (1989) Evidence for post-furnace formation of PCDD and PCDF implications form control. Chemosphere, Vol.18, pp.1051-1055

Griffin, R. D. (1986) A new theory of dioxin formation in municipal solid waste combustion. Chemosphere, Vol.15, pp.1987-1990

Hill, S.C., Smoot, L.D. (2000) Modeling of nitrogen oxides formation and destruction in combustion systems. Progr. in Energy and Combustion Sci. Vol.16, pp.417-458;

Huang, H., Buekens, A. (1995) On the mechanisms of dioxin formation in combustion processes. Chemosphere, Vol.31, pp.4099-4117

Hunsinger, H., Seifert, H., Jay, K. (2007) Control of PCDD/F formation under conditions of fluctuating combustion performance in MSWI. Organohalogen Compounds, Vol.69, pp.956-961

Jarosiński, J. (1996) Techniki czystego spalania (in polish). WNT, Warszawa;

Jay, K., Stieglitz, L. (1995) Identification and quantification of volatile organic components in emission of waste incineration plants. Chemosphere, Vol.30, pp.1249-1260

Lenoir, D., Wahrmeier, A., Schramm, K. W., Kaune, A., Zimmermann, R., Taylor, P. H., Sidhu, S. S. (1998) Thermal formation of polychlorinated dibenzodioxins and furans: investigation of relevant pathways. Environmental Engineering Science, Vol.15, pp.37-478

Lenoir, D., Wahrmeier, A., Sidhu, S. S., Taylor, P. H. (2001) Formation and inhibition of chloroaromatic micropollutants formed in incineration processes. Chemosphere, Vol.43, pp.107-114

Lindbauer, R. L., Wurst, F., Prey, Th. (1994) PCDD/F emission control for MSWI by SO_3 addition. Organohalogen Compounds, Vol.19, pp.355-359

Liow, M. Ch., Lee, W. J., Chen, S. J., Wang, L. Ch., Chung, Ch. H., Chen, J. H. (1997) Emission polycyclic aromatic hydrocarbons from medical waste incinerators. Journal of Aerosol Science, Vol.28, pp.549-550

Liu, K., Pan, W-P., Riley, J. T. (2000) A study of chlorine behavior In a simulated fluidized bed combustion system. Fuel, Vol.79, pp.1115-1124

Mascolo, G., Spinos,a L., Lotito, V., Mininni, G., Bagnuolo, G. (1997) Lab-scale evaluations on formation odf products of incomplete combustion in hazardous waste incineration: influence of process variables. Water Science and Technology, Vol.36, pp.219-226

Mastral, A. M., Callen, M. S., Garcia, T. (2000) Toxic organic emissions from coal combustion. Fuel Processing Technology, Vol.67, pp.1-10

Miller, J. A., Bowman, C. T. (1989) Mechanism and modeling of nitrogen chemistry in combustion. Progress in Energy and Combustion Science, Vol.15, pp.287-338

Modigell, M., Liebig, D. (1999) Modeling and simulating the waste incineration process. private communication

Nissen, W. R. (2002) Combustion and Incineration Processes. Marcel Dekker, New York,

Recknagel, H., Sprenger, E., Hönmann, W., Schramek, E. R. (1994) Taschenbuch für Heizung und Klimatechnik. - R. Oldenburg Verlag, München

Rigo, H. G., Handler, A. J. (1998) Is there a strong dioxin-chlorine link in commercial scale systems? Chemosphere, Vol.37, pp.2031-2046

Schmöckel, G., Streit, A. (1994) Emission organischer Stoffe bei der Holzfeuerung. Bayerisches Landsamt für Umweltschutz, Augsburg

Seeker, R. (2001) Current understanding of dioxin/furan formation and control. - private communication

Steele, R. C., Malte, P. C., Nicol, D. G., Kramlich, J. C. (1995) NO_x and N_2O in lean-premixed jet-stirred flames. Combustion and Flame, Vol.100, pp.440-449

Stieglitz, L., Vogg, H. (1987) On formation conditions of PCDD/PCDF in fly ash from municipal waste incinerators. Chemosphere, Vol.16, pp.1917-1922

Sukrut, S., Thipse, E., Dreizin, L. (2002) Metal partitioning in products of incineration of municipal solid waste. Chemosphere, Vol.46, pp.837-849

Thome-Kozmiensky, K. J. (1994) Thermische Abfallbehandlung. - EF-Verlag für Energieund Umwelttechnik, Berlin

Tosine, H. M., Clemen,t E. R., Ozvacic, V., Wong, G. (1985) Levels of PCDD/PCDF and other chlorinated organics in municipal refuse. Chemosphere, Vol.14, pp.821-827

Trenholm, A. (1998) Identification of PICs in hazardous waste combustion emissions. Waste Management, Vol.18, pp.485-492

Vogg, H., Stieglitz, L. (1986) Thermal behavior of PCDD/PCDF in fly ash form municipal incinerators. Chemosphere, Vol.15, pp.1373-1378

Wey, M.Y., Su, J.L., Yan, M.H., Wei, M.C. (1998) The concentration distribution of heavy metals under different operation conditions. The Science of the Total Environment, Vol.212, pp.183-193

Wielgosiński, G. (2009) Dioxin Emission from Thermal Processes and Methods of Emission Reduction. (in polish) - Polish Academy of Science, Lodz

Wienecke, J., Kruse, H., Huckfeldt, U., Eickhoff, W., Wassermann, O. (1995) Organic compounds in the flue gas of a hazardous incinerator. Chemosphere, Vol.30, pp.907913;

Wunsch, P., Matuschek, G., Kettrup, A. (1995) The influence of combustion bed temperature during waste incineration on slag quality. Thermochimica Acta, Vol.263, pp.95-100

Zeldovich Y. B. (1946) The Oxidation of Nitrogen in Combustion and Explosions. Acta Physcochimica, Academy of Sciences U.S.S.R., Vol.21, 577-628

Chapter-11

Multiobjective Optimization in Chemical Process Plant Design

1. Introduction: An Overview

There needs to be a methodical and effective way to solve multiobjective decision making problems. Process models like Aspen PlusTM and Aspen Dynamics are used a lot when the process or control system needs to be changed. But these models aren't made to look into other things, like safety and the environment. Because making decisions with multiple objectives is hard and can lead to conflicts, a combined optimization tool should be useful. This chapter goes into great detail about a way for solving this problem that is based on separate modules and its different stages.

2. Proposed methodology

The method is based on a number of standard, separate techniques. These methods have been changed as needed and then woven together to make a single plate shape. The main goal is to make sure that all screening and choice decisions made during the creation or modification of chemical process plants are made in the same way. This will help create a process that is better for the environment and safety while also being more cost-effective.The methodology (see Figure 1) consists of four layers/stages:

- Generation of alternatives and problem definition;
- Analysis of alternatives i.e. generation of relevant data for comparison of Environmental, economic and safety objectives
- Multiobjective decision analysis/ optimization
- Design evaluation stage i.e. decision making from the pareto-surface of non-inferior solution or ranking of alternatives

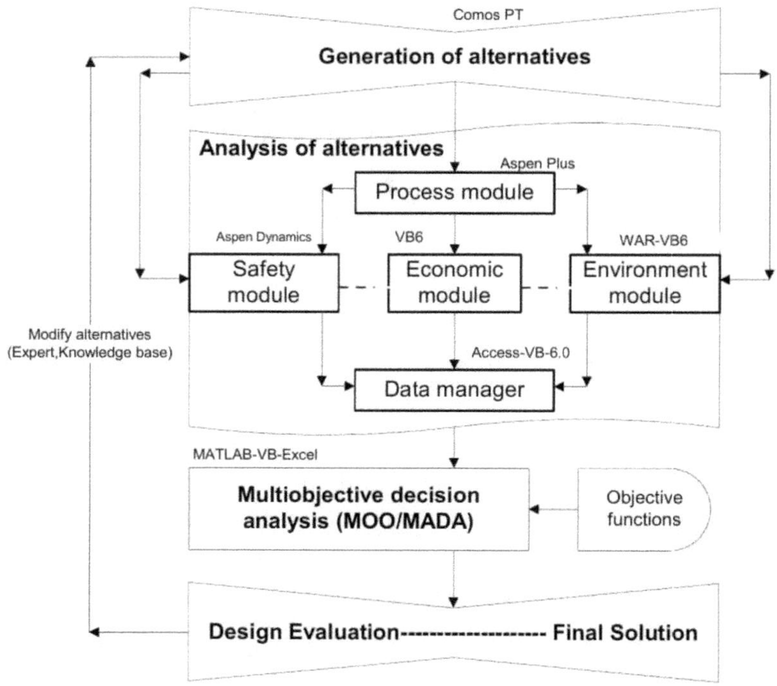

Fig. 1. Simplified block diagram of proposed methodology

2.1 Stage I: Generation of alternatives and problem definition

The first layer composed of following tasks:

Definition of the scope of the study,

Statement of key assumptions and the performance targets such as quality etc., Degree of freedom analysis, Identification of the key design, control, and manipulated variables, Definition of the system boundary,

Identification of constraints,

Choice of functional unit for all calculations,

Collection of relevant information about process and chemicals to be handled, Generation of different alternatives either based on suggestion from independent departments or using the individual objective modules from stage II.

Seader et al. (1999)[6] has described rules for selection of process variables in the book "Process design principles-synthesis, analysis and evaluation". The data and information about the process and chemicals involved such as thermodynamic and kinetic data can be found from journal articles, patents or handbooks. Current chemical prices can be obtained from market reports if not available in main plant documentation or company central data base. In addition to these sources, some data related to quantification of environmental impacts and material safety data sheets of chemical are also collected from commercial data bases so that an impact assessment and safety analysis can be performed in subsequent design steps. Commercial computer aided tool like Comos^{PT} can be used for plant documentation and to support stage-I of proposed methodology.

2.2 Stage II: Analysis of alternatives

This stage is composed of independent modules used to generate relevant information for evaluation of economic, safety and environmental performance objectives. These modules are:

- Process module
- Safety module [1,3]
- Economic module
- Environment module [2]
- and a data manager for managing the relevant information generated from these modules.

2.3 Process module

In the process module, an operation model of the process system has to be developed for evaluating alternatives. The configured simulation model has to be able to reproduce the selected results to an accepted degree of accuracy. This simulation model can be used for design and operation, revamping and debottlenecking of the process under study[7]. Three major integrated simulation systems widely used in the firms and companies for this purpose are Aspen technology (Aspen Plus, Aspen dynamics etc), Hyprotech (Hysys process, Hysys plant etc) and Simulation Sciences (Pro/II etc.). Aspen Plus ™12.1 is used in this work for development of simulation model and linked in a visual basic platform for integration with safety, economic and environment modules. The most important results available from the process simulation model are material and energy balance information for both streams and units, rating performance of units and tables and graphs of physical properties. A brief description of Aspen Plus ™ 12.1 and steps involved in development of the process simulation model is described here below.

Aspen Plus ™

Aspen Plus ™ supports both sequential modular and equation oriented computation strategy and allows the user to build and run a steady-state simulation model for a chemical process. It provides a flexible and productive engineering environment designed to maximize the results of engineering efforts, such as user interface mode manager, quick property analysis, rigorous and robust flowsheet modelling, interactive architecture, powerful model analysis tools and analysis and communication of results. Therefore, it lets the user to focus his/her energies on solving the engineering problems, not on how to use the software. It is not only good for process simulation but also allows to perform a wide range of other tasks such as estimating and regressing physical properties, generating custom graphical and tabular output results, sensitivity analysis, data-fitting plant data to simulation models, costing the plant, optimizing the process, and interfacing results to spreadsheets.

The development of a simulation model for a chemical process using Aspen Plus ™12.1 involves the following steps (see details in table 1):

1. Define the process flowsheet configuration by specifying

- Unit operations
- Process streams flowing between the units
- Unit operation models to describe each unit operation

2. Specify the chemical components,
3. Choose a thermodynamic model to represent the physical properties of the components and mixtures in the process,
4. Specify the component flow rates and thermodynamic conditions (i.e. temperature, pressure, or phase condition) of the feed streams,
5. Specify the operating conditions for the unit operations,

Table 1. Developmental process for an Aspen Plus ™ simulation model

Step	Used to
Defining the flowsheet	Break down the desired process into its parts: feed streams, unit operations, and product streams

Specifying stream properties and units	Calculate the temperature, pressure, vapor fraction, molecular weight, enthalpy, entropy and density for the simulation streams
Entering components	From a databank that is full of common components
Estimating property parameters	Property Constant Estimate System (PCES) can estimate many of the property parameters required by physical property models
Specifying streams	Streams connect unit operation blocks in a flowsheet and carry material and energy flows from one block to another. For all process feed streams, we must specify flowrate, composition, and thermodynamic condition
Unit operation blocks	We choose unit operation models for flowsheet blocks when we define our simulation flowsheet

2.4 Safety module

Safety module is based on combination of conventional standard risk analysis techniques and process disturbance simulation. This module not only generates relevant information related to safety aspects for multiobjective decision analysis but also used for safety/risk analysis and optimization. The purpose of this module is to determine risk from operational disturbances and to develop effective risk reductions. It can be divided into the following steps (Figure 2):

§ Simulation of process related malfunctions

i = { financial risk , environmental risk , human health risk }

Step 1: System description and objectives of analysis (before starting safety and risk analysis)

Step 2: Safety/risk analysis (identification of weak points via Extended HAZOP)

Step 3: Safety/risk assessment (categorization of risk via risk potential matrix (HAZOP decision matrix))

Step 4: Safety/risk system optimization

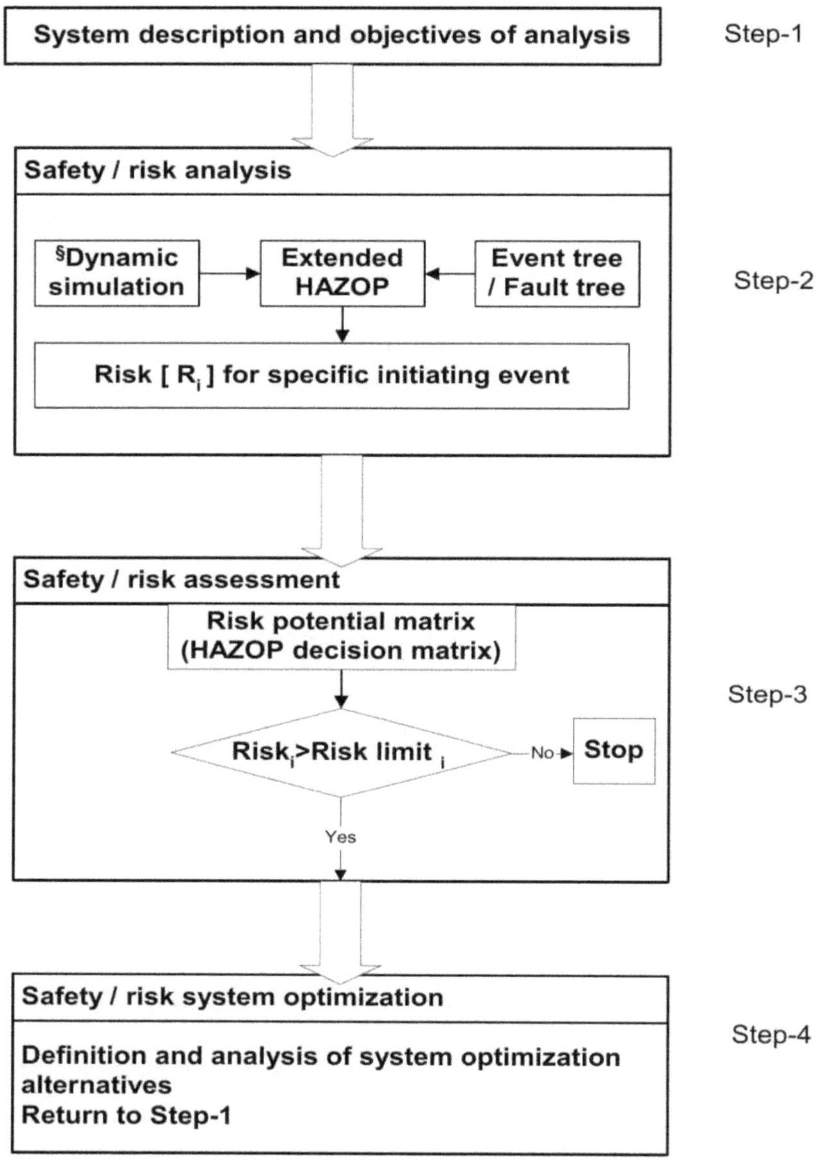

Fig. 2. Simplified block diagram of safety module

2.4.1 Step 1: (Before starting safety/risk analysis) - Description of system and objectives of analysis

For efficient safety/risk studies, the analyst must have an accurate description of the system to be investigated and a clear objective of the analysis study. Therefore, in this step the purpose, objectives, and scope of the study are clearly defined. The necessary information required for the study such as process flow diagrams, piping and instrumentation diagrams, plant layout schematics, material safety data sheets, equipment data sheets, operating instructions, start up and emergency shutdown procedures, and process limits, etc. is gathered from plant documentation. A team under a trained and experience leader with five to seven people including experts of the design and operation of the subject process may be formulated.

2.4.2 Step 2: Safety/risk analysis (Identification of weak points via extended HAZOP) Extended HAZOP

Our intention is to identify weak points due to disturbances in operation, which may or may not be hazardous, in order to improve safety, operability, and/or profit at the same time. Extended HAZOP (HAZOP supported by dynamic simulation, event tree and fault tree techniques and HAZOP decision matrix) is used not only for identification of weak points but also for generation and analysis of optimization proposals [8-11]. Extended HAZOP differs from the standard HAZOP approach in following aspects:

i. Use of dynamic simulation:

In Extended HAZOP, the analysis of the influence of disturbances (failures) on the behaviour of the process is based on shortcut or simplified hand calculations or dynamic simulation. Aspen dynamics is used for this purpose.

ii. Classification of risk related consequences:

Each established consequence (hazard) has to be expressed by a consequence class (C). The plant specific scoring (from 0 (lowest) to 8 (highest)) chart is given in Table 2 (a & b) based on principle consequence analysis. For classification of consequences based on principle release estimates, accident consequence analysis techniques (models for calculation of toxic, fire and explosion effects) and plant location data (capital investment, population density etc.) have to be considered.

Illustrative Example 1

Figure 3 shows the plant lay out considered for developing plant specific consequence scoring chart. The area around the plant is open fields (rural condition). As weather conditions changed around the year, so certain assumptions are made to results in worse case conditions for consequence analysis. These include weather conditions and wind speed that result in smallest value of dispersion coefficients. Therefore, stability "F" and wind speed as low as possible (1.5 m/s) is selected. It is assumed that 10 workers are present (working 24 h each day), which are not distributed uniformly, on the land in area (100 m x 100 m) around the column under study. Acetone is selected as representative fluid for consequence analysis.

Acetone vapours released from the vent line at a rate of 1616 kg/h due to loss of cooling medium. It is assumed that released vapours form a cloud for 30 minutes before.

Table 2. Scoring chart for Consequence Financial consequences [3.4]

Effects	Class	Financial loss (€)	Class related consequences: examples
Function impairment	*	< 10	: Product quality lowering (brief)
	1	$10^1 - 10^2$	: Product quality lowering
	2	$10^2 - 10^3$	: Product quality lowering (long term)
Functional Loss	3	$10^3 - 10^4$	: Production disturbance (brief) Soil contamination Safe dispersion of material release from vent line
	4	$10^4 - 10^5$	: Production disturbance Material release from the piping Pump damage (pressure impacts)
	5	$10^5 - 10^6$	: Production disturbance (long term) Jet fire as result of release of material from vent line Pool fire (from pump leakage)
Safety and Environmental pollution	6	$10^6 - 10^7$	Fireballs due to catastrophic rupture of vapour product line
	7	$10^7 - 10^8$	Vapour cloud explosion (ignoring domino effect)
	8	$>10^{8}$	Vapour cloud explosion along with domino effect

Effects	Class	Community	Class related consequences: examples
Function impairment	*	No effect on people	: Product quality lowering (brief)
	1	Nuisance effect	: Product quality lowering
	2	Minor irritation effect to people & local news	: Product quality lowering (long term)
Functional Loss	3	Moderate irritation effect to people and non	: Production disturbance (brief) Soil contamination Safe

		compliance to laws, local news	dispersion of material release from vent line
	4	Moderate irritation effects to people & environment, single injuries and regional news	: Production disturbance Material release from the piping Pump damage (pressure impacts)
	5	Significant effects to people and environment, > 1 injuries & regional news	: Production disturbance (long term) Jet fire as result of release of material from vent line Pool fire (from pump leakage)
Safety and Environmental pollution	6	Major effects to people and environement, multiple injuries, fatality likely, regional news	Fireballs due to catastrophic rupture of pipe or condenser (vapour product line)
	7	Severe effects to people and environment, fatality, regional news	Vapour cloud explosion (ignoring domino effect)
	8	Multiple fatalities and process shutdown certain, international news	Vapour cloud explosion along with domino effect

Environment and Health consequences ignited and leads to vapour cloud explosion. The physical effects of this scenario or event is calculated as:

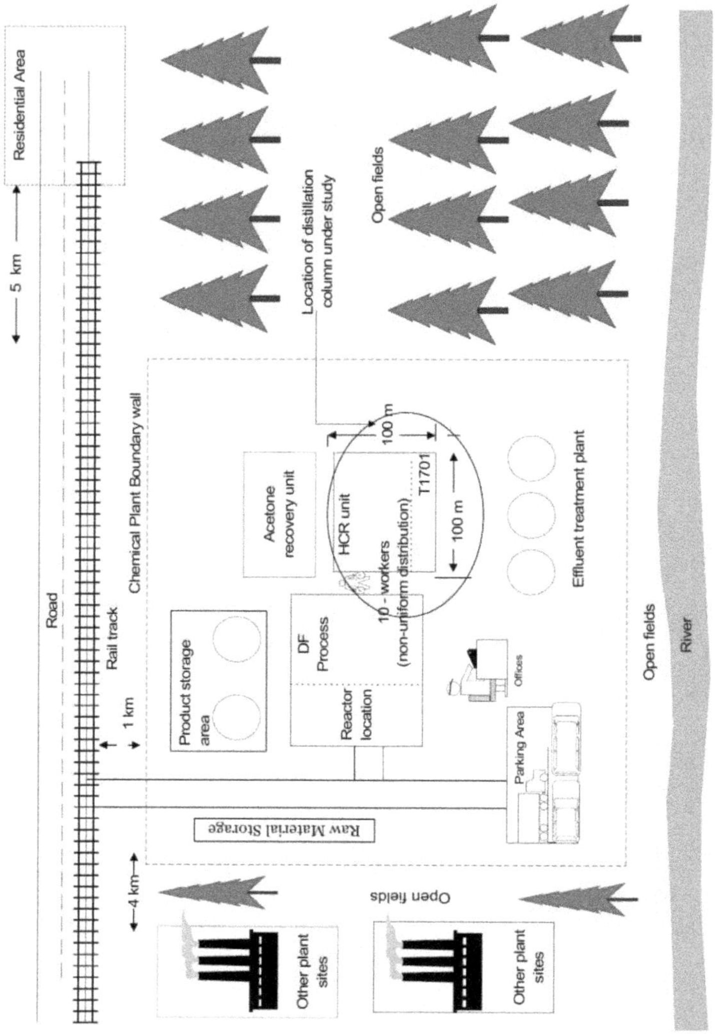

Fig. 3. Plant lay out for establishing consequence score chart

Weight of fuel in the cloud = M = 1616/2 = 808 kg

Then amount of TNT equivalent to the amount of this flammable material is

$$M_{TNT} = \alpha \cdot \frac{M \cdot H_c}{H_{TNT}}$$

Where α = explosion efficiency ~ 0.05 (Cameron 2005) H_c = heat of combustion of fuel ~ 3.03 × 104 kJ/kg for acetone

H_{TNT} = TNT blast energy ~ 5420 kJ/kg; so M_{TNT} = 225.25 kg

Then, using relation $Z = R/(M_{TNT})^{1/3}$ and figure 4, scaled distance and overpressure is estimated. Table 3 presents the results obtained.

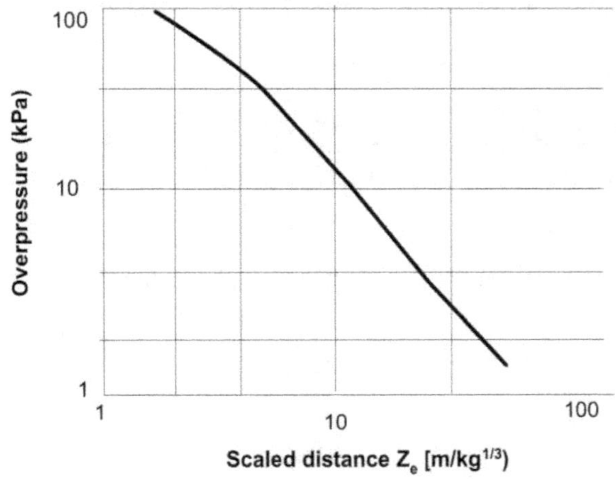

Fig. 4. Overpressure versus scaled distance for TNT explosions on flat surfaces (Tweeddale 2003, p. 115)

Table 3. Results of physical effects of vapour cloud explosion

Distance, R M	Scaled distance, Z m / kg1/3	Overpressure, Δp kPa
10	1.64	90
20	3.28	40
50	8.21	20
100	16.40	7

It is estimated that severe structural damage and 15% chance of fatality outdoors or 50% chance indoor will be experienced out to 20 m and almost complete destruction of all ordinary structures and 100% chance of fatality indoors to 10 m distance.[8] (see Cameron 2005, p. 268).

iii. Classification of frequencies of risk related consequences:

The frequency of occurring for each possible consequence (hazard) has to be expressed by a frequency class, called (F) according to the scoring chart for frequency (Table 3.4). Definition of frequency class may be supported by Event Tree and/or Fault tree analysis techniques or Layer of protection analysis (LOPA) or historical databases.

For establishing frequency class: Estimation / calculation of frequency of vapour cloud explosion and fatality of person because of release of material due to catastrophic rupture of distillation column.

Frequency of catastrophic rupture of column = 10 − 6 (Taken from table 4)

Table 4. Scoring chart for frequency [3.4]

Class	Frequency of occurring incident		
	Frequency $\frac{1}{y}$	Comprehension	Examples based on general data bases
9	$< 10^{-8}$	Very very small	Catastrophic rupture or leakage of pipe of diameter > 150 mm
8	$10^{-8} - 10^{-7}$	Very small	Catastrophic rupture of pipe of diameter ≤ 50 mm
7	$10^{-7} - 10^{-6}$	Small	Catastrophic rupture of fractionating system (excluding piping), storage tank rupture
6	$10^{-6} - 10^{-5}$	Less small	Pipe residual failure, 100 m full breach, Double wall tank leakage
5	$10^{-5} - 10^{-4}$	Moderate	Process vessel leakage of ≥ 1 mm diameter
4	$10^{-4} - 10^{-3}$	Less moderate	Pump leakage, Heat exchanger leakage
3	$10^{-3} - 10^{-2}$	Less high	Safety valve open spuriously, Large external fire
2	$10^{-2} - 10^{-1}$	High	Cooling water failure, BPCS instrument loop failure
1	$10^{-1} - 10^{0}$	Very high	Operator failure, Regulator failure , Solenoid valve failure
*	$>10^{0}$	Very very high	Power failure in developing countries, Operators failure under high stress

Probability of ignition of released material = 0.10 (CCPs 2000, Borysiewich 2004)
Probability of VCE if released material ignited = 0.01 (CCPs 2000, Borysiewich 2004)
Probability of fatality of a person exposed to overpressure of 40 kPa due to VCE = 0.20 (Tweeddale 2003, p. 117 Figure 5-14)

Then, frequency of vapour cloud explosion $= 10^{-6} \cdot 0.10 \cdot 0.01 = 10^{-9}$

So frequency class for this scenario = 9

Frequency of fatality of a person exposed to VCE = $10^{-9} \cdot 0.20 = 2.10^{-10}$

So frequency class for this scenario = 9

iv. Way of documenting the HAZOP results

The Extended HAZOP methodology worksheet for documenting the HAZOP team results is shown in Figure 5. Below consequence the physical effects and risk has to be documented first and next risk has to be classified using score charts (Table 3.2) related to financial, environment and health related consequences. The worst score of each risk has to be documented. For each risk related consequence, frequency class has also to be established.

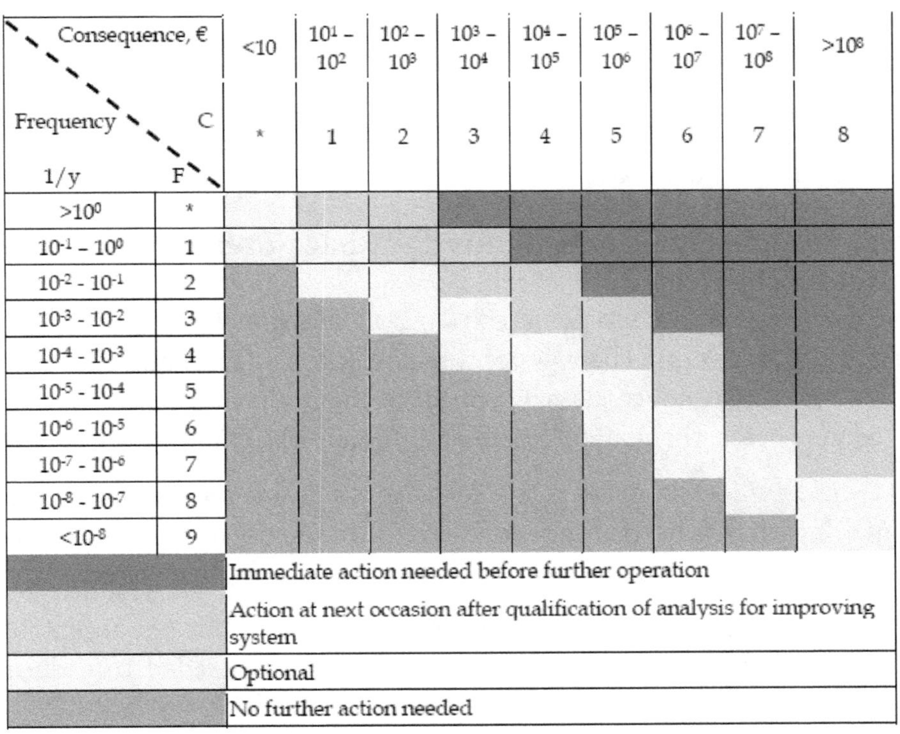

Fig. 5. Risk potential matrix (Extended HAZOP decision matrix)

Illustrative Example 2

Release of material to atmosphere from vent line or vapour line may disperse safely or has toxic effects or can lead to several outcomes such as flash fire, vapour cloud explosion and fire balls. So documenting consequence class in HAZOP work sheet, the score ' 8 ' of the most severe consequence will be documented.

Plant/P&ID : Equipment : Volume :			Process: Function:			Document: Page : Date :		
No	Guide word / Process Parameter	Detection / Safeguards	Possible causes	Consequences	FC	Recommended Actions	FC	Resp. / Ref.
				Physical effects: Risk related:				

Short cut calculations 2-Dynamic simulation 3-deterministic models 4- Event tree 5- Fault tree 6- Historic data base

2.4.3 Step 3: Safety/risk assessment - Risk potential matrix (HAZOP decision matrix)

Figure 6 shows the risk potential matrix (HAZOP decision matrix) used for order of magnitude ranking of events. The rows of the matrix consider frequency class, while the columns show the consequence class. Each cell in the matrix represents a risk category. For the decision process, the matrix is divided into four risk category levels.

Risk level I --- red area --- scenario in this level is intolerable and immediate action (pant or process modification) is needed to reduce that risk category or more detailed quantified analysis has to be carried out in order to find arguments for wrong preliminary decisions.

Risk level II --- grey area --- scenario in this level is tolerable but not acceptable for long period of time so action at next schedule maintenance is needed to reduce that risk category.

Risk level III --- yellow area --- scenario in this level is acceptable and any action to reduce that risk category is optional.

Risk level IV--- green area --- scenario in this level needs no action. Risk potential matrix (HAZOP decision matrix) may be also used for:

- Documentation of the status of the plant safety
- Selection and development of optimization proposals
- Importance of improvement
- Documentation of improvement achieved

The application of risk potential matrix (HAZOP decision matrix) in the Extended HAZOP is shown in figure 5. Arrows show the transformation of entries from the Extended HAZOP worksheet to the HAZOP decision matrix. The identity number (ID) of each scenario of the Extended HAZOP worksheet is placed in HAZOP decision matrix. Recommended actions for this scenario will be placed from Extended HAZOP sheet to the bottom of HAZOP decision matrix. First HAZOP decision matrix will shows the existing status and second HAZOP decision matrix shows the improved plant status after recommended actions.

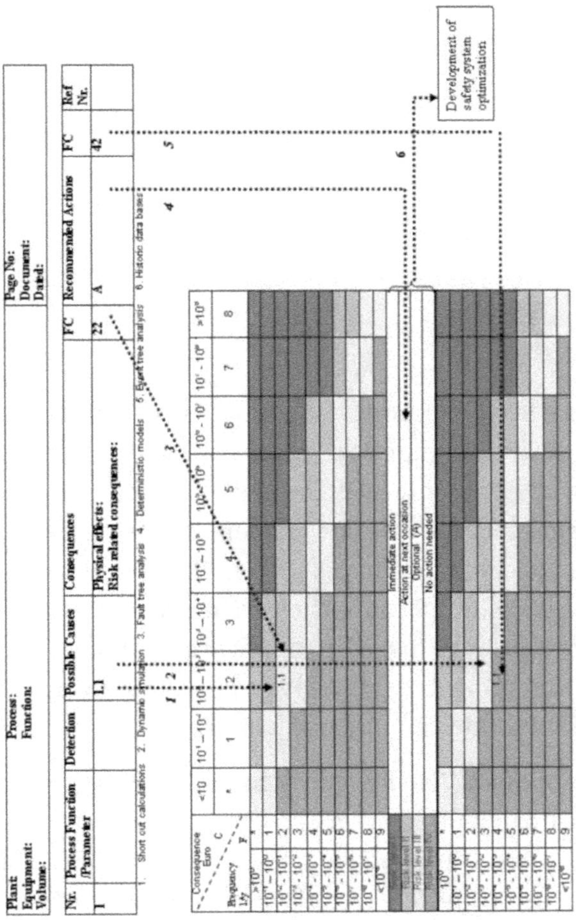

Fig. 6. Application of HAZOP decision matrix in Extended HAZOP

Similarly all results from Extended HAZOP worksheets are transferred to the HAZOP decision matrix. Keeping in view the risk target and depending on the scenario or recommended actions during the Extended HAZOP discussion, analysis team may reach a safety related modification proposal. Next, if safety/ risk optimization is in focus then weak points/scenarios with similar risk are clustered after analyzing HAZOP decision matrix and safety related optimization proposals are developed.

2.4.4 Step 4: Safety/risk system optimization (Development and analysis of optimization proposals)

In this step, safety related optimization proposals are generated and evaluated using dynamic simulation, Event tree analysis and/or Fault tree analysis. The optimization proposals can be developed at two levels:

- Simple optimization proposals e.g. addition of pressure alarm or change of location of sensor within the Extended HAZOP discussion
- Optimization proposals related to severe scenarios by evaluating risk potential matrix (HAZOP decision matrix)

The relevant information such as frequency and damage data will be transferred to economic module for safety related cost calculations and multiobjective decision making (if more than one alternatives developed).

2.5 Economic module

In all stages of design process, economic evaluation is crucial for the evaluation of process alternatives. Various objective functions are available in the literature of chemical engineering economics for economic evaluation of chemical processes. Some quite elegant objective functions, which incorporate the concept of the "time value of money", are net present value (NPV) and discounted cash flow. Business managers, accountants and economists prefer these methods because they are more accurate measures of profitability over an extended time period. However, application of these methods needs certain assumptions[12]. Total annualized cost (TAC) can be used as economic indicators/objective function for the evaluation of design alternatives and economic optimization.

Economic module developed in Visual Basic consists of two distinct sections. First section carries out standard cost calculations (i.e. Fixed capital investment (FC1) and operational cost (OC1)) and compute total annualized cost (TAC1) while second section carries out extended cost calculations i.e. process safety/risk related costs and computes the fixed capital investment related to safety system (FCISS), accident and incident damage related risk cost. Table 3.5 illustrates the difference of cost elements considered in standard practice of cost calculations of chemical process design and in this economic module. Figure 7 shows the simplified block diagram of economic module.

2.5.1 Standard cost calculations

Standard cost calculations involves fixed capital investment (FCI1) and operational cost (OC1). Fixed capital investment (FCI1) includes the cost of design and other engineering and construction supervision, all items of equipment and their installation, all piping, instrumentation and control systems, buildings and structures, and auxiliary facilities such as utilities, land and civil engineering work. Several capital cost estimate methods ranging from order of magnitude estimate (ratio estimate) to detailed estimate (contractors estimate) are used for the estimation of installed cost of the process units in the chemical plant.

Table 5. Elements of economic module and difference from standard cost calculations

Standard cost calculations	Economic module used in this work
- FCI1 = Fixed capital investment using either cost equations that have been derived by Ulrich or correlations developed by Guthrie depending on users choice **- OC1= Operating cost (including both direct (e.g. raw material, utilities etc.) and indirect costs (e.g. taxes, overhead cost etc)** **- TAC1 = total annualized cost = d · (FCI1) + OC1** **normally d is taken 0.15-25 but can also be computed using depreciation calculation methods**	- FCI1= Fixed capital investment using either cost equations that have been derived by Ulrich or correlations developed by Guthrie depending on users choice - FCI2= FCI1 + fixed capital investment related to safety system (FCISS) - OC1= Operating cost (including both direct (e.g. raw material, utilities etc.) and indirect costs (e.g. taxes, overhead cost etc) - TAC1 = total annualized cost = d · (FCI1) + OC1 - TAC2 = total annualized cost = d · (FCI2) + OC1 normally *d* is taken 0.15-25 but can also be computed using depreciation calculation methods
	Extended cost calculations - RC1 = risk cost 1= Asset risk cost + health risk cost + environmental risk cost

	- RC2 = risk cost2 = RC1 + production loss risk cost - RC3 = risk cost3 = process interruption cost - TRC = total risk cost = RC2 + RC3 - ECC = Extended costs

The most commonly used method that provides estimates within 20 – 30% of actual cost and widely used at design stage involve the usage of cost charts/correlations (Guthrie's article (1969) and book (1974), chapter 5 of Ulrich's 'A guide to chemical engineering process design and economics' (1984), 'Plant design and economics for chemical engineers' by Peters and Timmerhause (1991)) for estimating the purchase cost of major type of process equipment [13-15].

These cost charts / correlations were assembled in the 1960's or earlier and are projected to the date of installation using cost indices or escalation factors such as the chemical engineering plant cost index (published biweekly by chemical engineering magazine).

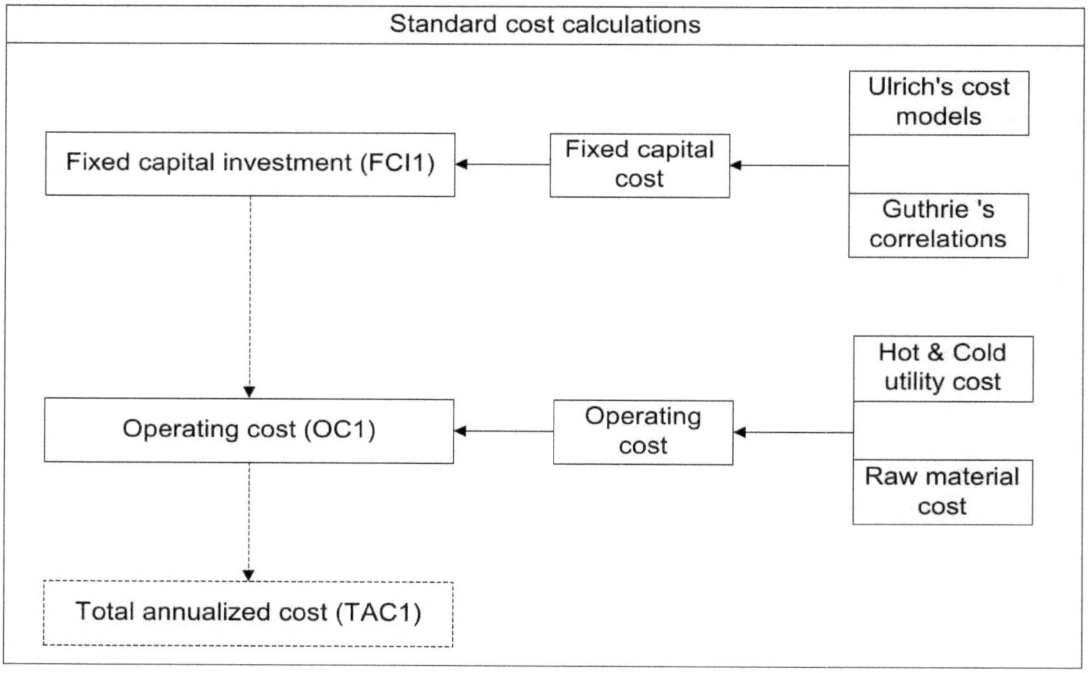

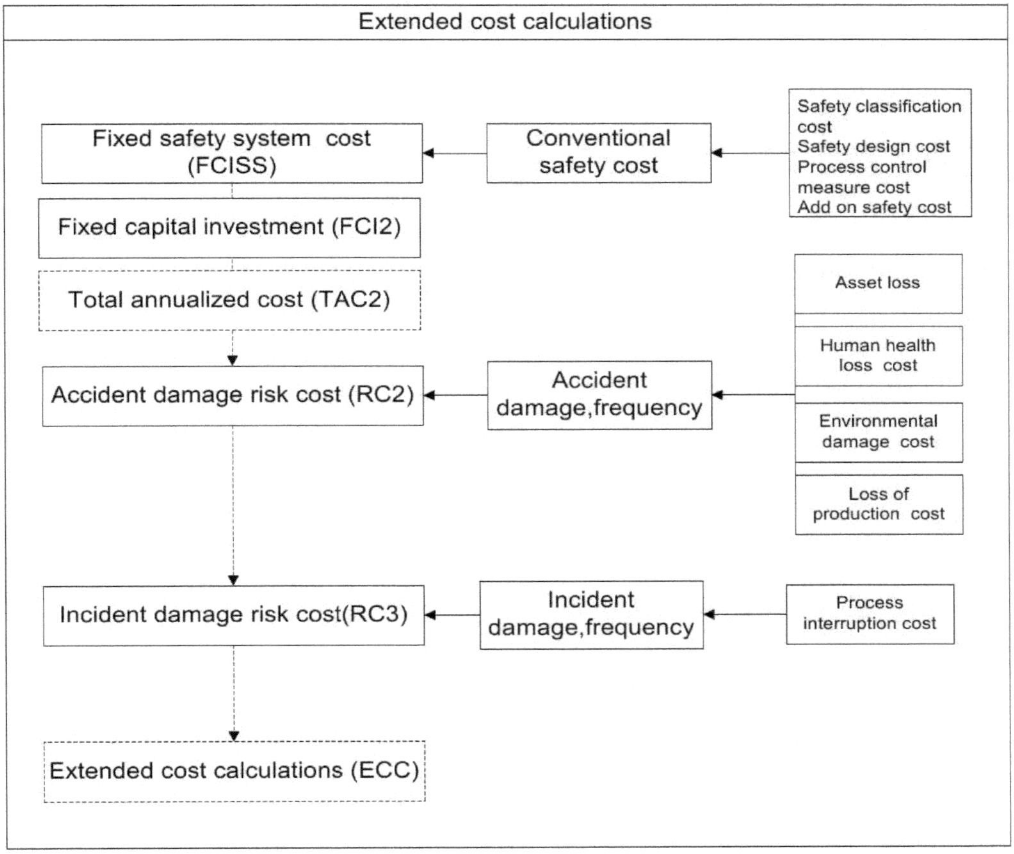

Fig. 7. Simplified block diagram of economic module

Note:

In Extended cost calculations, the costs such as insurance cost, market loss cost, loss of image and prestige cost should also be considered in addition. But in this module these costs are not included.

Marshall and Swift Index (also provided in chemical engineering magazine) and NelsonFabaar Index (from the oil and gas journal). For the comparison of process design alternatives, these study estimates for purchased cost of process units using cost charts or equations based on them are adequate. Given the purchase cost of a process unit, the installed cost is obtained by adding the cost of installation using factored-cost methods. For each piece of equipment Guthrie (1969,1974) provides factors to estimate the direct cost of labor, as well as, indirect costs involved in the installation procedure. The cost elements that are included in the estimation of fixed capital investment are shown in figure 8.

Direct costs:

- Equipment f.o.b* cost
- Equipment erection cost
- Piping
- Instrumentation
- Electrical
- Buildings, process
- Utilities
- Storages
- Site development
- Ancillary buildings

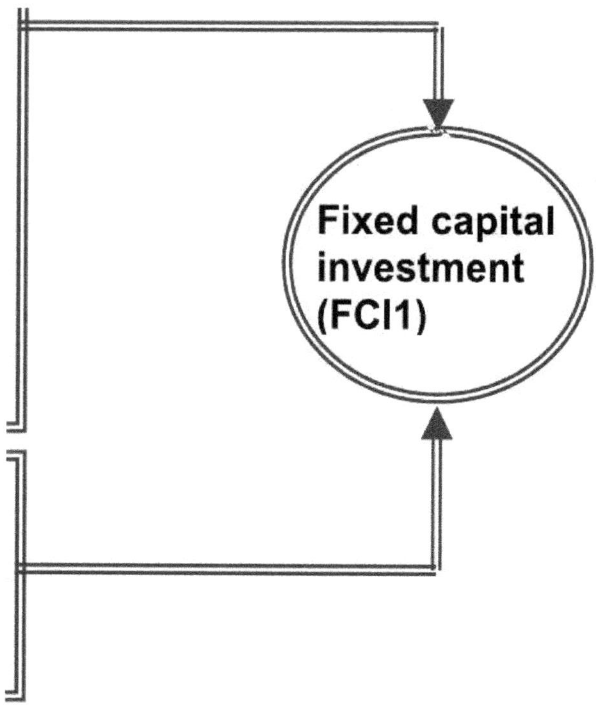

Fig. 8. Typical cost elements for fixed capital investment

Indirect costs:

- Design and Engineering
- Contractor's fee
- Contingency
  f.o.b ---- freight on board cost i.e. equipment purchase cost plus transport cost

The operating cost (OC1) of a chemical plant is divided into two groups:

- Fixed operating cost
- Variable operating cost

The elements in fixed operating cost includes maintenance cost, operating labor cost, laboratory cost, supervision cost, plant overheads, capital charges, taxes, insurance, licence fees and royalty payments while the variable operating cost consists of raw material costs, miscellaneous operating material costs, utilities (services) and shipping and packaging. However this division of operating cost is somewhat arbitrary and depends on the accounting practice of a particular organization. The typical cost elements included in operating cost "OC1" are shown in figure 9.

However, from the existing process optimization point of view, energy cost and raw material costs are more important and often considered.

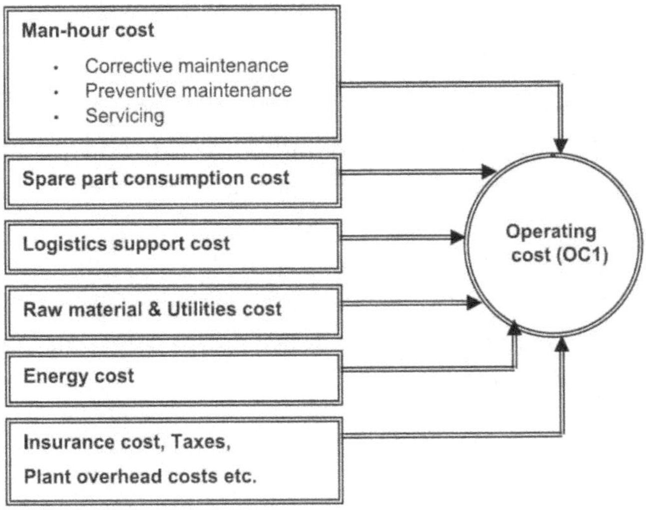

Fig. 9. Typical cost elements for operating cost (OC1)

Economic module developed in this thesis using Visual basic computes fixed capital expenditure using either cost equations that have been derived by Ulrich or correlations developed by Guthrie depending on users choice, The significant operating cost for process optimization i.e. energy consumption cost (heating and cooling utilities cost) can also be calculated by using this module. Once the FCI1 and OC1 are calculated, then total annualized cost is obtained using the following equation:

$$\text{Total annualized cost (TAC1)} = \mathrm{d} \cdot (\mathrm{FCI1}) + \mathrm{OC1} \tag{1}$$

Here d is depreciation or capital recovery factor and normally taken between 0.15-0.25 but can also be computed using depreciation calculation methods e.g. double declining balance method.

2.5.2 Extended cost calculations

The second section of the economic module (see Figure 7) carries out Extended cost calculations, which considers the fixed capital investment related to safety system, and risk cost due to accident and incident damage.

i. Fixed capital investment related to Safety system:

The fixed investment related to safety system is calculated by the following equation:

$$\mathrm{FCISS} = C_{SD} + \sum_{i=1}^{n} N_{SE,i} \cdot C_{SE,i} \tag{2}$$

Here, the first term C_{SD} is cost for safety design (i.e. cost related to safety classification, safety requirements and design specification, detailed design and engineering, factory acceptance test or pre-start up acceptance test and start up and correction). Table 3.6 gives the typical cost elements included in C_{SD} calculations. The second term $\sum_{i=1}^{n} N_{SE,i} \cdot C_{SE,i}$ is the sum of the purchase cost of safety equipment. Here CSE,i is the purchase cost of equipment " i " and NSE, i is the number (count) of that equipment. The costs for these devices are based on the recent detailed survey of available costs from various suppliers conducted by Khan and Annyotte (2004), however in the module the user has the possibility to enter the present market costs.

Table 6. Typical cost elements included in C_{SD} of safety system cost calculations
ii. Fixed capital investment (FCI2)

Safety classification cost e.g SIL determination cost	C_{SIL}
Safety requirements and design specifications (SRS) cost	C_{SRS}
Detailed design and engineering cost	C_{DE}
!Miscellaneous Cost:	C_{ME}
Initial training cost	C_{TC}
Factory acceptance test (FAT)/Installation/Pre-startup acceptance test (PSAT) cost	C_{FAT}
Startup and correction cost	C_{SCC}
$C_{SD} = C_{SIL} + C_{SRS} + C_{DE} + C_{ME} + C_{TC} + C_{FAT} + C_{SCC}$	

!power, wiring, junction boxes, operators interface cost

Then, extended fixed capital investment is calculated by adding FCISS to FCI1.

$$\text{FCI2} = \text{FCI1} + \text{FCISS} \tag{3}$$

iii. Total annualized cost (TAC2)

So, the extended total annualized cost will be calculated using extended fixed capital investment.

$$\text{TAC2} = \text{d} \cdot (\text{FCI2}) + \text{OC1} \tag{4}$$

Maintenance and repair cost of safety system should also be included in this calculation. But in this economic module these cost elements are not considered.
iv. Risk cost 1 (RC1)

Risk cost (RC1), which is the sum of property risk cost due to asset loss (PRC), health risk cost due to human health loss (HRC) and environmental risk cost due to environmental damage (ERC). The relations for calculation of these costs used in the module are:

Property risk cost due to asset lost (PRC) is the cost incurred due to lost of physical assets such as damage to property, loss of equipment due to accident/scenario and calculated by the equation below:

$$PRC = \sum_{i=1}^{n} \dot{F}_{A,i} \cdot A_{D,i} \cdot C_{A,i} \cdot t_{op} + \sum_{j=1}^{n} \dot{F}_{I,j} \cdot C_{D,j} \cdot t_{op} \tag{5}$$

$\dot{F}_{A,i}$ is frequency of occurring the hazardous accident, $A_{D,i}$ is damage area due to that accident, $C_{A,i}$ is the asset cost per unit area, t_{op} is total operation time, $\dot{F}_{I,j}$ is incident occurring frequency and $C_{D,j}$ is incident damage cost. Health risk cost due to human health lost (HRC) is the cost of fatality and/or injury due to the accident scenario under study.

$$HRC = \sum_{i=1}^{n} \dot{F}_{A,i} \cdot N_{\text{Peop, eff}} \cdot C_{H,\text{ life}} \cdot t_{op} \tag{6}$$

Here, $N_{\text{peop, eff}}$ is the number of person affected due to accident and is equal to $N_{\text{peop,eff}} = pop \cdot \wp$. Where POP is the population around the area of accident and $\wp$ is the population distribution factor ($\wp$ is 1 if population is uniform distributed (maximum value) and $\wp$ is 0.2 if population is localized and away from the area of accident (minimum value)) and $C_{H,\text{life}}$ is dollar value of human life or health. Though attempts to put value on human life have caused criticism and it changes from place to place. But a value for this can be obtained by dividing the annual gross national product by the annual number of births or by estimating how much money the person would have earned if not killed by the accident (Tweeddale 2003). A value for cost of loss of lives, marginal cost to avert the fatality, for the highest category of involuntariness risk 14 × 106$ is used in this work (Passman, H.J. et al. 2003).

Environmental risk cost due to environmental damage is the cost incurred due to environmental damage.

$$ERC = \sum_{i=1}^{n} \dot{F}_{ED} \cdot A_{ED,i} \cdot C_{ED,i} \cdot t_{op} \tag{7}$$

Where $A_{ED,i}$ is the environmental damage area due to scenario "i", $\dot{F}_{ED}$ is the frequency of release of material to environment and $C_{ED,i}$ is the environmental damage cost per unit area. so the sum of these three risk costs gives:

$$RC1 = PRC + HRC + ERC \tag{8}$$

v. Risk cost 2 (RC2)

Risk cost 2 (RC2), which is the sum of risk cost1 (RC1) and production loss risk cost (PLRC), accounts for accident damage risk cost. Here, production loss risk cost due to asset damage (PLRC) accounts for the cost due to the production loss because of accident and given by:

$$PLRC = \sum_{i=1}^{n} \dot{F}_{A,i} \cdot t_d \cdot \dot{C}_p \cdot t_{op} \tag{9}$$

Where PLRC is the production loss risk cost, td is the time lost due to accident and $\dot{C}_p$ is the production loss value in \$/h. Thus Risk cost 2 (RC2) is

$$RC2 = RC1 + PLRC$$

vi. Risk cost 3 (RC3)

Risk cost 3 (RC3), which is sum of process interruption cost due the spurious trip of the safety system and process interruption cost because of safe shut down to avoid from accident, accounts for incident damage risk cost and calculated as follow:

$$RC3 = \left(\sum_{i=1}^{n} \dot{F}_S^{\mathrm{trip}} \cdot t_{\mathrm{trip}} + \dot{F}_R^{\mathrm{trip}} \cdot t_{dR} \right) \cdot \dot{C}_p \cdot t_{op} \tag{10}$$

Here, $\dot{F}_S^{\mathrm{trip}}$ is spurious trip frequency, $\dot{F}_R^{\mathrm{trip}}$ is safe shut down frequency when trip system demand arises, ttrip is down time due to spurious trip and t_{dR} is down time to safe shut down when trip system demand arises.

vii. Total Risk cost (TRC)

Total Risk cost (TRC) is the sum of all risk costs:

$$TRC = RC2 + RC3 \tag{11}$$

Total risk cost can be annualized by dividing it with total operation time (t_{op}):

$$TAC_{\mathrm{risk}} = TRC/t_{\mathrm{op}}$$

viii. Extended Cost (ECC)

Extended cost calculations (ECC) is Life cycle related cost and calculated as follow :

$$ECC = FCI2 + PVC \tag{12}$$

Here, PVC is present value of the annual costs (OC1, TAC $_{\mathrm{risk}}$) and calculated as follow:

$$PVC = \left(OC1 + TAC_{\text{risk}} - \text{Insurance cos } t\right) \cdot \frac{1-(1+R)^{-t_{ly}}}{R}$$

R is the present interest rate and t_{ly} is the number of years (predicted life of system). Besides, the cost elements mentioned above in Extended cost calculation section, the other elements such as warranty/insurance cost, lost of image and prestige cost, market lost cost should also be considered but quantification of these elements is still almost impossible.

2.6 Environment module

Environment module consists of four steps and introduced an environmental performance index (EPI1) for evaluation of environmental performance and environmental pollution index (EPI2) as environmental objective to be integrated along with economics. The environmental performance index (EPI1) is calculated by combining total PEI based on WAR algorithm[16,17], resource depletion, energy conservation and fugitive emission rates while environmental pollution index (EPI2) is calculated by combining total PEI based on WAR algorithm and fugitive emissions because in this case other factors like resource depletion and energy consumption will be integrated in economic module or objective function. The Analytic hierarchy process (AHP) is used as multicriteria decision analysis tool for combining these different impacts and determination of weighting factors of individual impact categories in total PEI and later on in environmental performance index (EPI1) and environmental pollution index (EPI2) calculations. The module is developed using Microsoft Visual Basic 6.0 and WAR GUI (WAR graphical user interface) is integrated in the user plate form. The steps are:

Step I : Problem definition and data gathering

Step II : Individual impact categories calculation

Step III: Determination of weighting factors

Step IV: Environmental performance index calculation

Figure 10 shows the simplified block diagram of environment module and tasks to be performed.

2.6.1 Step I: Problem definition and data gathering

The primary task in step 1 is problem framing and scope definition. Information such as material and energy balance information, process conditions, process technology and nature of used materials/chemicals should be retrieved from process module. Process flow diagram is to be re-examined for identification of additional waste and emission streams. Collect additional data and information for environment evaluation to fill gaps. As sources of emissions such as fugitive emission sources, venting of equipment, periodic equipment cleaning, incomplete separations etc. are often missing in process so process is analyzed to identify these sources.

2.6.2 Step II: Individual impact categories calculation (Potential environmental impact calculations based on WAR algorithm)

The software WAR GUI (waste reduction algorithm graphical user interface) from the US Environmental Protection Agency is used to calculate individual potential environmental impacts. The generalized formula based on WAR algorithm for calculating individual PEI is given in equation 13.

$$PEI_L = \left(\dot{M}_b \cdot \sum_{k}^{\text{Comps}} x_{kb} \cdot \psi_{kL} + \dot{Q}_r \cdot \psi_L^E \right) / \dot{M}_p \text{ [Impact /kg product]} \qquad (13)$$

Where PEI_L is the potential environmental impact of category L, $\dot{M}_b$ is mass flow rate of base (effluent) stream, x_{kb} is the mass fraction of component k in the base stream, ψ_{kL} is the normalized impact score of chemical k for category L, $\dot{Q}_r$ is energy rate supplied for separation and ψ_L^E is the normalized impact score of category L due to energy. The sensitivity analysis results of individual potential environmental impact with respect to optimization variables should also be performed.

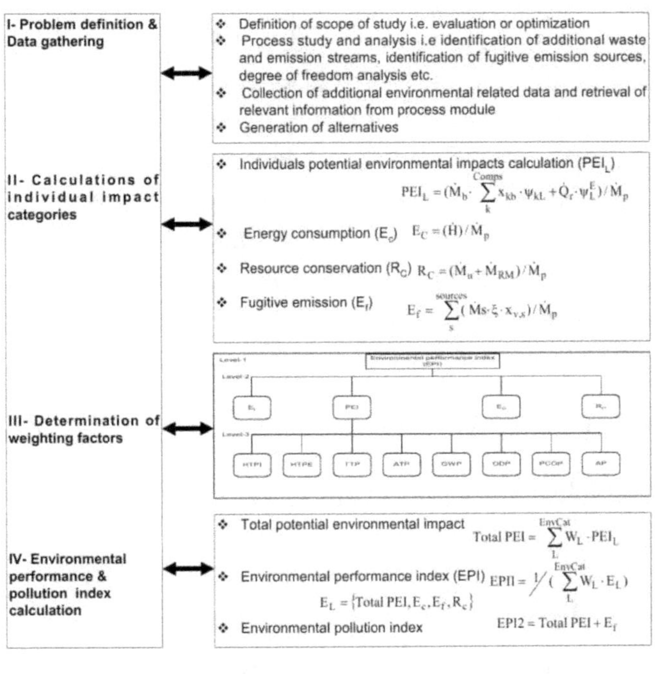

Fig. 10. Simplified block diagram of environment module

Energy consumption factor (E_C)

Energy consumption factor refers the total amount of energy consumed in the process per unit of product and is calculated as follow:

$$E_C = (\dot{H})/\dot{M}_p \text{ [kJ/kg product]} \tag{14}$$

Here $\dot{H} = \dot{M}_{\text{steam}} \cdot \hat{h}_{\text{steam}} + E_E$ where $\dot{M}_{\text{steam}}$ is the mass flow rate of steam [kg/h], $\hat{h}_{\text{steam}}$ is the enthalpy of steam per kg[KJ/kg], E_E is electrical energy consumed per unit time [KJ/h] and $\dot{M}_p$ is product rate [kg/h]. The sensitivity analysis of this factor with respect to optimization variables should also be performed.

Resource conservation factor (R_C)

The resource consumption refers all needed raw materials and utilities used and given by:

$$R_C = (\dot{M}_u + \dot{M}_{RM})/\dot{M}_p \text{ [kg/kg product]} \tag{15}$$

Where R_C is the resource conservation factor, $\dot{M}_U$ is utilities consumption rate, $\dot{M}_{RM}$ is raw material consumption rate.

Fugitive emission factor (E_f)

Fugitive emissions are unplanned or unmanaged, continuous or intermittent releases from unsealed sources such as storage tank vents, valves, pump seals, flanges, compressors, sampling connections, open ended lines etc and any other non point air emissions. These sources are large in number and difficult to identify. These emission rates depends on factors such as the age and quality of components, specific inspection and maintenance procedures, equipment design and standards of installation, specific process temperatures and pressures, number and type of sources and operational management commitment[18]. However, four basic approaches for estimating emissions from equipment leaks in a specific processing unit, in order of increasing refinement, in use are:

- Average emission factor approach
- Screening ranges approach
- EPA correlation approach
- Unit-specific correlation approach

All these approaches require some data collection, data analysis and/or statistical evaluation. On the other hand, using fundamental design / engineering calculations for accurate fugitive emission estimations for each source present in the process industry are difficult due to:

- large number and type of fugitive emission sources
- dependence of emission rates on other factors along with design and operating conditions e.g. installation standards, inspection and maintenance procedure etc.

As focus in this work is to integrate fuggitve emissions into environmental performance evaluation and optimization objectives so average emission factor approach giving a bit over estimates are used. Average emission factors for estimating fugitive emissions from fugitive sources found in synthetic organic chemical manufacturing industries operations (SOCMI) obtained from the US Environmental Protection Agency L & E Databases are used. The relation used in this work for calculation of fugitive emissions is:

$$E_f = \sum_{s}^{\text{sources}} (\dot{M}s \cdot \xi \cdot x_{v,s})/\dot{M}_p [\text{ kg/kg product }] \tag{16}$$

Here E_f is fugitive emission factor per unit of product, $\dot{M}_s$ mass flow rate through the source ' s ', ξ is average emission factor and $x_{v,s}$ is mass fraction of volatile component through source 's' and $\dot{M}_p$ is product rate. It is assumed $x_{v,s}$ for the process fluids through fugitive sources such as pump seals, valves, flanges and connection is equal to 1 , i.e. fluids are composed entirely of volatile compounds.

2.6.3 Step III: Determination of weighting factors (Application of multicriteria decision analysis technique)

The integration of these individual impact categories into one index is a hierarchical multicriteria decision analysis problem. The analytic hierarchy process (AHP) is used for this purpose[19] and a computer programme for it is developed in VB 6.0. In this stage, first a hierarchical structure of the problem, which is structured hierarchically similar to a flow chart, is constructed. The overall objective is placed at the top while the criteria and subcriteria are placed below. For example, as shown in figure 11, the overall objective Environmental performance index (EPI1) is placed at the top (level 1), then below (level 2) are criterias Total PEI, Ef , Ec and RC and after this (level 3) sub-criterias as HTPI, HTPE, TTP, ATP, GWP, ODP, PCOP and AP. After this using the numerical scale given in table 2.6, two pairwise comparison matrices (see Table 7 and 8) are constructed for determination of weights for aggregation of individual impact categories of WAR to total PEI and for determination of weights of total PEI, Ef, Ec and RC to Environmental performance index (EPI1).

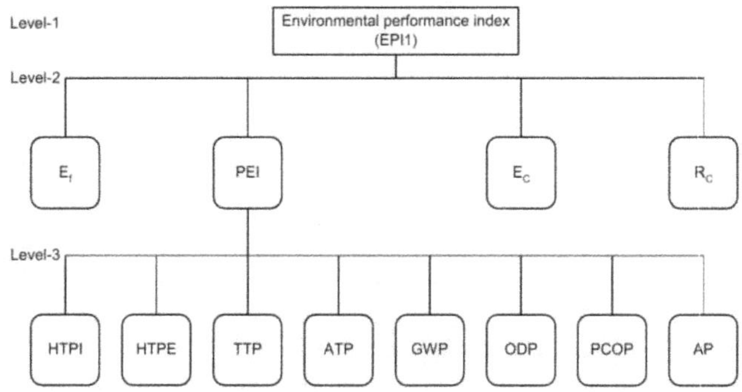

Fig. 11. Hierarchical structuring of multicriteria decision analysis problem for integrating individual environmental impacts

The right hand upper diagonal information in both matrices is to be provided by the decision maker giving the relative importance of the two criteria using the numerical scale of table 2.6 while the left hand lower diagonal is the reciprocal of the right hand upper diagonal. Once these pair wise comparison matrices are constructed, then developed computer programme using the AHP method, determines the weighting of individual impact categories. The level of inconsistency of decision makers input is checked by consistency ratio before giving the output. Consistency ratio less than 0.1 is good and for ratios greater than 0.1, the input to pair wise matrix should be re-evaluated.

Table 7. Pairwise comparison matrix for individual impact categories at level 3

Pairwise comparison matrix								
	HTPI	HTPE	TTP	ATP	GWP	ODP	PCOP	AP
HTPI	1	A12	A13	A14	A15	A16	A17	A18
HTPE		1	A23	A24	A25	A26	A27	A28
TTP			1	A34	A35	A36	A37	A38
ATP				1	A45	A46	A47	A48
GWP					1	A56	A57	A58
ODP						1	A67	A68
PCOP							1	A78
AP								1
W_L	W1	W2	W3	W4	W5	W6	W7	W8

Table 8. Pairwise comparison matrix for individual impact categories at level 2

Pairwise comparison matrix				
	PEI	R_C	E_C	E_f
PEI	1	A12	A13	A14
R_C		1	A23	A24
E_C			1	A34
E_f				1
W_L	W1	W2	W3	W4

2.6.4 Step IV: Environmental performance & pollution index calculation

In the final step, first Total PEI is determined by multiplying each impact category values with its relevant weighting factor WL as given below:

$$\text{Total PEI} = \sum_{L}^{\text{EnvCat}} W_L \cdot PEI_L \tag{17}$$

After calculating Total PEI, Environmental performance index (EPI1) is determined for each alternative by multiplying the values of Total PEI, Ef, EC and RC with its relevant weighting factor WL (table 8) as given below:

$$EPI1 = 1/\left(\sum_{L}^{\text{EnvCat}} W_L \cdot E_L\right) \tag{18}$$

Where $E_L = \{\ \text{Total PEI}, E_c, E_f, R_c\}$

and environmental pollution index (EPI2) is calculated as follow

$$EPI2 = \text{Total } PEI + E_f \tag{19}$$

The higher value of environmental performance index (EPI1) shows that the process is environmentally better and vice versa. While the higher value of environmental pollution index (EPI2) shows that the environmental performance of process is worse.

2.7 Data manager

The relevant information generated from process module, safety module and environment module for each alternative is transferred to data manager. This information is used to formulate process diagnostic tables and multiobjective decision-making problem formulation. These tables consist of mass input/output table, energy input/output table, capital and utility annual expense summary, environmental impact summary and frequency of occurance of an event and their consequence categories and safety cost.

2.8 Stage III: Multiobjective decision analysis/ optimization

The purpose of this layer/stage is to set up multiobjective decision making/optimization among these conflicting objectives. The aim is to find out the trade-off surface for each alternative and /or complete ranking of alternatives. The calculation loop used for it is shown in figure 12.

In each independent performance module i.e. economic, environment and safety module, relevant information is generated and transferred automatically or manually to data manager for each alternative generated or under study. Before transferring the values of performance objective functions, each objective function is optimized within their independent module such as:

- Process/Economic optimization of each alternative is carried out using SQP optimization algorithm build within Aspen Plus ™.
- The lower and upper limits for Environmental objective functions are calculated using environmental module from the material and energy balance information from process model.
- Safety/risk aspects are optimized in the safety module and information such as hazard occurance frequency, safety cost data (fixed safety system cost, accident and/or incident damage risk cost) for each alternative is transferred to the data manger.

Depending on the case under study or objectives of the study, graphical tool box of MatLab and/or multiobjective optimization technique (goal programming) or multiattribute decision analysis technique (PROMETHEE and/or AHP) is used for multiobjective decision analysis. The Data Manager is linked with MatLab 7.0 via Excel link Toolbox in the

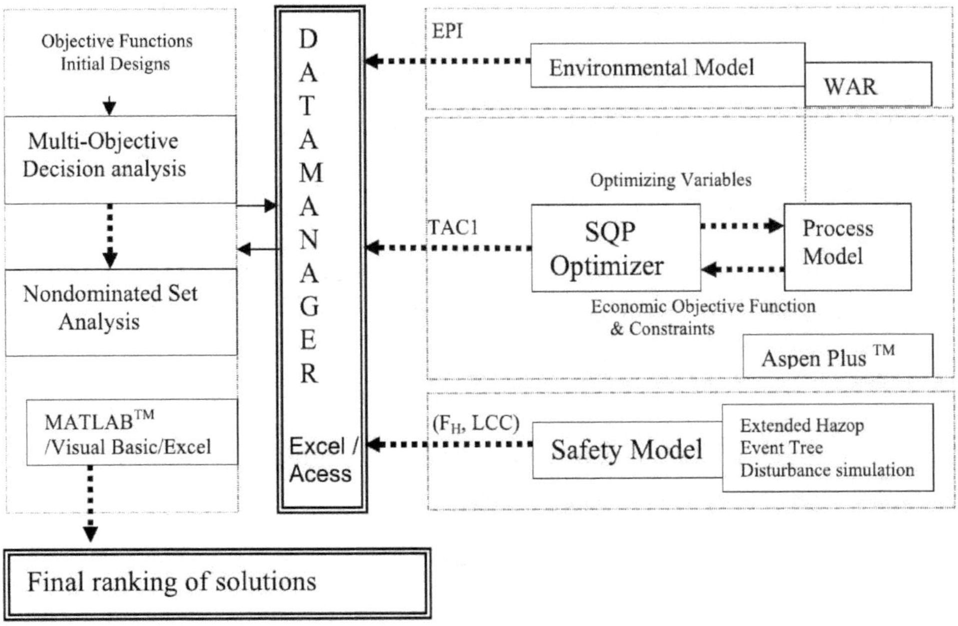

Fig. 12. Calculation loop for multiobjective optimization

integrated interface. Aspen plusTM is linked in the integrated interface via Visual basic 6.0 and Microsoft Excel. AHP technique is also programmed for the cases under study in Visual basic 6.0. The computer realization of these links in the integrated interface is explained in chapter four.

2.9 Stage IV: Design evaluation

This layer's objective is to identify the optimal option and/or establish a comprehensive rating of the alternatives under consideration, depending on the outcomes of the third layer of the created approach. The Pareto method (non-dominated analysis) or PROMETHEE is employed for this purpose.

3. References

[1] Ramzan, N., Naveed, S., Feroze, N. and Witt, W.; "Multicriteria decision analysis for safety and economic achievement using PROMETHEE: A case study" Process Safety Progress (A Journal of Americal Institute of Chemical Engineering), 28(1), 68-83 (2009).

[2] Ramzan, N., Degenkolbe, S. and Witt, W.; "Evaluating and improving environmental performance of HC's recovery system: A case study of distillation unit", Chemical

Engineering Journal (Journal published by Elsevier), 40(1-3), 201-213(2008). ISSN: 1385-8947.

[3] Ramzan, N., Compart, F. and Witt, W.; "Methodology for the generation and evaluation of safety system alternatives based on extended Hazop" Process Safety Progress (A Journal of Americal Institute of Chemical Engineering), 26(1),35-42 (2007).

[4] Ramzan, N., Compart, F. and Witt, W.; "Application of extended Hazop and event tree analysis for investigating operational failures and safety optimization of distillation column unit" Process Safety Progress (A Journal of Americal Institute of Chemical Engineering), 26(3),248-257 (2007).

[5] Ramzan, N., Witt, W.; "Multiobjective optimization in distillation unit: A case study", The Canadian Journal of Chemical Engineering, 84(5), 604-613(2006).

[6] Seader et al. (1999), Process design principles-Synthesis, Analysis, and Evaluation, John Wiley & Sons Inc. New York, 338-370.

[7] Nawaz, Z. Mahmood, Z. (2006) Importance Modeling, Simulation and Optimization in Chemical process Design, The Pakistan Engineer (Journal of Institute of Engineers Pakistan), 38-39.

[8] Cameron, I., Raman, R. (2005), Process Systems Risk Management, Elsevier Academic Press, NY, ISBN 0-12-156932-2

[9] Crowl, D.A., Louvar, J.F. (1999), Chemical Process Safety: Fundamentals with applications," Prentice Hall, New York.

[10] Kletz, T.A. (1997), Hazop-past and future, Reliability engineering and system safety, 55, 263-266.

[11] Lees, F.P. (1996), Loss prevention in CPI, Butterworth's, London, UK.

[12] Peters, M. S. and K.D. Timmerhaus (1991), Plant Design and Economics for Chemical Engineers, Ed.2nd, McGraw-Hill, New York, 90-145.

[13] Douglas, J. (1998), Conceptual design of chemical processes, McGraw Hill Inc.

[14] Guthrie, K.M. (1969), Data and techniques for preliminary capital cost estimating, Chem. Eng., 114-1421.

[15] Guthrie, K.M. (1974), Process plant estimating, evaluation and control, Craftsman, Solano Beach, CA.

[16] Cabezas, H. Bare, C. & Mallick, K. (1999), Pollution prevention with chemical process simulators: the generalized waste reduction (WAR) algorithm-full version. Computers and Chemical Engineering, 23,623-634.

[17] Cabezas, H. Bare, C. & Mallick, K. (1997), Pollution prevention with chemical process simulators: the generalized waste reduction (WAR) algorithm, Computers and Chemical Engineering, 21s, s305-s310.

[18] Dimian, A. C. (2003), Integrated design and simulation of chemical processes, 1st Eds., Elsevier Netherlands, 1-30,113-134.

[19] Dev, P.K. (2004), Analytic hierarchy process helps evaluate project in Indian oil pipelines industry, International journal of operation and production management, 24(6), 588-604

Chapter-12

Computational Fluid Dynamics Modelling of Fluidized Bed with Embedded Tubes

1. Introduction: An Overview

Fluidized beds are extensively utilized in combustion and chemical sectors. Immersed tubes are typically employed to promote heat transfer or regulate temperature in fluidized beds. Tubes exposed to solid particle impact may experience significant erosion wear. A multitude of studies has focused on erosion in tubes submerged in fluidized beds and the many affecting elements (cf. Lyczkowski and Bouillard, 2002). According to Achim et al. (2002), the factors can be categorized into particle properties, mechanical design, and operating circumstances. Some previous experimental studies have focused on bubble and particle behaviors (Kobayashi et al., 2000, Ozawa et al., 2002), tube attrition, erosion or wastage (Bouillard and Lyczkowski, 1991; Lee and Wang, 1995; Fan et al., 1998; Wiman, 1994), heat transfer (Wong and Seville, 2006, Wiman and Almstedt, 1997) and gas flow regimes (Wang et. al, 2002).

Previous numerical studies were also performed using different CFD codes. Recently He et al. (2009, 2004), using the K-FIX code adapted to body fitted coordinates investigated the hydrodynamics of bubbling fluidized beds with one to four immersed tubes. The erosion rates predicted using the monolayer kinetic energy dissipation model were compared against the experimental values of Wiman (1994) for the two tube arrangement. The numerical values were three magnitudes lower than the experimental ones. Also employing a eulerian-eulerian model and the GEMINI numerical code, Gustavsson and Almstedt (2000, 1999) performed numerical computations and comparison against experimental results (Enwald et. al., 1999). The authors observed a satisfactory qualitative concordance between the experimental and numerical erosion results, and the contributions to erosion from various fluid dynamics events surrounding the tube were delineated.

This work reexamines the behavior of submerged tubes in a gas fluidized bed with various tube configurations using the Eulerian-Eulerian two-fluid model and the MFIX code. The objective of the numerical simulations is to compare and examine impacts not previously explored in the aforementioned references.

2. Two fluid and erosion model

The mathematical model is based on the assumption that the phases can be mathematically described as interpenetrating continua; the point variables are averaged over a region that is large compared with the particle spacing but much smaller than the flow domain (see Anderson, 1967). A short summary of the equations solved by the numerical code in this study are presented next. Refer to Benyahia et al. (2006) and Syamlal et al. (1993) for more detailment.

The continuity equations for the fluid and solid phase are given by :

$$\frac{\partial}{\partial t}(\varepsilon_f \rho_f) + \nabla \cdot (\varepsilon_f \rho_f \vec{v}_f) = 0 \quad (1)$$

$$\frac{\partial}{\partial t}(\varepsilon_s \rho_s) + \nabla \cdot (\varepsilon_s \rho_s \vec{v}_s) = 0 \quad (2)$$

In the previous equations $\varepsilon_f, \varepsilon_s, \rho_f, \rho_s, \vec{v}_f$ and $\vec{v}_s$ are the volumetric fraction, density and velocity field for the fluid and solids phases. The momentum equations for the fluid and solid phases are given by:

$$\frac{\partial}{\partial t}(\varepsilon_f \rho_f \vec{v}_f) + \nabla \cdot (\varepsilon_f \rho_f \vec{v}_f \vec{v}_f) = \nabla \cdot \bar{\bar{S}}_f + \varepsilon_f \rho_f \vec{g} - \vec{I}_{fs} \quad (3)$$

$$\frac{\partial}{\partial t}(\varepsilon_s \rho_s \vec{v}_s) + \nabla \cdot (\varepsilon_s \rho_s \vec{v}_s \vec{v}_s) = \nabla \cdot \bar{\bar{S}}_s + \varepsilon_s \rho_s \vec{g} + \vec{I}_{fs} \quad (4)$$

$\bar{\bar{S}}_f \bar{\bar{S}}_s$ are the stress tensors for the fluid and solid phase. It is assumed newtonian behavior for the fluid and solid phases, i.e.,

$$\bar{\bar{S}} = (-P + \lambda \nabla \cdot \vec{v})\bar{\bar{I}} + 2\mu S_{ij} \equiv -p\bar{\bar{I}} + \bar{\tau}\, S_{ij} = \frac{1}{2}[\nabla\vec{v} + (\nabla\vec{v})^T] - \frac{1}{3}\nabla \cdot \vec{v} \quad (5a, b)$$

In the above equation P, λ, μ are the pressure, bulk and dynamic viscosity, respectively. In addition, the solid phase behavior is divided between a plastic regime (also named as slow shearing frictional regime) and a viscous regime (also named as rapidly shearing regime). The constitutive relations for the plastic regime are related to the soil mechanics theory. Here they are representated as:

$$p_s^p = f_1(\varepsilon^*, \varepsilon_f).\, \mu_s^p = f_2(\varepsilon^*, \varepsilon_f, \phi) \qquad (6)$$

In the above equation ε^* is the packed bed void fraction and ϕ is the angle of internal friction.

A detailing of functions f_1 to f_4 and f_9 can be obtained in Benyahia (2008). On the other hand, the viscous regime behavior for the solid phase is ruled by two gas kinetic theory related parameters (e, Θ).

$$p_s^v = f_3(\varepsilon_s, \rho_s, d_p, \Theta, e)\ \mu_s^v = f_4(\varepsilon_s, \rho_s, d_p, \Theta^{1/2}, e) \qquad (7)$$

The solid stress model outlined by Eqs. (6) and (7) will be quoted here as the standard model. Additionally, a general formulation for the solids phase stress tensor that admits a transition between the two regimes is given by :

$$\bar{\bar{S}}_s = \begin{cases} \phi(\varepsilon_f)\bar{\bar{S}}_s^v + [1 - \phi(\varepsilon_f)]\bar{\bar{S}}_s^p & \text{if } \varepsilon_f < \varepsilon^* + \delta \\ \bar{\bar{S}}_s^v & \text{if } \varepsilon_f \geq \varepsilon^* + \delta \end{cases} \qquad (8)$$

According to Pannala et al.(2009), two diferent formulations for the weighting parameter " ϕ " can be employed :

$$\phi(\varepsilon) = \frac{1}{1 + v^{\frac{\varepsilon - \varepsilon^*}{2\delta\varepsilon^*}}}\ \phi(\varepsilon) = \frac{\text{Tanh}\left(\frac{\pi(\varepsilon - \varepsilon^*)}{\delta\varepsilon^*}\right) + 1}{2} \qquad (9a, b)$$

In the above equation the void fraction range δ and the shape factor v are smaller values less than unity. It must be emphasized that when δ goes to zero and ϕ equals to unity, the "switch" model as proposed by Syamlal et al. (1993) based on the Schaeffer (1987) can be recovered. The models based on eqs. (9a) and (9b) will be referred in the numerical simulations as BLEND S and BLEND T, respectivelly.

On the other hand, the Srivastava and Sundaresan (2003), also called "Princeton model", can be placed on the basis of Eq. (9) Also in equations (4) and (5) $\vec{I}_{fs}$ is the momentum interaction term between the solid and fluid phases, given by

$$\vec{I}_{fs} = -\varepsilon_s \nabla P_f - \beta(\vec{v}_s - \vec{v}_f) \qquad (10)$$

There is a number of correlations for the drag coefficient β (Eqs. 11 to 16). The first of the correlations for the drag coefficient is based on Wen and Yu (1966) work. The Gidaspow drag coefficient is a combination between the Wen Yu correlation and the correlation from Ergun (1952). The Gidaspow blended drag correlation allows controlling the transition from the Wen and Yu, and Ergun based correlations. In this correlation the χ blending function was originally proposed by Lathowers and Bellan (2000) and the value of parameter C controls the degree of transition. From Eq. (14), the correlation proposed by Syamlal and O'Brien (1993) carries the advantage of adjustable parameters C_1 and d_1 for different minimum fluidization conditions. The correlations given in Eq. (15) and Eq. (16) are based on Lattice-Boltzmann simulations. For detailments of these last drag correlations refer to the works by Benyahia et al. (2006) and Wang et al. (2010).

$$\beta_{\text{Wen-Yu}} = \frac{3}{4} C_D \frac{\rho_f \varepsilon_f \varepsilon_s \left|\vec{v}_f - \vec{v}_s\right|}{d_p} \varepsilon_f^{-2.65}$$

$$C_D = \begin{cases} \dfrac{24}{\text{Re}}\left(1 + 0.15\,\text{Re}^{0.687}\right) & \text{Re} < 1000 \\ 0.44 & \text{Re} \geq 1000 \end{cases} \qquad \text{Re} = \frac{\rho_f \varepsilon_f \left|\vec{v}_f - \vec{v}_s\right| d_p}{\mu_f} \tag{11}$$

$$\beta_{\text{Gidaspow}} = \begin{cases} \beta_{\text{Wen-Yu}} & \varepsilon_f > 0.8 \\ \beta_{\text{Ergun}} = \dfrac{150\,\varepsilon_s (1-\varepsilon_s)\mu_f}{\varepsilon_f d_p^2} + \dfrac{1.75\,\rho_f \varepsilon_s \left|\vec{v}_f - \vec{v}_s\right|}{d_p} & \varepsilon_f \leq 0.8 \end{cases} \tag{12}$$

$$\beta_{\text{Gidaspow-blended}} = \chi\, \beta_{\text{Wen-Yu}} + (1-\chi)\beta_{\text{Ergun}} \qquad \chi = \frac{\tan^{-1}\left(C\,(\varepsilon_f - 0.8)\right)}{\pi} + 0.5 \tag{13}$$

$$\beta_{\text{Syamlal-OBrien}} = \frac{3}{4}\frac{\rho_f \varepsilon_f \varepsilon_s}{V_r^2 d_p}\left(0.63 + 4.8\sqrt{\frac{V_r}{\text{Re}}}\right)^2 \left|\vec{v}_f - \vec{v}_s\right|$$

$$V_r = 0.5A - 0.03\,\text{Re} + 0.5 \times \sqrt{(0.06\,\text{Re})^2 + 0.12\,\text{Re}\,(2B - A) + A^2} \tag{14}$$

$$A = \varepsilon_f^{4.14} \qquad B = \begin{cases} C_1 \varepsilon_f^{1.28} & \varepsilon_f \leq 0.85 \\ \varepsilon_f^{d_1} & \varepsilon_f > 0.85 \end{cases}$$

$$\beta_{\text{Hill-Koch-Ladd}} = 18\mu_f (1-\varepsilon_s)^2 \varepsilon_s \frac{F}{d_p^2} \qquad F = f_9(F_0, F_1, F_2, F_3) \tag{15}$$

$$\beta_{\text{Beestra}} = 180\frac{\mu_f \varepsilon_s^2}{d_p^2} + 18\frac{\mu_f \varepsilon_f^3 \varepsilon_s \left(1 + 1.5\sqrt{\varepsilon_s}\right)}{d_p^2} + 0.31\frac{\mu_f \varepsilon_s\,\text{Re}}{\varepsilon_f d_p^2}\frac{\left[\varepsilon_f^{-1} + 3\varepsilon_f \varepsilon_s + 8.4\,\text{Re}^{-0.343}\right]}{\left[1 + 10^{3\varepsilon_s}\,\text{Re}^{-0.5-2\varepsilon_s}\right]} \tag{16}$$

For closing the model, a transport equation for the granular energy Θ provides a way of determine the pressure and viscosity for the solid phase during the viscous regime. Equation (17) is a transport equation for the granular energy Θ. Its solution provides a way of determine the pressure and viscosity for the solid phase during the viscous regime. The terms $\kappa_s \gamma$ and ϕ_{gs} are the granular energy conductivity, dissipation and exchange, respectively.

$$\frac{3}{2}\left[\frac{\partial}{\partial t}\varepsilon_s \rho_s \Theta + \nabla \cdot \rho_s \vec{v}_s \Theta\right] = \bar{\bar{S}}_s : \nabla \vec{v}_s - \nabla \cdot (\kappa_s \nabla \Theta) - \gamma + \phi_{gs} \qquad (17)$$

$$\kappa_s = f_5\left(\varepsilon_s, \rho_s, d_p, \Theta^{1/2}, e, \beta\right) \qquad (17)$$

$$\gamma = f_6\left(\varepsilon_s, \rho_s, d_p, \Theta^{3/2}, e\right) \qquad (18)$$

$$\phi_{gs} = f_7\left(\varepsilon_s, \rho_s, d_p, \Theta, \left|\vec{v}_f - \vec{v}_s\right|, \beta\right) \qquad (18)$$

In the algebraic approach, instead solving the full equation (17), the granular energy is obtained by equating the first term on the right hand side with the dissipation term.

The model where Eqs. (5) to (8) and (17) are solved is the kinetic theory model, termed here as KTGF. Conversely, in the constant solids viscosity model (CVM) the solids pressure is defined as in Eq. (6) and the solids viscosity in either plastic and viscous regimes is set equal to a constant.

For erosion calculations in this work we use the monolayer energy dissipation model (Lyczkowski and Bouillard, 2002). In that model the kinetic energy dissipation rate for the solids phase in the vicinity of stationary immersed surfaces is related to erosion rate in m/s by multiplication with an appropriate constant. This constant is function of surface hardness, elasticity of collision and diameter of particles hitting the surface. The kinetic energy dissipation rate Φ_s in W/m^3 for the solids phase is given by :

$$\Phi_s = \begin{bmatrix} \bar{\varepsilon}_s & \vec{v}_s : & \vec{v}_s + \beta \\ & & \frac{v_s}{2} \end{bmatrix} \qquad (19)$$

3. Numerical method

MFIX (Multiphase Flow with Interphase eXchanges) is an open source CFD code developed at the National Energy Technology Laboratory (NETL) for describing the hydrodynamics, heat transfer and chemical reactions in fluid-solids systems. It has been used for describing bubbling and circulating fluidized beds, spouted beds and gasifiers. MFIX calculations give transient data on the three-dimensional distribution of pressure, velocity, temperature, and species mass fractions.

The hydrodynamic model is solved using the finite volume approach with discretization on a staggered grid. A second order accurate discretization scheme was used and superbee scheme was adopted for discretization of the convective fluxes at cell faces for all equations in this work. With the governing equations discretized, a sequential iterative solver is used to calculate the field variables at each time step. The main numerical algorithm is an extension of SIMPLE. Modifications to this algorithm in MFIX include a partial elimination algorithm to reduce the strong coupling between the two phases due to the interphase transfer terms. Also, MFIX makes use of a solids volume fraction correction step instead of a solids pressure correction step which is thought to assist convergence in loosely packed regions. Finally, an adaptive time step is used to minimize computation time. See Syamlal (1998) for more details. The immersed obstacles were implemented using the cut-cell technique available in the code (Dietiker, 2009)

The numerical runs were based on experiments of Wiman (1994, 1997) in an air pressurized bed with horizontal tubes for four different tube-bank geometries. The T2 and I4 are portrayed in Fig. 1 whereas the S4 and S4D tubes geometry are given in Figure 3. Figure 2 details the domain for the numerical simulations and circumferential angle coordinate for erosion measurement. In Fig. 2a is detailed the mesh for the T2 arrangement This nonuniform stretched grid, follows Cebeci et al. (2005) approach, has fine spacing close to the surface of the tubes and coarse spacing away from the surface. For all the other arrangements the mesh was uniform as depicted in Fig. 2b. The mesh employed for the bed without tube, and the I4 arrangement was 60×260 cells, and for the S4 and S4D arrangement 60×340. The bed was operated at room temperature (24 C) at pressures between 0.1 and 1.6 MPa and at two different excess velocities: $U_{f1} - U_{mf} = 0.2$ m/s and $U_{f1} - U_{mf} = 0.6$ m/s. Here, U_{f1} is the superficial fluidization velocity based on the free bed crosssection. The voidage at minimum fluidization was 0.46 and the minimum fluidization velocities was 0.42,0.31 and 0.18 m/s, for 0.1,0.4 and 1.6 MPa pressures, correspondingly. The particle diameter and density were 700μ m and 2600 kg/m^3. The bubble parameters obtained from simulation were based on methodology described in Almstedt (1987), using numerical probes in the domain, centered at (0.15,0.55) and separated from 15 mm . The target tube for erosion measurements is the one centered at (0.18,0.55) for all the tube arrangements. More information about the experiments can be accessed from Wiman (1994,1997).

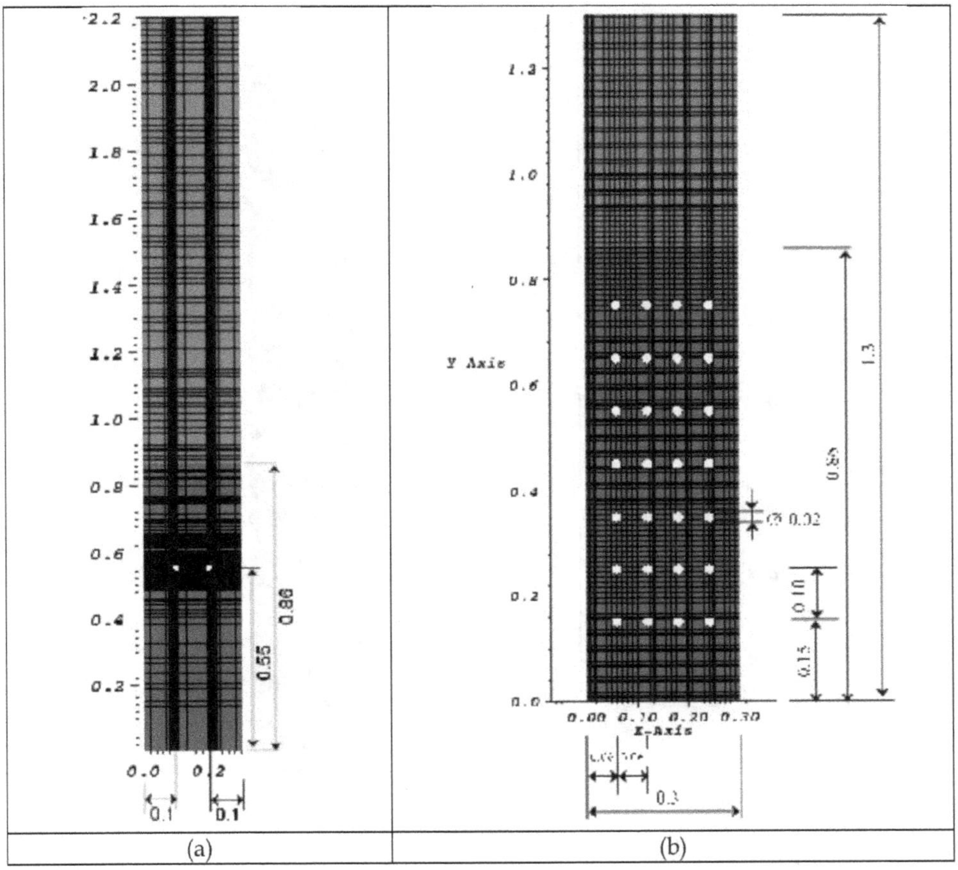

Fig. 1. Geometry and mesh for T2 and I4 tube arrangement.

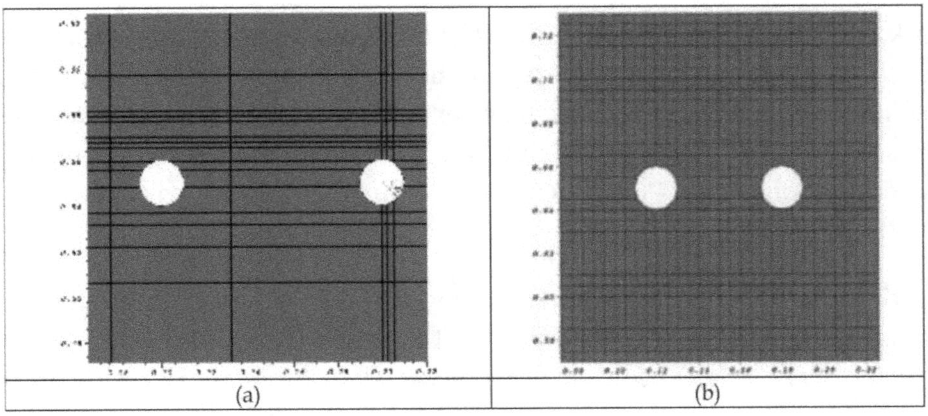

Fig. 2. Mesh detailment for (a) T2 arrangement (b) all the others arrangement

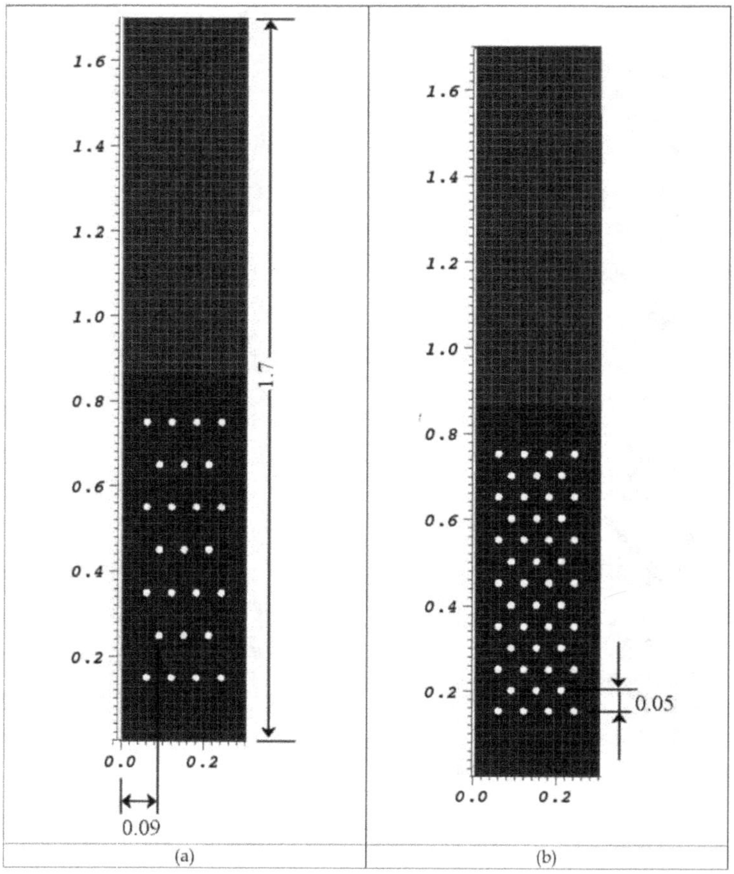

Fig. 3. Geometry and mesh for (a) S4 and (b) S4D tube arrangement

In this work, the parameters for controlling the numerical solution (e.g., under-relaxation, sweep direction, linear equation solvers, number of iterations, residual tolerances) were kept as their default code values. Moreover, for setting up the mathematical model, when not otherwise specified the code default values were used. The computer used in the numerical simulations was a PC with OpenSuse linux and Intel Quad Core processor. The simulation time was 20 s .

For generating the numerical results, e employed the parameters listed above, referred here as baseline simulation. In addition, for the baseline simulation we employed the SyamlalO'Brien drag model, the standard solid stress model, and slip and non-slip condition for solid and gas phase, correspondingly. The previous set of models will be referred in the result's section as baseline simulation models.

4. Results and discussion

Figure 4 is a sampling plot showing the instantaneous solids velocities and gas volumetric fraction fields following two different bubbles passage around the obstacle. Analysis of Fig. (4) shows the bubble splitting mechanism taking place and the characteristic time scale for the bubble passage. After the bubble passage, the solid wake has higher solid velocity magnitude around the obstacle. The solids upward movement following the bubble wake and wall downward movement is kept unchanged. Generally, for beds without internal obstacles, bubble size increases with bed height, particle size and superficial velocity. Analysis of CFD results shows that the presence of tubes was found to alter such general trends for bubble growth.

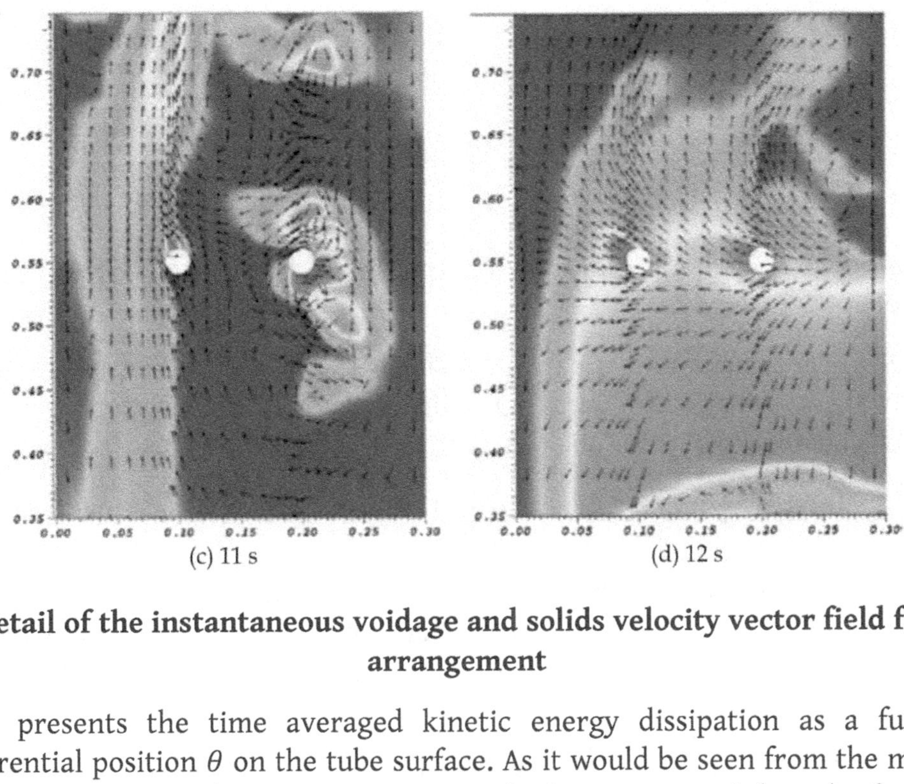

(c) 11 s (d) 12 s

Fig. 4. Detail of the instantaneous voidage and solids velocity vector field for the T2 arrangement

Figure 5 presents the time averaged kinetic energy dissipation as a function of circumferential position θ on the tube surface. As it would be seen from the majority of results the most severe dissipation occurs on the lower parts of the tube for an angle corresponding to 90 degrees. This fact is in agreement with experimental measured values of erosion from Wiman (1994). Analysis of results shows that, except for the BVK drag model, all the drag models predict the highest dissipation rate occurring at 90 degrees. The value predicted using the Syamlal-O'Brien drag model is the highest. There is a noticeable difference in the value of dissipation value between the Gidaspow and Gidaspow blend drag models, whereas the results by the Wen-Yu drag locate in the intermediate range. By his turn, the results by HYS and Koch-Hill models locates near the Wen-Yu values.

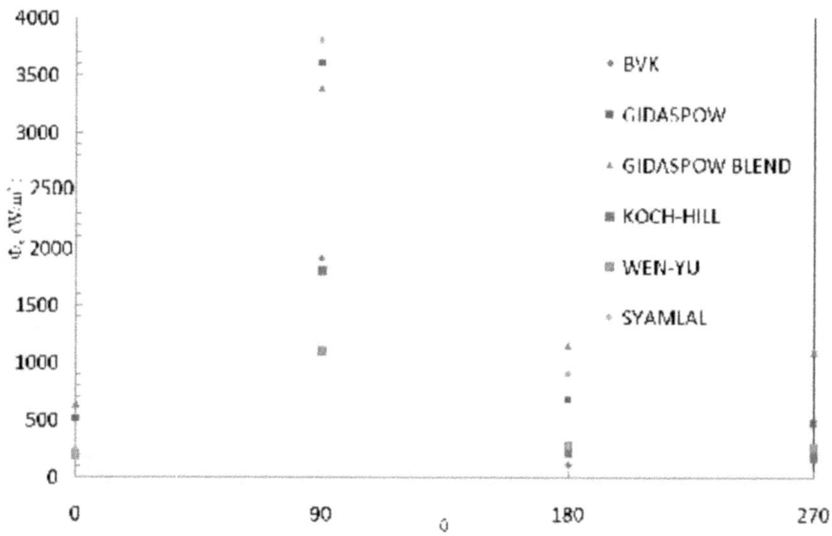

Fig. 5. Time averaged kinetic energy dissipation predicted for different gas-solid drag models - T2 arrangement

Also, analysis of transient simulated results shows the highest values of kinetic energy dissipation rate occur when the particle fraction suddenly changes from a low to a high value, which corresponds to the tube being hit by the wake of a bubble.

Figure 6 presents the result of kinetic energy dissipation for different solid stress models. As shown the peak values of the dissipation rate are close to 180 degrees for the model with (baseline) and without the blending function discussed in section 2 . On the other hand, the maximum value predicted by the constant solids viscosity model (NO KTGF) locates around 180 degrees, and it is superior to the predicted by the solids kinetic energy theory.

Figure 7 presents the result of kinetic energy dissipation for different tube surface slip conditions. The baseline case considers the free slip condition for the solids and non-slip condition for the gas. The peak values do not change when considering the solids with the non-slip condition. On the other hand, when considering slip conditions for both phases the value decreased. The above results, suggests the free slip condition for the gas phase on the surface of the tube plays an important role on the value of the kinetic energy dissipation.

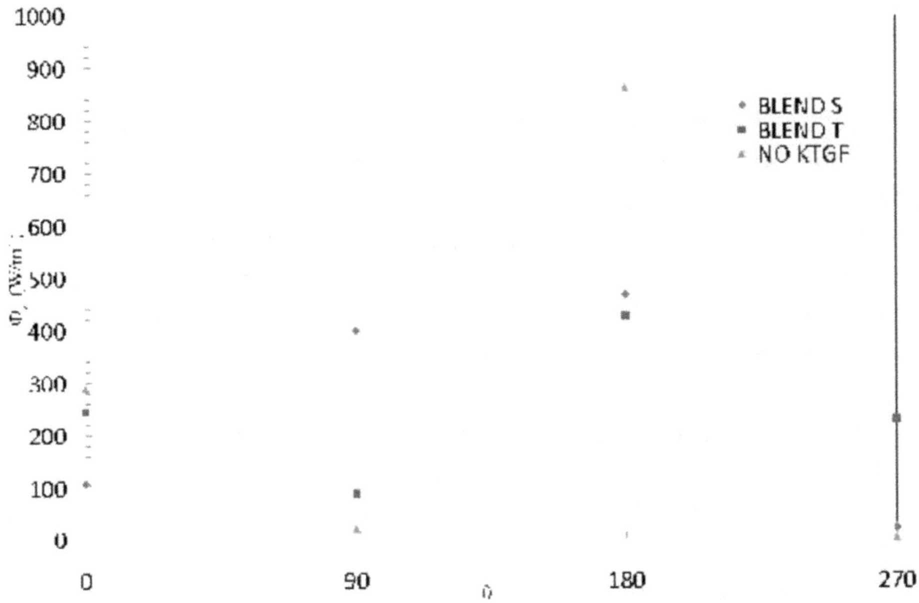

Fig. 6. Time averaged kinetic energy dissipation predicted for different solid stress models

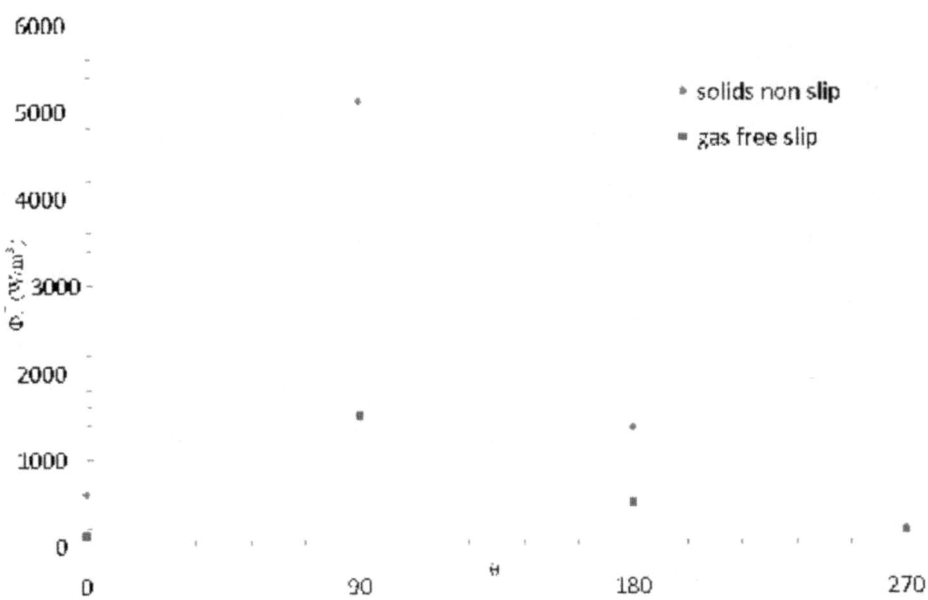

Fig. 7. Time averaged kinetic energy dissipation predicted for different tube surface slip conditions

Figures 8 and 9 shows a comparison between the numerically predicted time averaged values of the kinetic energy dissipation rate, using the baseline simulation models discussed in section 2, and those based on the experimental values for the T2 arrangement in the work of Gustavson and Almstedt (2000). A comparison of the numerical and experimental results for the two different operational pressures (c.f. Figs 8 and 9) shows some degree of discordance. However, the lack of minutely agreement with experimental values, the results can be compared for recognizing similar drifts. For instance, from the Fig 8, the higher simulated values of the dissipation rate occurring in the bottom position of the tube, i.e. for $\theta < 180^{\circ}$, is in agreement with the experimental counterpart. Similarly, the increase of dissipation rate with increased operational pressure for the simulated results is in agreement with the experimental values. Regarding the erosion and baseline models used for the simulation, some remarks towards better agreement with experimental values can also be done. According to the monolayer erosion model and its discussion above Eq. (19) some degree of uncertainty is associated to the multiplying constant, as the exact value of elasticity of collision is not known. It is also expected, that adjustments in the baseline simulation models, such as those outlined in Figs. (5) to (7) would produce better agreement.

Figure 10 is a sampling plot showing the instantaneous gas volumetric fraction fields for different tube arrangements. Analysis of Fig. (10) shows the influence of immersed obstacles on the bubble splitting mechanism taking place and the bubble passage pattern. Above the tube bank, the bubble appears to grow to size similar to the without tube geometry. For the geometry with tubes the bubble encompasses the obstacles but not at the full width of the bed. The interaction is stronger for the denser tube geometry.

Figures 11 and 12 shows a comparison between the numerically predicted values of bubble frequency, using the baseline simulation models discussed in section 2, and those based on the experimental measurements from the works of Almstedt (1987) and Wiman (1995). As shown in the Fig. 10, the calculated values of N_b are underestimated at higher pressures, while at low pressures, there is a quite good agreement between calculated and experimental results. This conclusion holds true both for the I4 and for the S4D tube arrangement. As in the experimental results there are no noticeable differences for the two tubes arrangements.

Figures 13 and 14 shows a comparison for the bubble frequency for the bed without tube and with the S4 arrangement. As shown in Fig. 13 for the bed without tubes the trend points to a frequency agreement between 0.4 and 0.6 MPa . Up this range the values differences increases as pressures increases up to 1.6 MPa . For the S4 arrangement the trend is similar to the I4 and S4D arrangement. The experimental results depicted in Figs.

11 to 14 suggest that the frequency increases with pressure both with and without tubes. For numerical results this holds true only for low pressures, i.e., 0.1 and 0.4 MPa . The numerical results suggest a maxima occurring between 0.4 and 1.6 MPa . On the other hand, the numerical results corroborate the experimental trend that the mean frequency is higher for the bed with tubes than for the freely bubbling bed.

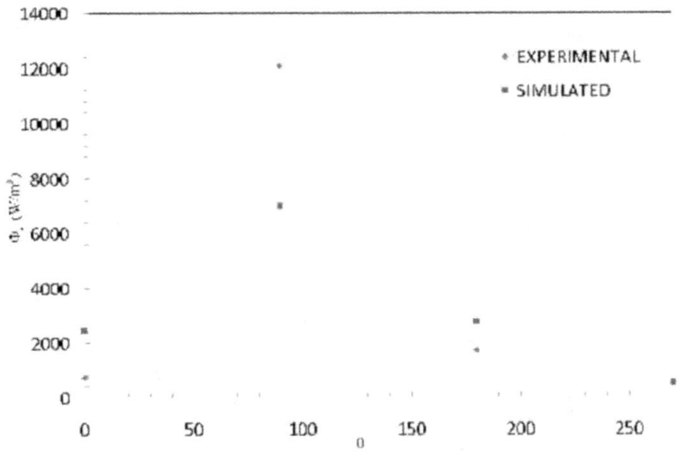

Fig. 8. Simulated kinetic energy dissipation at various circumferential angular positions in the surface of the tube and values from experimental results by Gustavsson and Almstedt, 2000 . Operational pressures : 0.8 MPa ;

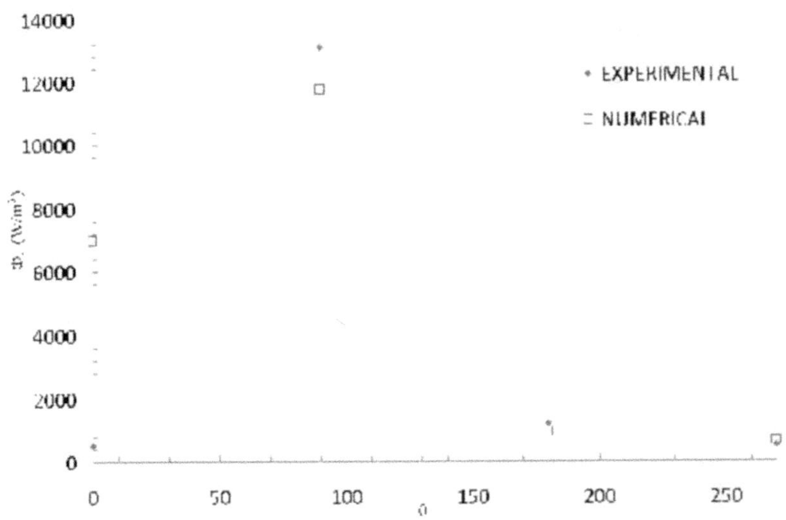

Fig. 9. Simulated kinetic energy dissipation at various circumferential angular positions in the surface of the tube and values from experimental results by Gustavsson and Almstedt, 2000 . Operational pressures : 1.6 MPa

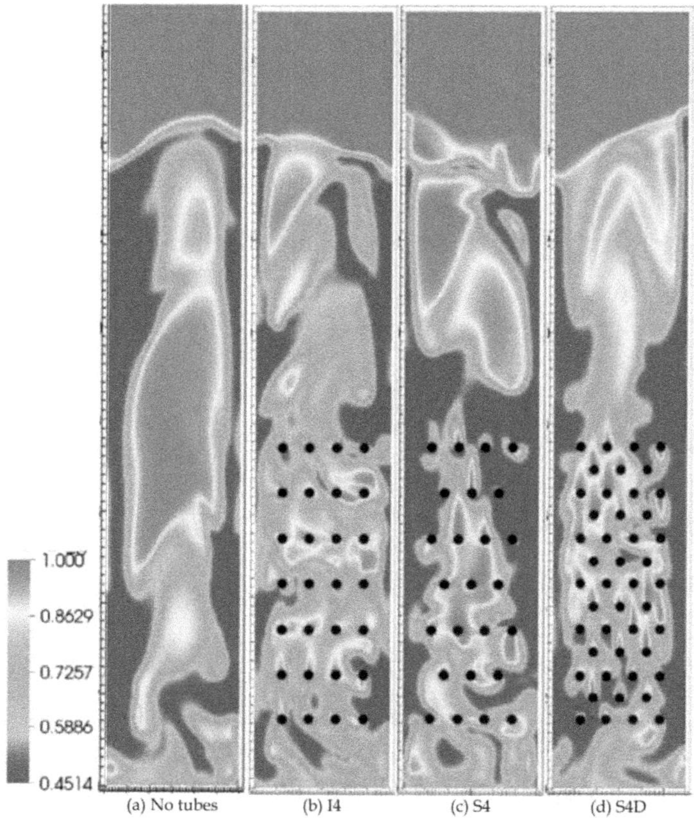

Fig. 10. Snapshots of voidage field at 4 s for different tube arrangements. P = 0.4MPa, Uf = 0.6 m/s

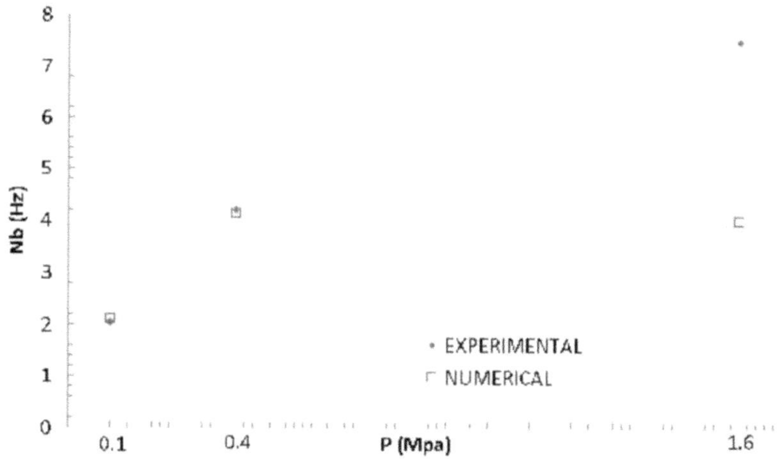

Fig. 11. Nb versus pressure, numerical X experimental: I4 arrangement, U_f = 0.6 m/s

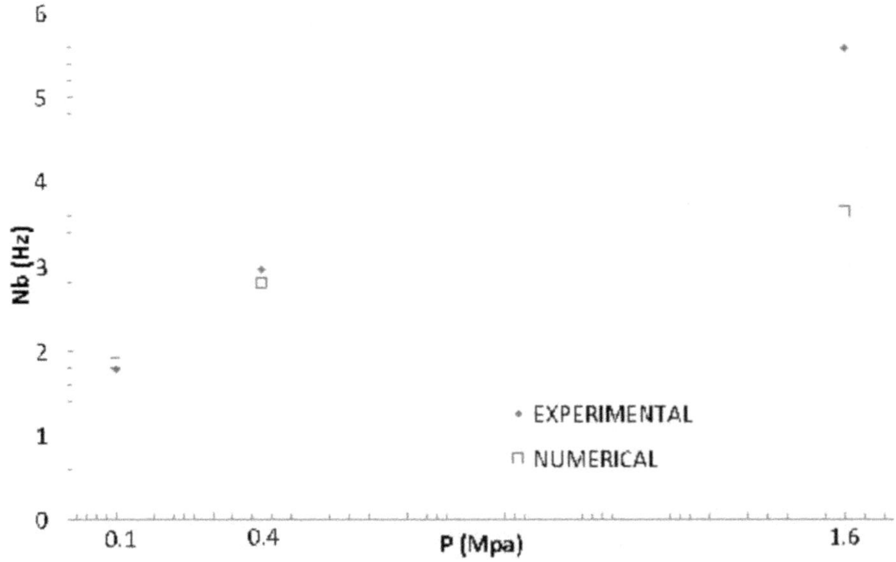

Fig. 12. Nb versus pressure, numerical X experimental: S 4 D arrangement, $U_f = 0.6$ m/s

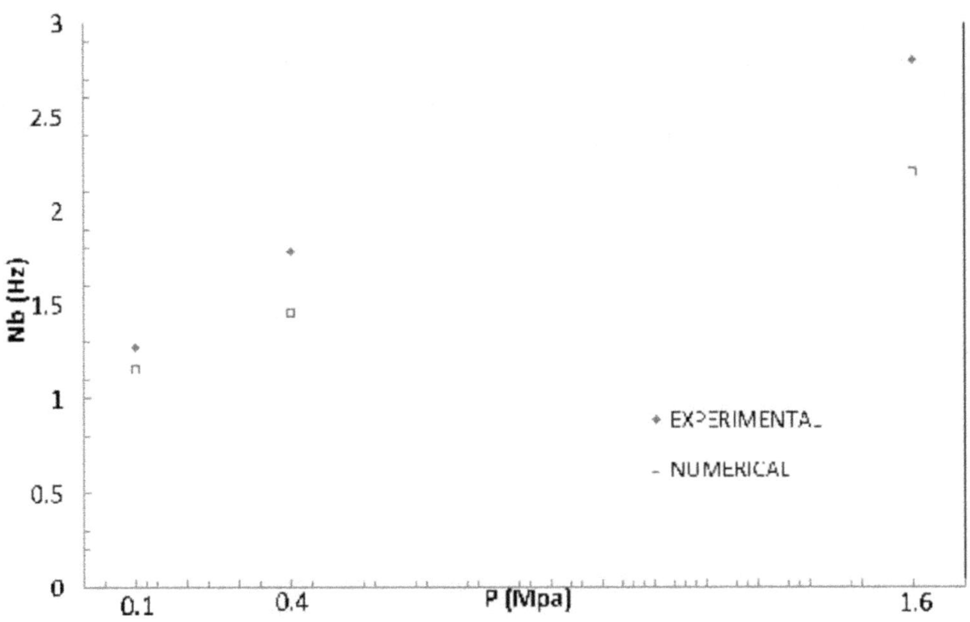

Fig. 13. Nb versus pressure, numerical X experimental: no tubes, $U_f = 0.6$ m/s

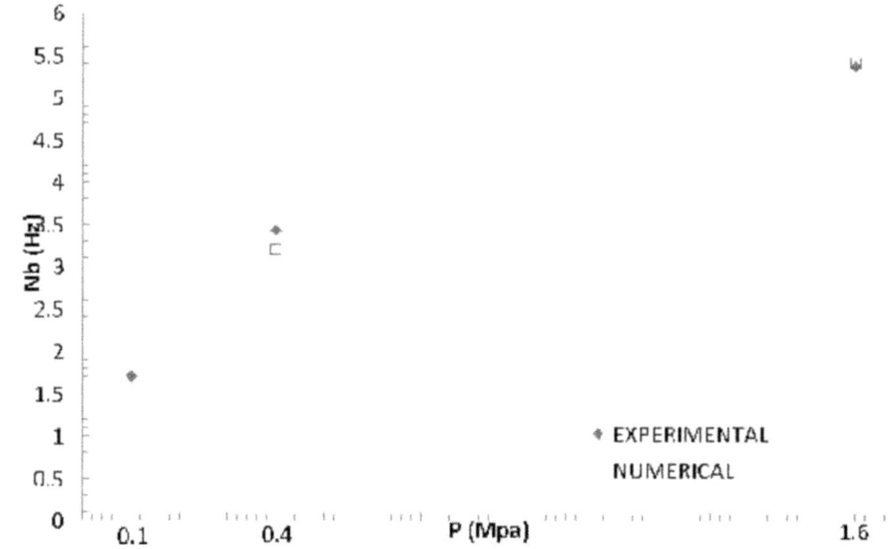

Fig. 14. Nb versus pressure, numerical X experimental: S 4 arrangement, $U_f = 0.6\ m/s$

Figure 15 to 18 shows a comparison for the mean bubble velocity. In all cases, the simulated results are underestimated in relation to the experiments. However, the trend observed for the experimental results with a maxima around 0.4 MPa is verified for the numerical results for all the tube arrangements. For the bed without tubes the experimental increase trend of Vb with pressure is valid for pressures higher than 0.4 MPa .

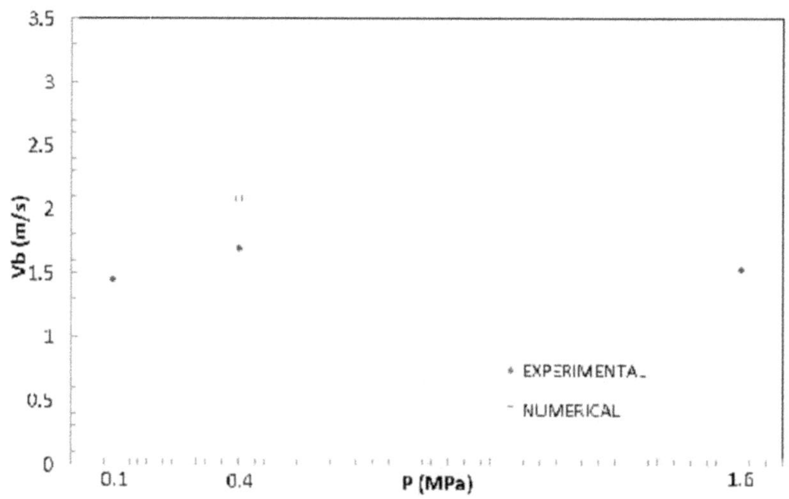

Fig. 15. Vb versus pressure, numerical X experimental: : I4 arrangement, $U_f = 0.6\ m/s$

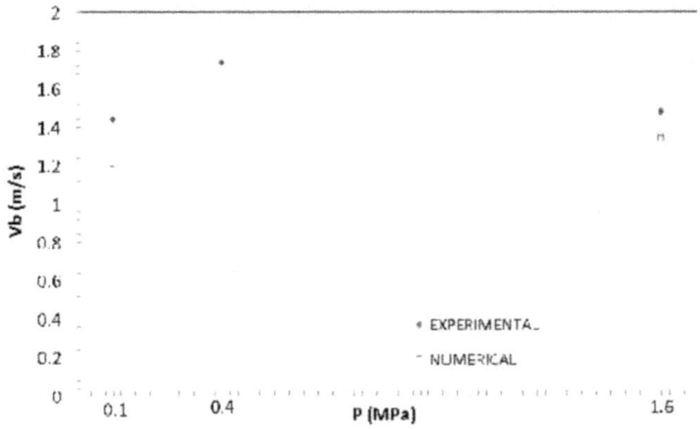

Fig. 16. Vb versus pressure, numerical X experimental: S4D **arrangement,** $U_f = 0.6$ m/s

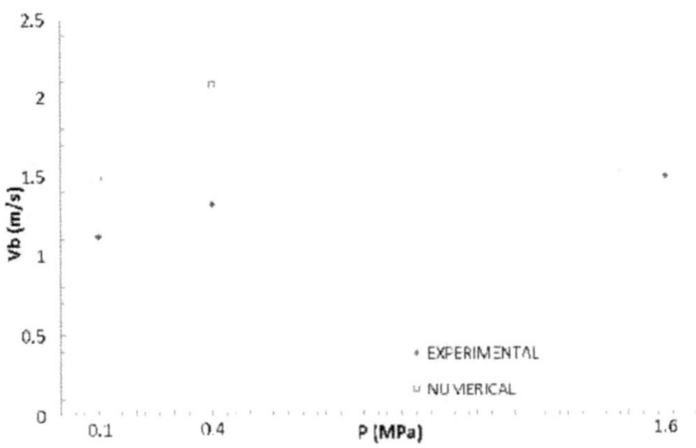

Fig. 17. Vb versus pressure, numerical X experimental: : no tubes, $U_f = 0.6$ m/s

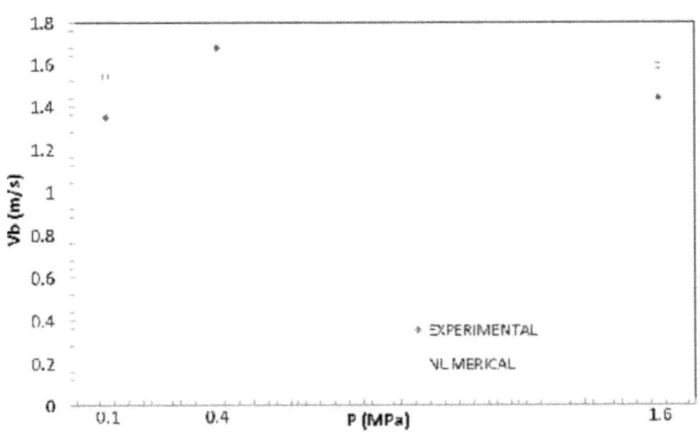

Fig. 18. Vb versus pressure, numerical X experimental: : S 4 arrangement, $U_f = 0.6$ m/s

Figures 19 and 20 shows a comparison for the bubble frequency for $U_f = 0.2$ m/s. Comparison with results in Figs. 12 and 14 shows that the numerical results, although smaller are closer to the experimental. The tendency holds true for both S4 and S4D arrangements. Comparison with the numerical results for the S4D and S4 arrangements.

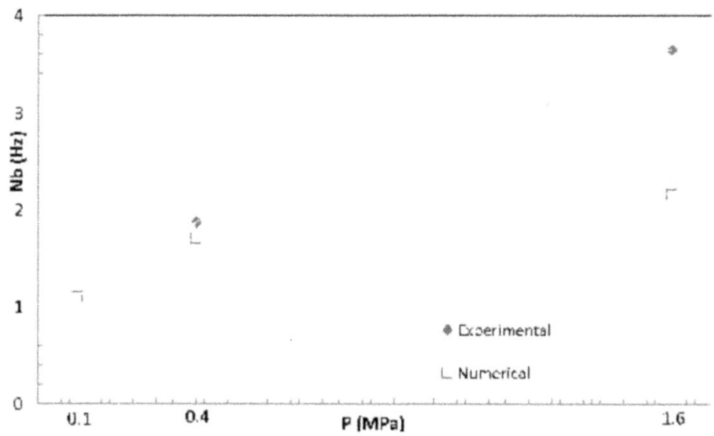

Fig. 19. Nb versus pressure, numerical X experimental: : S 4 arrangement, $U_f = 0.2$ m/s

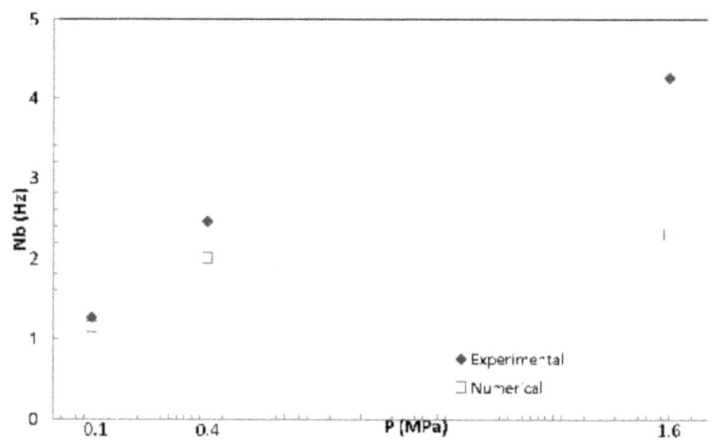

Fig. 20. Nb versus pressure, numerical *X* **experimental: S4D arrangement,** $U_f = 0.2$ m/s **given in Fig. 6 and 7, also shows that Nb increases with increasing excess velocity. The last, is the same drift verified for the experimental values.**

Figures 21 and 22 present the results for the time averaged kinetic energy dissipation as a function of circumferential position θ on the tube surface for the S4 arrangement for two distinct pressures. As it would be seen the highest experimental results are between 90 and 240 degrees, while for the numerical are between 30 and 150 .

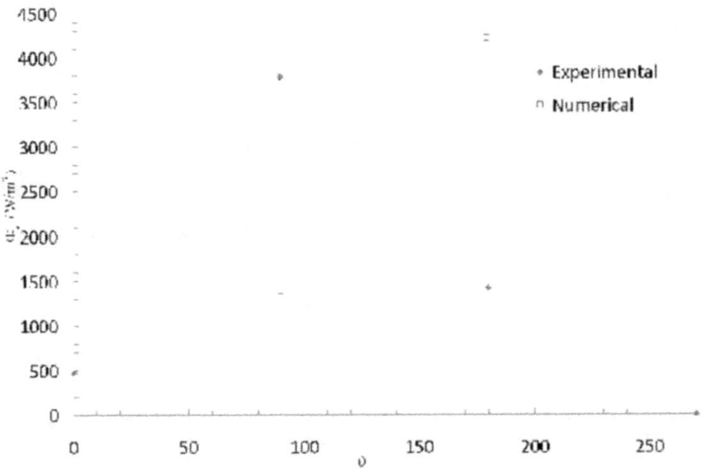

Fig. 21. Time averaged kinetic energy dissipation predicted for different tube arrangements at two different operating pressures: : S4 arrangement, P = 0.1MPa

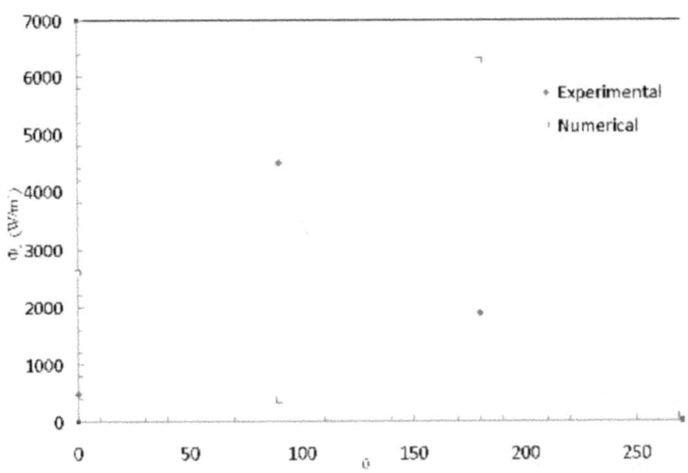

Fig. 22. Time averaged kinetic energy dissipation predicted for different tube arrangements at two different operating pressures : : S4 arrangement, P = 1_6MPa

The numerical values are over predicted in the range 0 to 100 degrees and above 240 degrees. Within the range of 100 to 240 degrees, the numerical values are inferior to the experimental results. Comparable results are confirmed for the 1.6 MPa pressure, with the experimental curves exhibiting less sensitivity to pressure fluctuations. The numerical values exhibit greater sensitivity to pressure, albeit within the same scale of magnitude.

5. Concise Summary

This study numerically explored the hydrodynamics of two-dimensional beds with immersed tubes. The simulations were conducted using an experimental setup featuring various tube bank designs, operating pressures, and gas excess velocities. The aim of this study was twofold: to examine impacts not previously addressed in the literature and to assess the viability of the MFIX code for this type of research. The simulation results were presented in terms of the averaged rate of kinetic energy dissipation of solids and bubble characteristics, including frequency and mean velocity. A comparison of the numerical frequency values with experimental data demonstrates strong concordance at low pressures. Additionally, analogous trends were seen, such as an increased frequency corresponding to higher excess velocities. The computational estimates for bubble velocity align more closely with experimental data at elevated pressures. A maximum bubble velocity for intermediate pressure is discernible in the bed with tubes; however, this could not be corroborated by the numerical data. Our findings indicate substantial effects on the expected dissipation rates and, subsequently, on the erosion rate when utilizing various drag models. The dissipation rate is affected by the application of blending functions for the transition between the plastic and viscous regimes of solid flow or by the implementation of a constant viscosity model. The results are highly sensitive to the slip condition of the gas phase at the tube's surface. Under a free slip condition for the gas phase, the minimal dissipation values are achieved. In his analysis, the range of angles yielding maximum values for the kinetic dissipation energy of solids is displaced in comparison to the experimental results. Ultimately, observations can also be made to enhance concordance with experimental results for the energy dissipation model. According to the monolayer erosion model and the discussion above Eq. (19), there exists a degree of ambiguity regarding the multiplication constant, as the precise value of collision elasticity remains unknown.

6. References

Achim, D., Easton, A. K., Schwarz, M. P., Witt, P.J., Zakhari, A., 2002, "Tube erosion modelling in a fluidised bed", Applied Mathematical Modelling, Vol 26, pp. 191201.

Almstedt, A. E., 1987, A study of bubble behaviour and gas distribution in pressurized fluidized beds burning coal. Thesis for the degree of Licentiate of Engineering, Chalmers University of Technology, Goteborg. Sweden.

Anderson, T. B., 1967, "A fluid mechanical description of fluidized beds: Equations of motion", Industrial Engineering Chemical Fundamentals, Vol 6, pp. 527-539.

Benyahia, S., 2008, "Validation study of two continuum granular frictional flow theories", Industrial Engineering Chemical Research, 47, 8926-8932.

Benyahia, S., Syamlal, M., O'Brien, T. J., "Summary of MFIX Equations 2005-4", 1 March 2006, Available from http://www.mfix.org/documentation/MfixEquations2005-4-1.pdf.

Benyahia, S., Syamlal, M., O'Brien, T. J., 2006, "Extension of Hill-Koch-Ladd drag correlation over all ranges of Reynolds number and solids volume fraction", Powder Technology, 162, 166-174.

Bouillard, J. X., Lyczkowski, R. W., 1991, "On the erosion of heat exchanger tube banks in fluidized-bed combustor", Powder Technology, Vol 68, pp. 37-51.

Cebeci, T., Shao, J. P., Karyeke, F., Laurendeau, E., 2005, Computational Fluid Dynamics for Engineers, Horizon Publishing Inc, USA, 402p.

Dietiker, J., "Cartesian Grid User Guide", 4 September 2009, Available from https://mfix.netl.doe.gov/documentation/Cartesian_grid_user_guide.pdf.

Enwald, H., Peirano, E., Almstedt, A. E., Leckner, B., 1999, "Simulation of the fluid dynamics of a bubbling fluidized bed: Experimental validation of the two-fluid model and evaluation of a parallel multiblock solver", Chemical Engineering Science, 54, 311-328.

Ergun, S., 1952, "Fluid-flow through packed columns", Chemical Engineering Progress, Vol 48, n. 2, pp. 91-94.

Fan, J. R., Sun, P., Chen, L. H., Cen, K. F., 1998, "Numerical investigation of a new protection method of the tube erosion by particle impingement", Wear, Vol 223, pp. 50-57.

Gustavsson, M., Almstedt, A. E., 1999, "Numerical simulation of fluid dynamics in fluidized beds with horizontal heat exchanger tubes", Chemical Engineering Science, 55, 857866.

Gustavsson, M., Almstedt, A. E., 2000, "Two-fluid modelling of cooling-tube erosion in a fluidized bed", Chemical Engineering Science, 55, 867-879.

He, Y. R., Lu, H. L., Sun, Q. Q., Yang, L. D., Zhao, Y. H., Gidaspow, D., Bouillard, J., 2004, "Hydrodynamics of gas-solid flow around immersed tubes in bubbling fluidized beds", Powder Technology, 145, pp. 88-105.

He, Y. R., Zhan, W., Zhao, Y., Lu, H. Schlaberg, I., 2009, "Prediction on immersed tubes erosion using two-fluid model in a bubbling fluidized bed", Chemical Engineering Science, 64, pp. 3072-3082.

Kobayashi, N., Yamazaki, R., Mori, S., 2000, "A study on the behaviour of bubbles and solids in bubbling fluidized beds", Powder Technology, Vol 113, pp. 327-344.

Lathowers, D., Bellan, J., 2000, "Modeling of dense gas-solid reactive mixtures applied to biomass pyrolysis in a fluidized bed", Proceedings of the 2000 U.S. DOE Hydrogen Program Review, NREL/CP-570-28890. USA.

Lee, S. W., Wang, B. Q., 1995, "Effect of particle-tube collision frequency on material wastage on in-bed tubes in the bubbling fluidized bed combustor", Wear, Vol 184, pp. 223-229.

Lyczkowski, R. W., Bouillard, J. X., 2002, "State-of-the-art review of erosion modeling in fluid/solid systems", Progress in Energy and Combustion Science, Vol 28, pp. 543602.

Ozawa, M., Umekawa, H., Furui, S., Hayashi, K., Takenaka, N., 2002, "Bubble behavior and void fraction fluctuation in vertical tube banks immersed in a gas-solid fluidizedbed model", Experimental Thermal and Fluid Science, Vol 26, pp. 643-652.

Pannala, S., Daw, C. S., Finney, C. E. A., Benyahia, S., Syamlal, M., ÓBrien, T. J., "Modelling the collisional-plastic stress transition for bin discharge of granular material", 2009, Powders and Grains 2009 - Proceeding of the 6th International Conference on Micromechanics of Granular Media, pp. 657-660.

Schaeffer, D. G., 1987, "Instability in the evolution equations describing incompressible granular flow", Journal Differential Equations, v. 66, pp. 19-50.

Siravastava, A., Sundaresan, S., 2003, "Analysis of a frictional-kinetic model for gas-particle flow", Powder Technology, v. 129, pp. 72-85.

Syamlal, M., 1998, "MFIX Documentation, Numerical Techniques", Technical Note, DOE/MC-31346-5824, NTIS/DE98002029, National Technical Information Service, Springfield, VA, USA.

Syamlal, M., Rogers, W. A., O'Brien, T. J., 1993, "MFIX Documentation, Theory Guide", Technical Note, DOE/METC-94/1004, NTIS/DE94000087, National Technical Information Service, Springfield, VA, USA.

Wang, J., van der Hoef, M. A., Kuipers, J. A. M., 2010,"CFD study of the minimum bubbling velocity of Geldart A particles in gas-fluidized beds", Chemical Engineering Science, 65, pp. 3772-3785.

Wen, C. Y., Yu, Y. H., 1966, "Mechanics of Fluidization", Chemical Engineering Progress Symposium Series, Vol 62, n. 62, pp. 100-111.

Wiman, J., 1994, An experimental study of hydrodynamics and tube erosion in a pressurized fluidized with horizontal tubes. Thesis for the degree of Licentiate of Engineering, Chalmers University of Technology, Goteborg. Sweden.

Wiman, J., Almstedt, A. E., 1997, "Hydrodynamics, erosion and heat transfer in a pressurized fluidized bed: influence of pressure, fluidization velocity, particle size and tube bank geometry", Chemical Engineering Science, Vol 52, pp. 2677-2695.

Wong, Y. S., Seville, J. P. K., 2006, "Single-particle motion and heat transfer in fluidized beds", AIChe Journal, Vol 52, pp. 4099-4109.

Chapter-13

Design Optimization of Multi-Effect Evaporation for Elevated BPE Systems

1. Introduction: An Overview

Historically, optimization was exclusively examined for conventional flow patterns such as forward and backward feed. Simple criteria were employed to differentiate between them, focusing on the viscosity and the temperature of the initial dilute solution (T_F). Consequently, forward feed was typically preferred for the evaporation of low-viscosity hot solutions with a temperature $T_P>T_p$, where T_P represents the target temperature of the end product. This significantly reduces the liquid heating load. Consequently, reverse feed was advised for highly viscous cold liquors.

Furthermore, several contributions to the optimum synthesis of multiple effect evaporator systems (MEES) have been reported to date. Most prior publications concentrated on the analysis rather than the synthesis of evaporation systems. It is commonly considered that critical structural factors, such as vapor and liquid flow patterns and the number of impacts, are established data, despite their significant impact on the performance of a MEES.

Nishitani and Kunugita (1979) initially introduced a multiobjective problem formulation to ascertain the optimal flow pattern of a multiple-effect evaporator system. Nevertheless, they did not take into account stream mixing or splitting. Furthermore, the solution approach executed the simulation of the MEES for all conceivable flow patterns sequentially. Hillebrand and Westerberg (1988) recently devised a straightforward model to directly calculate the utility consumption for multiple-effect evaporator systems that exchange sensible heat with external streams. Westerberg and Hillebrand (1988) introduced the term "heat shunt" to heuristically determine the optimal liquid flow pattern. However, significant assumptions such as a constant boiling point elevation, absence of liquid bypassing, and little heat of mixing considerably constrain the applicability of their findings.

A rigorous mathematical model previously created (Irahola & Cerdá, 1994) has been employed to address the issue at hand. It contemplates the potential for optimizing the desired variables. This has enabled the model to address multiple problem types, including simulation, optimization, optimal synthesis, and constrained partial reengineering of multi-effect evaporation systems (MEES) (Irahola, 2008). The generated mathematical model is a Mixed-Integer Nonlinear Programming (MINLP) model, solved utilizing commercial tools. The methodology was effectively implemented in three industrial challenges. The best configuration, influenced by the feed and product temperature, employs a unique liquid flow pattern that frequently deviates from traditional forward and backward feed, resulting in significant savings (Irahola & Cerdá, 1996). It should be noted that: the bifurcation of live steam flow may serve as a superior alternative to the conventional steam cascade; the optimal fixed cost curve is not invariably monotonically increasing; the accurate allocation of the areas of the MEES effects, suitable operating conditions, and the judicious selection of liquid and vapor flow patterns are critical determinants in the optimal design of the MEES. The primary drawback of the proposed method for solving the defined mathematical model (MINLP type) for the optimal synthesis of the MEES is the existence of multiple local optima, which complicates the attainment of a global optimum.

2. Evaporation of a sodium hydroxide aqueous solution

Among the solutions of industrial interest that present a high increase in boiling point are sodium hydroxide solutions (caustic soda). The concentration of these substances by evaporation, presents significant disadvantages due to the characteristics of the caustic solutions, namely:

- Have a high boiling point elevation (BPE) which implies a great loss in the temperature difference available.
- Concentrated solutions are highly viscous, which severely reduces the rate of heat transfer in natural circulation evaporators.
- They can have detrimental effects on steel, causing what is called caustic fragility. In addition, they may require removal of large amounts of salt when the solution is concentrated.

Since the transfer of heat (U) of liquor film coefficient, depending on the speed of the caustic solution through tubes (among other variables), usually, seeks a high speed in order to obtain a large coefficient (Kern, 1999).

According to the literature, it has taken as standard for the concentration of caustic soda, a evaporation system of two or three effects operate in backward feeed (Kern, 1999). In this study, found that the structure in counterflow or backward feed, obviously presents a high performance, but is not the best. In order to confirm what was said, is going to solve a problem.

3. Optimal synthesis of a multi-effect evaporation system for the concentration of caustic soda

3.1 Problem

Find the optimal MEES to concentrate 30040lb/h(13626 kg/h) of an aqueous solution of sodium hydroxide from 10 to 50% by weight. The type of used evaporator is long vertical tube with forced circulation. Available in the plant: live steam boiler to 63.69 Psia (4.48 kg/cm2). The allowable minimum absolute pressure in an effect is of 1.942 Psia (0.1365 kg/cm2) (Geankoplis, 1983).

3.2 Mathematical model

The scope of the rigorous mathematical model is limited by the following assumptions adopted in the formulation:

In each effect, the vapor and liquid phases are in equilibrium.

b. A solid phase never arises in any effect.

c. The impact of the hydrostatic head on the liquid boiling point is neglected.

d. There is no leakage or entrainment.

e. Heat losses from any effect need not be considered.

f. The steam always condenses completely.

g. Subcooling of the condensate is very small.

h. Flow of noncondensables is negligible.

i. The concentrated final product is withdrawn from a single effect which in turn does not transfer liquid to any other one.

j. If necessary, you can use a heat exchanger or condenser so that the product go out to the preset temperature (Tp).

k. Not consider any type of pump between the effects.

To solve the mathematical model and find the optimal design should be available before, the mathematical expressions for all dependent variables: enthalpy of steam (H), enthalpy of solution of soda caustic (h), latent heat of vaporization of the water (λ), overall heat transfer coefficient (U), temperature of the solution in the evaporator (T) and costs of forced circulation evaporator, barometric condenser multijet, surface condenser and heat exchanger. In general, useful information is available in graphics, which presented various authors cited in the bibliography, but there are no equations of those curves. These situations and other problems are resolved below. According to Standiford (1963), in forced circulation evaporators, film transfer coefficient (h) on the side of the liquid, can be calculated from the conventional Dittus-Boelter equation for forced circulation when there is no boiling.

$$\frac{hD}{k} = 0.0023\left(\frac{DG}{\mu}\right)^{0.8}\left(\mu\frac{Cp}{k}\right)^{0.4} \tag{1}$$

If there is this equation for two points and combine both equations, you can find the functionality of $\mathbf{h}_1$ with respect to another point (2) as:

$$h_1 = h_2\left(\frac{\mu_2}{\mu_1}\right)^{0.4}\left(\frac{C_{p1}}{C_{p2}}\right)^{0.4} \tag{2}$$

As the overall heat transfer coefficient U is practically determined by the film coefficient h fluid side, the above equation can be used to obtain a correlation for U. In Geankoplis for T = 105°F and X = 0.5, data is U = 400 (Btu/ hft °F). T and X is obtained from a graph (Horvath, 1985) $\mu_2 = 22.84$ centipoise. Then, for five values of concentration and six temperature values are obtained from graphics (Horvath, 1985) the values of Cp_1 and μ_1. With the data obtained can be calculated according to the above expression, the overall heat transfer coefficient U_1 with reference to U_2. Thus, U is plotted vs. X (Fig. 1) and U vs. T (Fig. 2). Finally, using the triple X, T, U_i, can be found by regression, the functionality of U in terms of concentration and temperature:

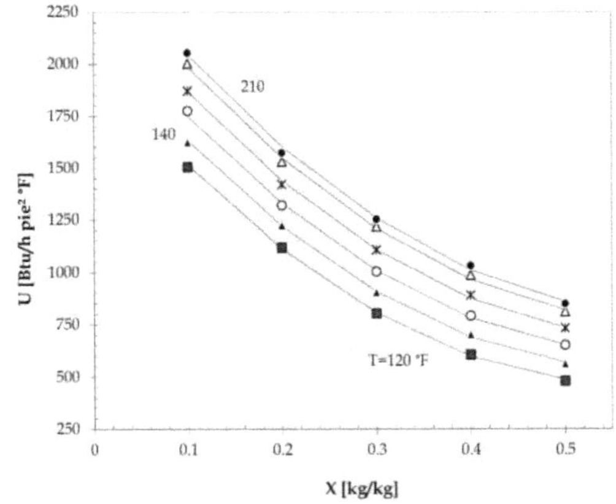

Fig. 1. Overall heat transfer coefficient correlation as a function of solute concentration.

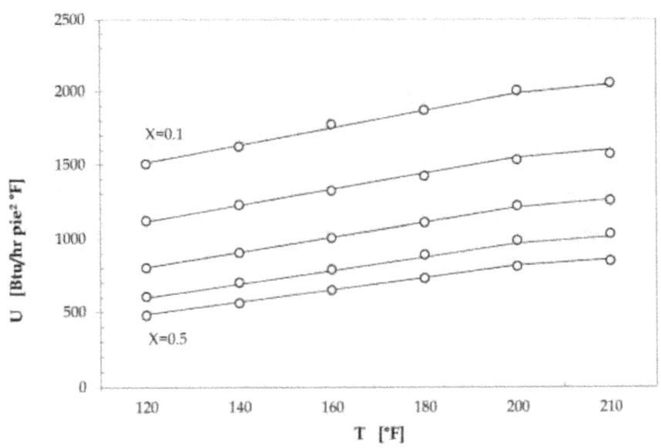

Fig. 2. Overall heat transfer coefficient correlation as a function of temperature.

$$U = 1254.780865 - 4954.6700\,X + 4832.059524\,X^2 + 6.321549\,T - 4.30974\,XT\left[\frac{Btu}{h\,ft_2\,°F}\right] \quad (3)$$

where: X : mass fraction and T[=]°F

The regression function is chosen to achieve maximum correspondence with the data, but at the same time trying to maintain, if possible, the simplicity. Nevertheless, we could not avoid the bilinear term. Figures 1 and 2, you can appreciate the good fit of the correlation found for U.

Also, to represent the enthalpy of solution, depending on the concentration and temperature, it became a non-linear regression with data from Foust (1980) (Fig. 3):

$$h = -10.250000 - 319.591837\,X + 939.795918\,X^2 + 0.963929\,T - 0.335714\,XT \left[\frac{Btu}{lb}\right] \quad (4)$$

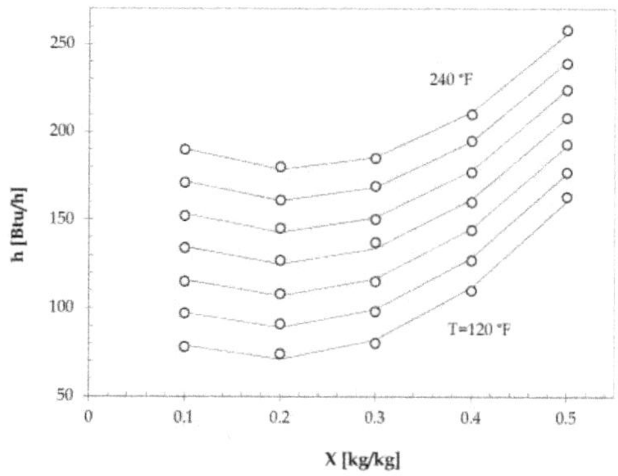

Fig. 3. Enthalpy of solution

The functional expression of the cost of barometric condenser multijet (C_{CBM}) is obtained by regression analysis of the chart presented by Peters (1991). Again, searching for mathematical expressions that minimize non-convexities of the mathematical model, we found that a quadratic expression represents excellent, the cost curve in the range of work desired (Fig. 4) (1992):

$$C_{CBM} = 4{,}753104 + 2{,}480885w + 0{,}281818w^2(10^3USD) \quad (5)$$

Where: w[=]gpm. The flow rate of cooling water (w) required in the condenser is directly given by (Kern, 1999):

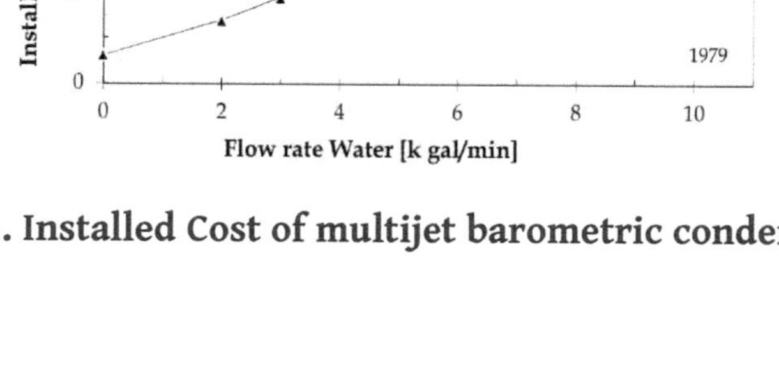

Fig. 4. Installed Cost of multijet barometric condenser.

$$w = \frac{Q}{500(T_s - T_w - t_a)}(\text{Gpm}) \tag{6}$$

where:

Q = heat load, Btu / h

T_s = temperature of saturated steam, ° − F

T_w = temperature of cooling water, ° − F

$t_a = 15° - \text{F}$ = degree of approximation at T_s

The cost of forced circulation evaporator is obtained based on information reported by Maloney, 2008:

$$CE_{FC} = 2420.5A^{0.7121}U\$S \tag{7}$$

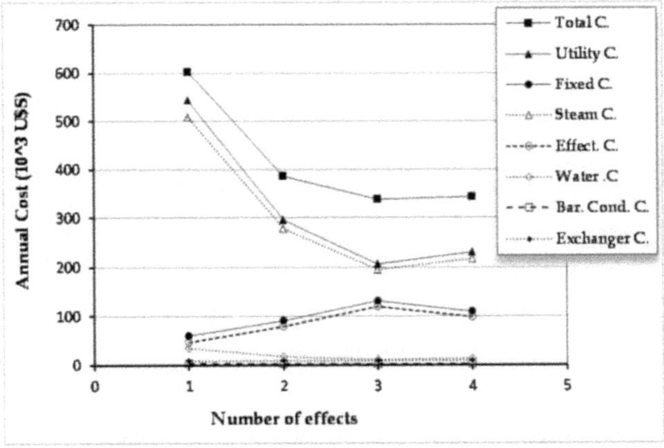

Fig. 5. Evolution of total operating costs.

The information of the data used in the resolution of the problem is presented in table 1. Also, the thermodynamic properties of the remaining functions and costs of services and equipment are presented.

In order to consider probable situations that could be presented in the industry, will study the cases in which the temperature of the weak solution (feed) is equal, higher or lower than the temperature of the strong solution (product). In the three cases, only change the values of the inlet of the weak solution temperature. The rest of the parametric conditions remain fixed.

Table 1. Data for example and, functional expression of thermodynamic properties and the costs of utilities and equipment.

Feed and product conditions					
Item		Feed Feed			
Flow rate lb/h (Kg/h)				60080	(2725.2)
Weight fraction		0.1		0.5	
Temperature °F(°C)	Case I	180	(82.2)	130	(54.4)
	Case II	130	(54.4)		
	Case III	80	(26.7)		
Operation conditions					
		Temperature °F (°C)			
Steam		296.6 (147.0)			
Cool water		(min) 89.6 (32.0) (max) 107.6 (42.0)			
Effect		(min) 125	(51.7) (max) 294.6 (145.9)		
		Minimum allowable temperature difference °F (°C)			
Condensers		18.0 (10)			
Heat Exchangers					
Thermodynamic properties					
Specific vapor enthalpy		$H_i = 1075 + 0.3466Tv_i$ [Btu/lb]			
Solvent Latent Heat of Vaporization		$\lambda_i = 1104 - 0.65Tv_i$[Btu/lb]			
Operating temperature at effect i		$T_i = (1 + 0.1419526X_i)Tv_i - 9.419608X_i + 271.3627X_i{}^2$[°F]			
Costs of utilities and equipment					
Steam		2.92210^{-6} (USD/Btu) 2.104E − 2			
Cool water (305 K)		1.95210^{-7} (USD/Btu) 1.405E − 3			
Surface condenser		$1092.83A^{0.65}$ (USD)			
Heat exchanger		$1144.16A^{0.65}$ (USD)			

4. Analysis and discussion

4.1 Case I. Feed temperature higher than the temperature of the product ($T_P > T_P$)

4.1.1 Comparative analysis of the optimal solution found

Adopted $T_F = 180°F(82.2°C)$. Coinciding with the generally accepted criterion of optimality for the evaporation of caustic soda, has been found that the optimum number of effects of the MEES is equal to three (Fig. 5). However, a mixed structure {2,1,3} has been found in the path of the current liquid instead of backward feed. The feed stream enters to the effect 2 and then continuous countercurrent to the effect 1. Then go out and circulates in forward feed to the effect 3 . (mixed liquid flowpattern). In this new structure that is presented (Fig. 6), we see significant increases in boiling point of the solution: 10.5°C and 40°C, in effects 1 and 3 respectively.

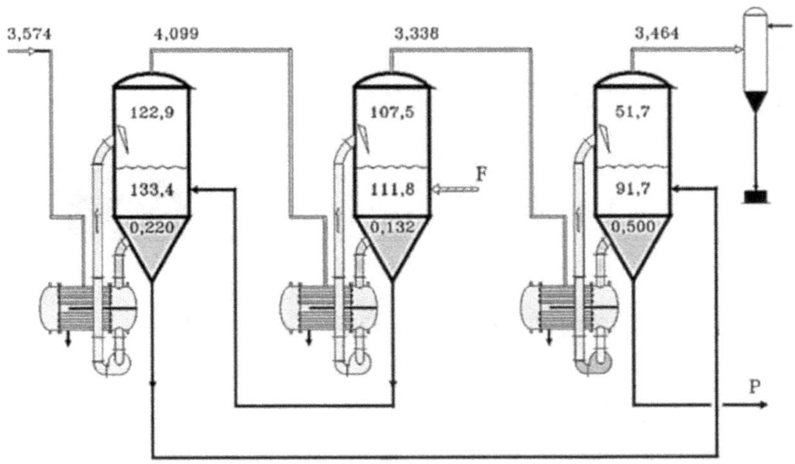

Fig. 6. Optimal configuration three-effect MEES. (TAC=337835 USD)

The optimal solution will be the one with the lowest total annual cost (TAC). From this point of view, the classical structure proposed as an optimum solution: evaporation system of three effects of equal area arranged backward feed (BFA), is 4.31% more expensive than the optimal solution (SO) (Fig. 7). More, even if it is allowed to in the structure backward feed, the effects have distinct areas (BF), do not get a better result that the optimal solution found. The difference in cost is 4.03%. In the figure 7, we also present results for a forward feed evaporation system. This structure, in its classic form forward feed with effects of equal area (FFA), is 7.91% more expensive. Which could corroborate, in some way, because in the past the BFA structure was preferred. If is allowed that the effects have different areas in the structure forward feed (FF) is very interesting the result obtained. The correct distribution of areas has led to a decrease in the total cost. The decrease is so great that, now, the FF structure is better than any of the above structures in backward feed. Its total annual cost (TAC) is only 2.95% greater than the of the optimum solution found (Fig. 7). In general, it appears that whatever the structure of MEES, the operating cost is significantly greater than the fixed cost (Fig. 8). It is approximately 65% of the total cost.

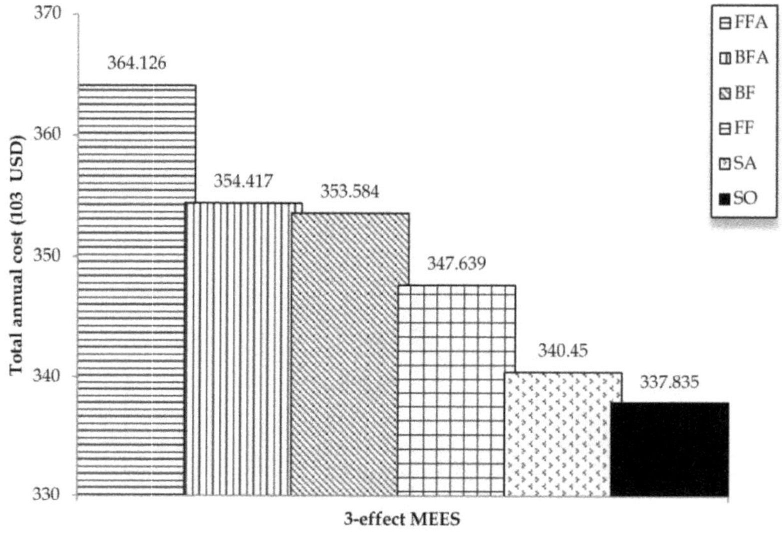

Fig. 7. Comparison with Typical Flowpatterns.

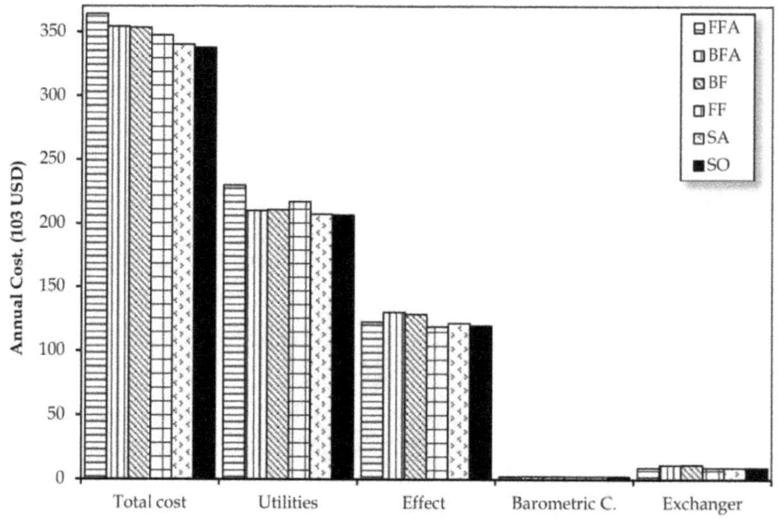

Fig. 8. Relative incidence of Operating and fixed Costs

4.1.2 Impact of the flow pattern

In this case, the trajectory of the liquid stream is the determining factor in the performance of a given MEES. The benefit achieved is even greater than obtained by allowing the effects having different areas with each other. That said, is based on the result of the structure SA. This has the same flow pattern that the optimal solution, but the effects of evaporation system are of equal area. The increase in cost is only 0.8% compared to MEES optimum (Fig. 7). From the practical point of view, the alternative SA may be the best option.

4.1.3 Profiles of the structural and parametric variables

As it will be seen later, only in this case it can be seen some regularity in the curves of the structural and parametric variables. Furthermore, after reaching the optimal point generally the next curve is anomalous with respect to the preceding ones. In the last effect occurs the maximum concentration jump (ΔX) of the solution (Fig. 9). At the same time, it has the maximum area as shown in the curves of the 1 to 4 -effect optimal MEES (Fig. 10). The flow rate of steam produced in the effects is approximately the same. However, this does not apply to MEES with greater number of effects than the optimal. (Fig. 11). The profile of the temperature to the optimum MEES of a different number of effects, does not have a regular aspect. However, the maximum temperature jump occurs in the last effect (Fig. 12).

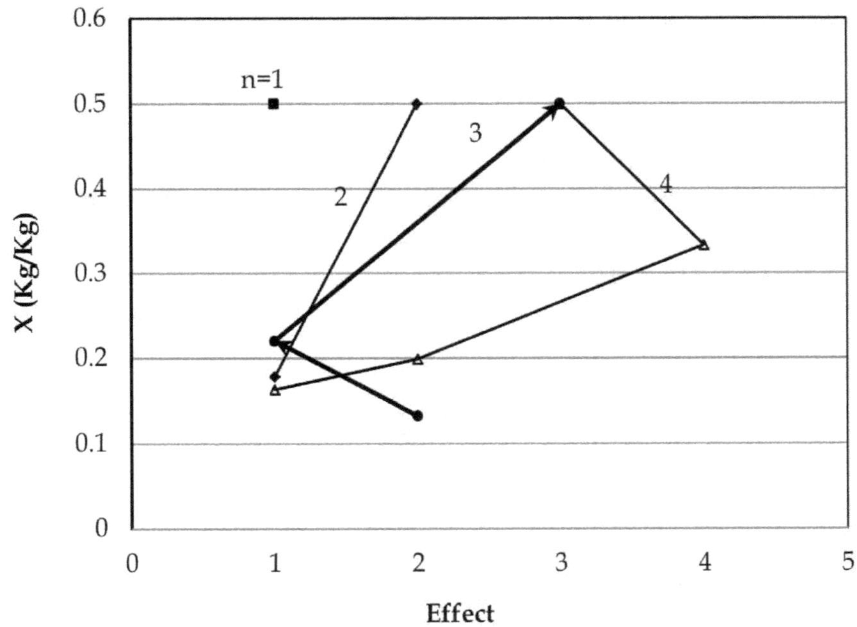

Fig. 9. Concentration profiles in the 1 to 4 -effect MEES. (n : number of effects)

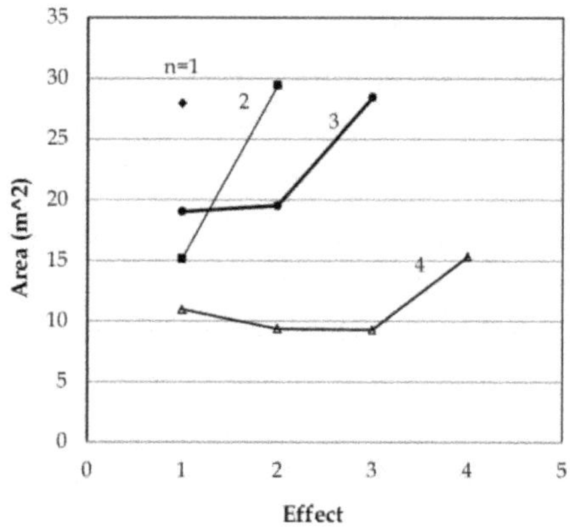

Fig. 10. Area profiles in the 1 to 4 -effect MEES (n : number of effects).

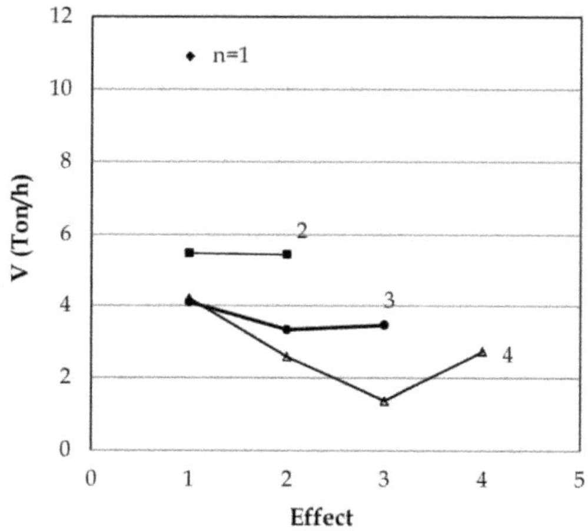

Fig. 11. Flow rate profiles in the 1 to 4 -effect MEES (n : number of effects).

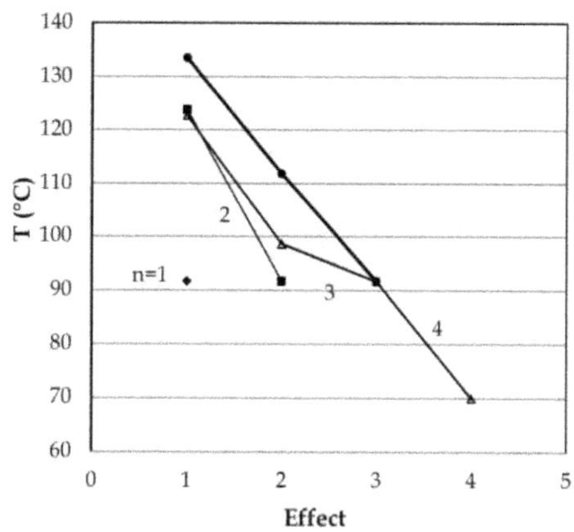

Fig. 12. Temperature profiles in the 1 to 4 -effect MEES (n : number of effects).

4.2 Case II. Feed temperature equal than the temperature of the product ($T_F = T_P$)

4.2.1 Impact of the splitting of the live steam flow rate on the optimal solution

The result obtained when TF = TP, is different, not only to the found for the case I, but also with respect to the classical position. Found structure is highly innovative and simple in its conception.

If you look at the evolution of the total annual cost (TAC) curve of the 1, 2, 3, 4 and 5 effect Optimal MEES, it was found that the four-effect MEES is that of lower cost (optimal quasiglobal) (Fig. 13). The flow pattern is backward feed and in this aspect, this result coincides with the classical motion, but not with the number of effects: proposed here a four-effect MEES, instead of three. However, this new proposal, would not be really the best alternative, if it was not associated to the new steam flow pattern proposed (Fig. 14). It emphasizes, splitting in the live steam flow pattern, it enters parallel to the effects 1 and 2; the by-passing effect 2 by the vapor stream from effect 1 and finally, the mixing of vapor streams from effects 1 and 2 for heating effect 3 .

Against, this new trajectory of the flow of steam, first doubt that occurs, is the performance of this configuration against the unifilar cascade of high thermodynamic efficiency.

4.2.2 Comparative study of proposed flow pattern with respect to traditional configurations

For the purpose of explaining the improvement achieved, compares the structure backward feed with effects of different areas (BF) and the optimal solution (SO). Both structures have equal number of effects and same trajectory of the liquid flow and only differ in the

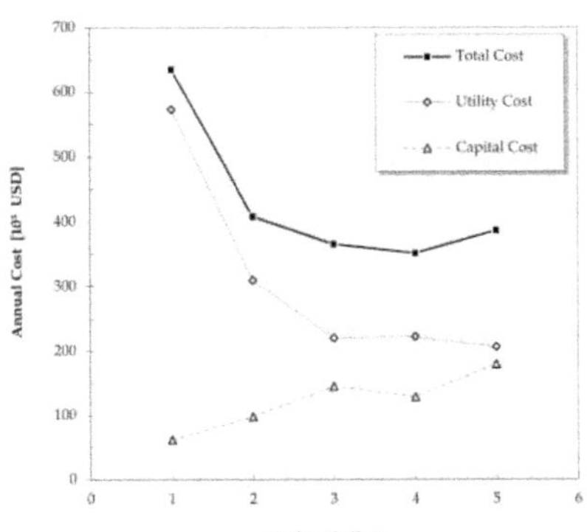

Fig. 13. Relative incidence of Operating and fixed Costs.

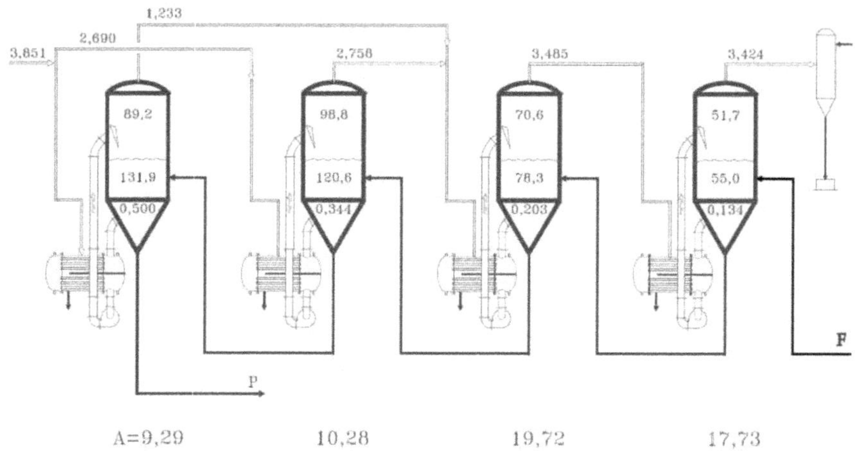

Fig. 14. Optimal configuration four-effect MEES. (TAC=351082 USD)

trajectory of the flow of steam. Two comparisons were made: one relating to the cost of auxiliary services and the other with respect to the cost of the effects, which is almost all of the fixed cost.

4.2.3 Energy efficiency and fixed cost of the traditional structure

The BF structure has a higher efficiency since the cost of auxiliary services is 174963 USD, 21.30% less than the cost for the SO. On the other hand, the cost of the effects is 193008 USD, i.e. 167,8% of the respective cost observed in SO . The net result of the comparison of the total costs, indicates that the BF structure is 8.9% more expensive than SO . With these results, following the classical position we can say, that BF 4 effects is not better than the optimum found (SO), because the MEES should be a structure BFA 3 effects, not four, which was used to compare. Therefore, will be then verified the validity of this rule, for the case study.

4.2.4 Difference in the number of effects due to non-traditional flow patterns

The optimal number of effects found by the mathematical model does not coincide with the optimal number for BFA and FFA traditional structures or even structures BF and FF . This explains why in the absence of a mathematical model to explore the multiple alternatives of design, the best answer to the problem was until now, a countercurrent system.

It was found that the optimum number of effects to structures backward feed and forward feed is 3 . However, the developed model proposes the structure SO of 4 effects as the best solution. Therefore, to verify the quality of it, is advisable to compare the best results found for each structure.

4.2.5 The optimal solution compared to traditional structures

The total cost of the BFA MEES is 4.2% more than the optimal. Therefore, the structure and number of effects, traditionally proposed do not seem to be the most appropriate. Then, one might think that if you remove the restriction of equal area of the effects, could be improved, significantly, the current result. The results show that the BF structure of three effects is 4.1% more expensive than SO (Fig. 15). However, despite the difference in the number of effects, is convenient to analyze in more detail these recent results.

Structurally, BF and SO differ only in the flow pattern of steam. The cost of the auxiliary services of BF is 1.0% lower than the SO. On the other hand, the fixed cost is 13.4% greater determining to SO submit one minor TAC (Fig. 16).

4.2.6 Profiles of the structural and parametric variables

This case is characterized because the profiles of the process variables, for the various intermediate optimal MEES, they have no similarity among themselves. In particular, notes that the optimal solution presents the most discordant curve with respect to the others.

The temperature profile is irregular with temperature differences between effects nonuniform, being the most important jump located between 2 and 3 effect, following the drop of temperature effects 3 and 4, both heated with secondary steam. (Fig. 17). The greater temperature difference between the heating steam and the solution to evaporate, occurs in the effect 2 of the four effect MEES (Fig. 18), through the use of live steam in the effect.

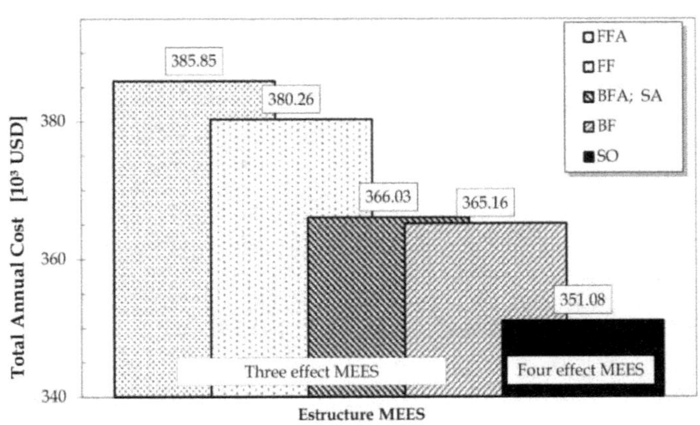

Fig. 15. Comparison with Typical Flowpatterns.

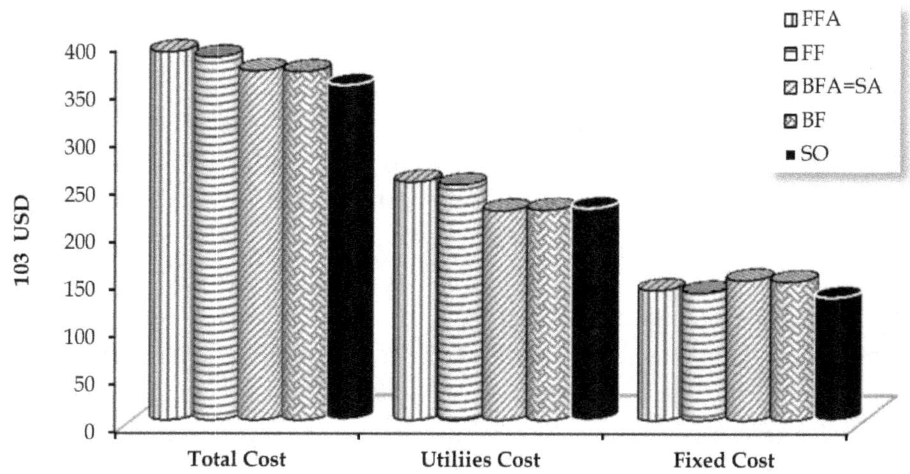

Fig. 16. Comparison with Typical Flowpatterns (Case II).

Thermal jumps that are achieved with the optimum structure are higher that in the triple and quintuple-effect MEES. In addition, it should be noted that the thermal jump in effect 2 is almost doubled with respect to other effects.

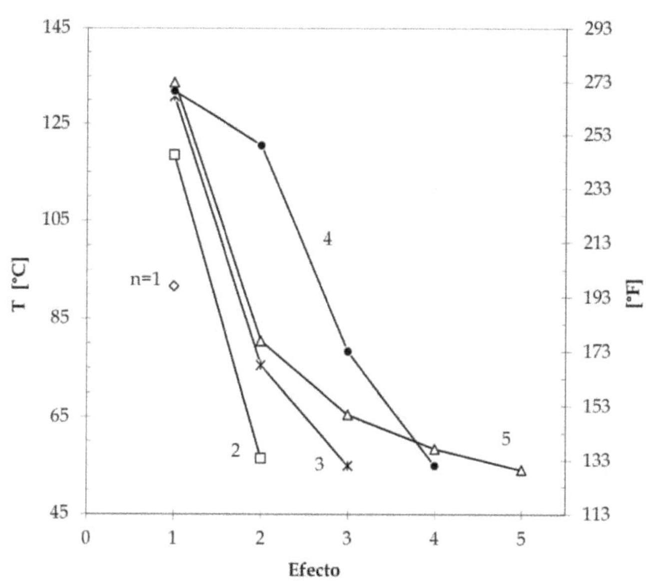

Fig. 17. Temperature profile optimal MEES for n = 1 to 5 effects.

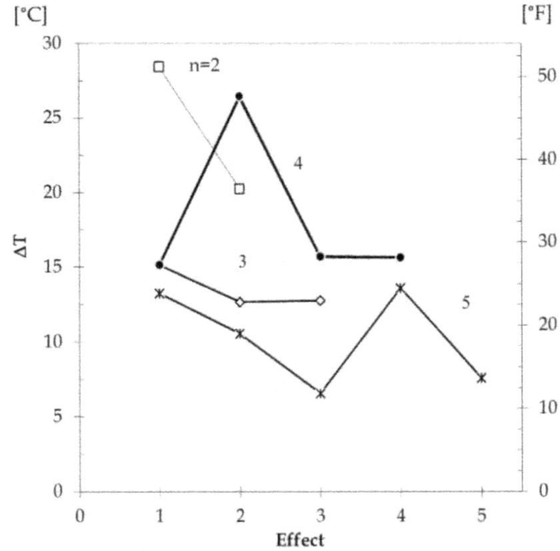

Fig. 18. Temperature difference profiles between the condensation and evaporation chambers in each effect for 2, 3, 4 and 5 effects.

Similarly, the profile of the area for each MEES, is far from being uniform, with the highest values located in the lower thermal effects. (Fig. 19). Should be mentioned, that the problem had been resolved for a fixed range area ($100,1000ft^2$).

The flow rate of solvent evaporated in each effect is approximately the same in the double and triple-effect MEES (Fig. 20), where the chosen structure is backward feed. But in the optimal solution and after this, the values of vapor flow rate are far from each other. Although it can be seen that, in SO the curve is regular and decreasing with temperature effects, i.e. the greater evaporated flow rate occurs in the effect 4.

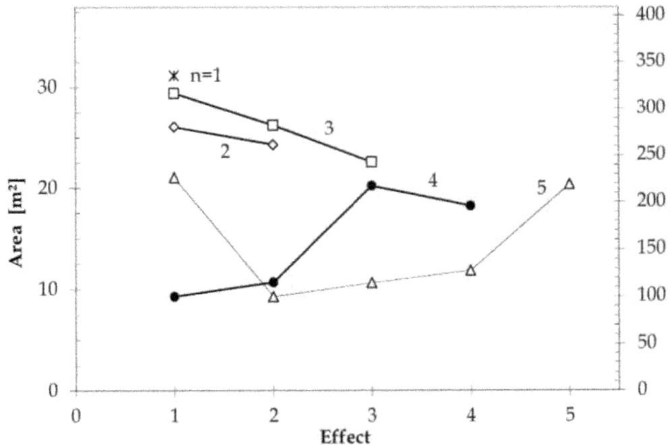

Fig. 19. Area profiles in the 1 to 5 effect MEES (n: number of effects)

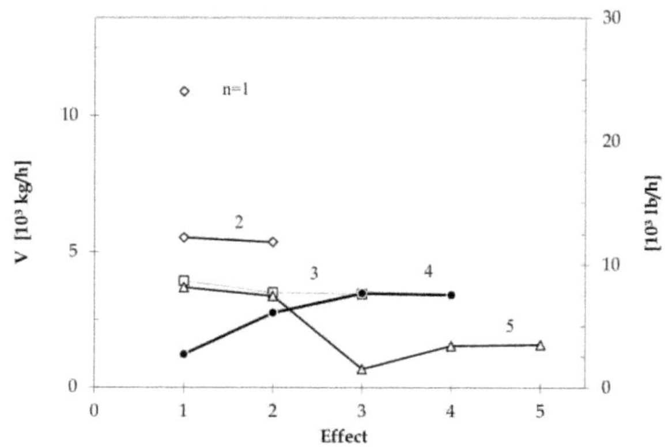

Fig. 20. Flow rate profiles in the 1 to 5 -effect MEES (n : number of effects).

The profile of the global coefficient U is similar: lower in effect 1 and higher in the effect 4 . It is important to clarify that, the values of the flow rate of steam produced in each effect are not directly indicative parameters. Yes it is a relative measure, for example, the percentage of the current liquid evaporates. Thus, the percentage amount of solvent evaporated, with respect to the flow rate of solvent that enters each effect, is maximum in the second effect (41%), followed in decreasing order, the third effect (34%), the first (31%) and finally the fourth (25%).

The concentration curve of the solution is monotone increasing, considering the path of the liquid stream (Fig. 21). The biggest jump in percentage concentration occurs in the second effect (69%), then in descending order, the third effect (51%), the first (45%) and the fourth (34%).

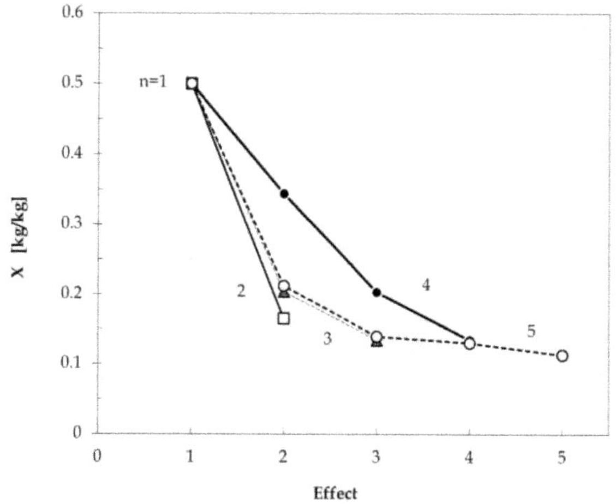

Fig. 21. Concentration profiles in the 1 to 5-effect MEES

The recent analysis of incremental concentration, the carried out for the steam produced in the effects, and the biggest jump thermal observed in effect 2 of the Optimal MEES show the importance of this effect. Its presence is the root cause of the improvement achieved in SO. This is achieved, thanks to the optimal design of the MEES, which allows an adequate relationship of the variables that define the system.

4.3 Case III. Temperature of the weak solution lower than strong solution ($T_F < T_P$)

4.3.1 Uniqueness of the cost curves

This case presents great similarity with the Case II. The optimal effect is the same (4) and intermediate structures found for $n \leq 4$ are almost identical. The best configuration for the liquid stream, given by {4,3,2,1}, as in Case II, use live steam to heat the first two effects.

Evolution of cost curves show a singularity with respect to the previously analyzed cases. Now, to increase the number of effects to reduce the TAC, the consumption of live steam begins to increase (Fig. 22) rather than continue to decline with the addition of a new effect, and the cost of the effects decreasing, rather than continue to increase. Isn't traditional behavior: a curve monotone decreasing for the cost of the live steam and one monotone increasing for the cost of the effects. Here is not complied with this scheme because of the significant reduction of the U coefficient and the high increase of the boiling point (BPE) in the effects of greater concentration. It is now "pays" with live steam part of the savings in the areas of effects.

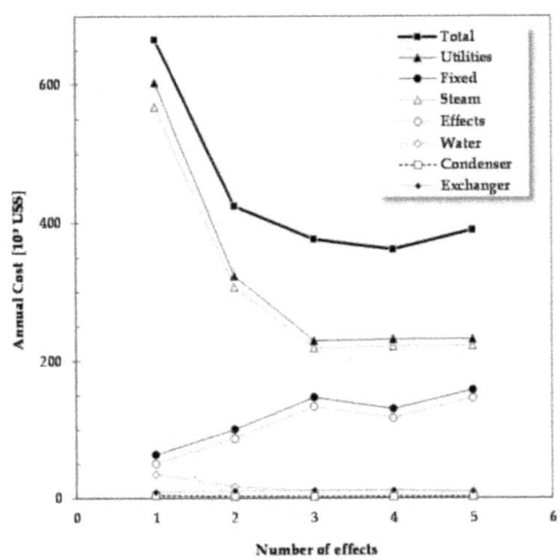

Fig. 22. Evolution of the TAC and its component terms with the number of effects (Case III).

4.3.2 Structure and distribution of temperature, concentration and heat transfer area

A structural analysis allows us to appreciate that the cuadruple-effect MEES results have added an effect, between the first and the second of the triple-effect MEES (Figs. 23 and 24, Table 2.).

Looking to reduce the total cost, the new effect requires one of the lowest values of area of heat transfer (Fig. 25). For this purpose, it operates with a large temperature difference; almost double that for the remaining effects (Fig. 26). On the other hand, contrary to expectations, their presence causes a slight increase in the consumption of live steam and allows at the same time, an increase in the thermal jump in the last two effects of, approximately, 6°F(3.3°C).

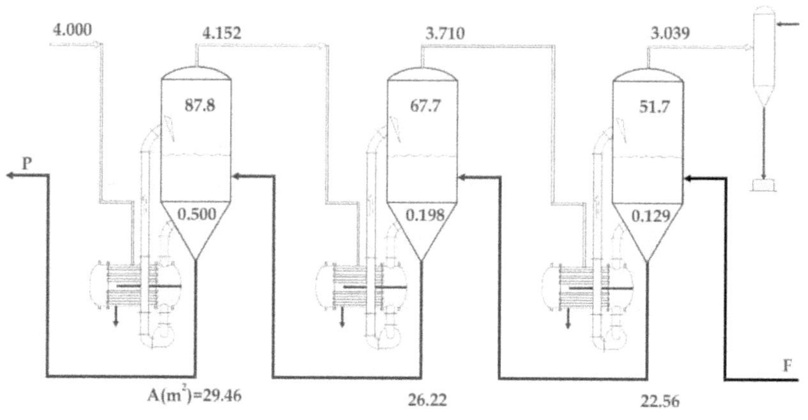

Fig. 23. Optimal Structure for three effects. (Local optimun).

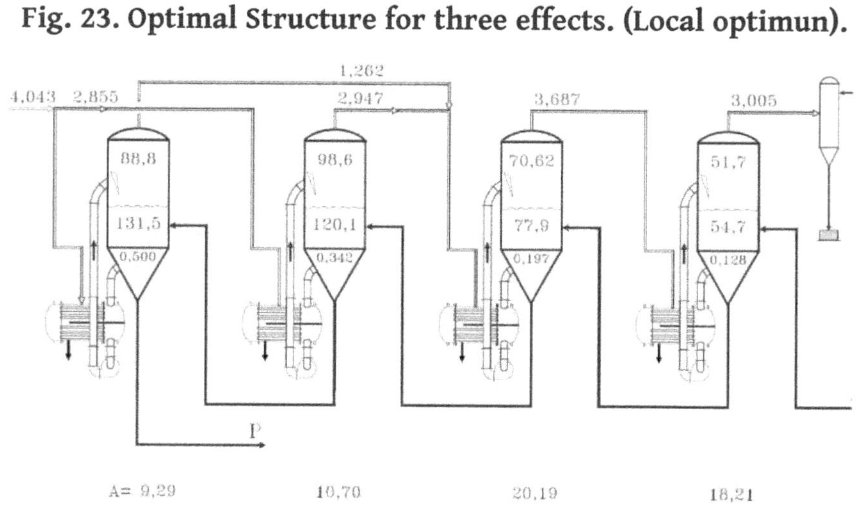

Fig. 24. Optimal Structure for four effects. (Optimal Solution).

Table 2. Results of alternative evaporation systems. (Vs: Live steam flow rate. V: Secondary Steam Flow rate)

Alternative Optimal Solutions (Case III)							
Equal area effects MEES: SA							
Effect	**T**	**Tv**	**Vs**	**L**	**V**	**X**	**F**
(i)	[°C]		[10^3 kg/h]			[kg/kg]	[m^2]
1	137.99	94.92	4.105	0	3.232	0.500	
2	128.93	107.28	0	0	2.979	0.340	15.33
3	80.62	73.39	0	0	3.665	0.195	
4	54.72	51.67	0	13.626	2.971	0.128	
Backward feed MEES. Optimal area effects: BF							
Effect	T	Tv	Vs	L	V	X	F
1	130.40	87.83	4.000	0	3.232	0.500	29.47
2	74.99	67.71	0	0	3.710	0.198	26.22
3	54.76	51.67	0	13.626	3.039	0.129	22.57
Backward feed MEES. Equal area effects: BFA							
Effect	T	Tv	Vs	L	V	x	F
1	128.39	85.95	3.984	0	4.126	0.500	
2	73.12	65.84	0	0	3.719	0.199	26.54
3	54.77	51.67	0	13.626	3.055	0.129	
Forward feed MEES. Optimal area effects: FF							
Effect	T	Tv	Vs	L	V	X	F
1	132.77	127.95	4.822	13.626	3.517	0.135	20.37
2	116.53	107.02	0	0	3.714	0.213	18.96
3	91.67	51.67	0	0	3.671	0.500	32.69
Forward feed MEES. Equal area effects: FFA							
Effect	T	Tv	Vs	L	V	X	F
1	135.12	130.25	4.870	13.626	3.530	0.135	
2	121.62	111.99	0	0	3.691	0.213	24.40
3	91.67	51.67	0	0	3.680	0.500	

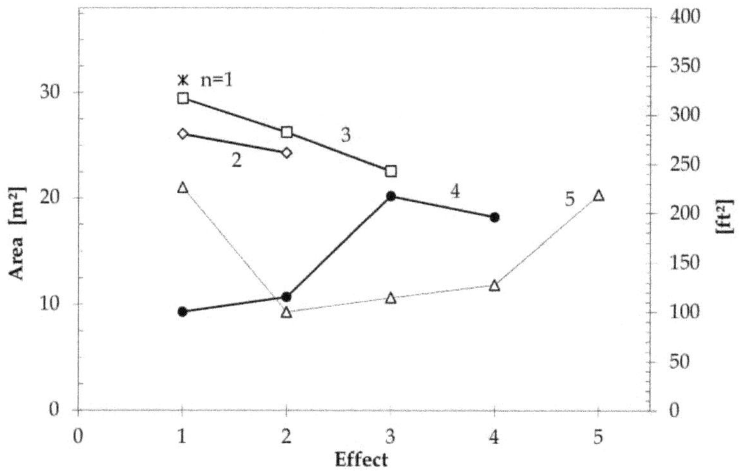

Fig. 25. Area profiles in the 1 to 5 -effect MEES.

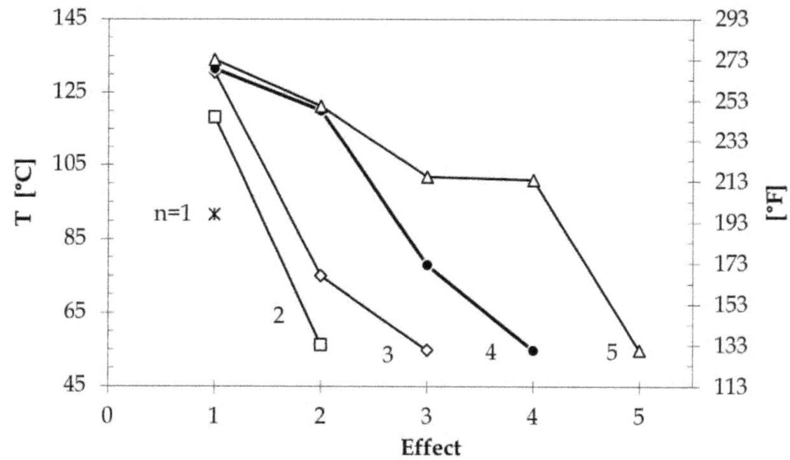

Fig. 26. Temperature profiles in the 1 to 5 -effect MEES.

The net result is a drastic decrease in the area of thermal transfer, as you can see through the cost of the effects (Fig. 27). Moreover, the distribution of the heat transfer area curve changes dramatically, from being monotone decreasing for a triple-effect MEES to have a nonmonotonic behavior in the Optimal MEES. Presents a maximum in the effect 3 and a minimum in the effect 1 , where the product is removed and in which the coefficient U takes its smallest value (Fig. 25). This was achieved by lowering the flow rate of the solvent evaporated in the first two effects, especially in the first (Fig. 28).

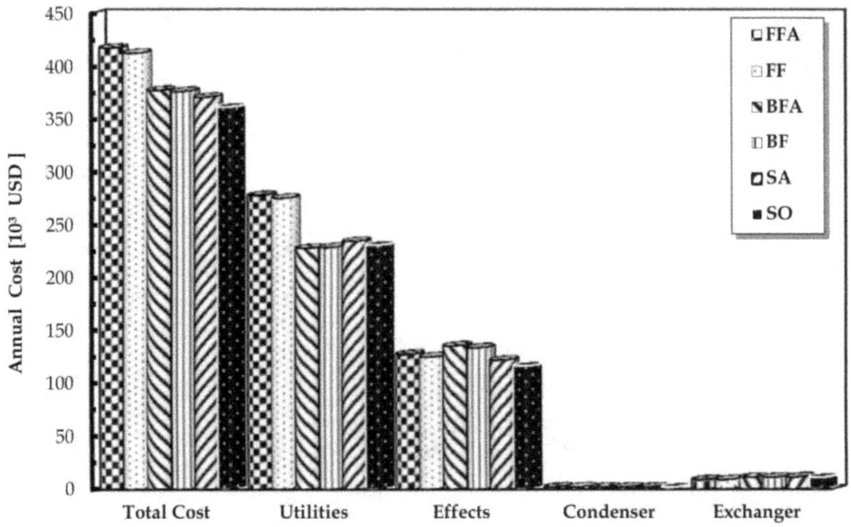

Fig. 27. Impact of the liquid flow pattern and the distribution of heat transfer area in different items of the total annual cost of the MEES (Case III).

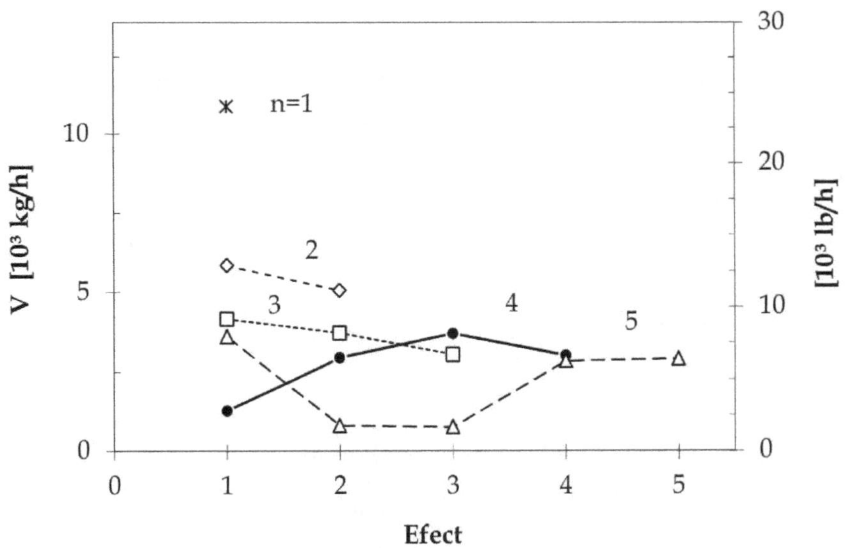

Fig. 28. Flow rate steam profiles in the 1 to 5 effect MEES.

Profile of concentration of the solution in the optimal MESS shows a non-uniform increase of concentration along the evaporator train (Fig. 29). Thus, following the path of the current liquid is seen a percentage increase of 28% in the effect 4,53% in the effect 3,73% in effect 2 and 46% in the first effect. Similarly as stated in Case II, the largest concentration jump occurs in effect 2, as opposed to what was observed in Case I, where the largest jump occurs in the last effect, where in addition, the product leaves.

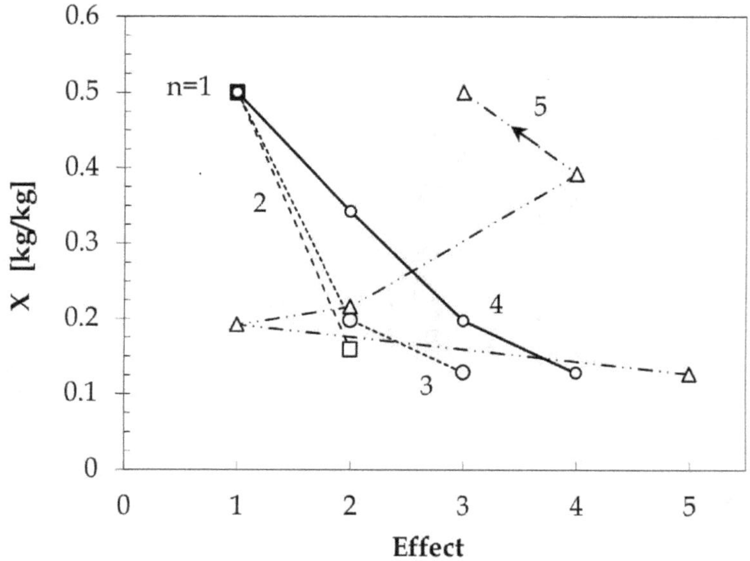

Fig. 29. Concentration profiles in the 1 to 5 effect MEES.

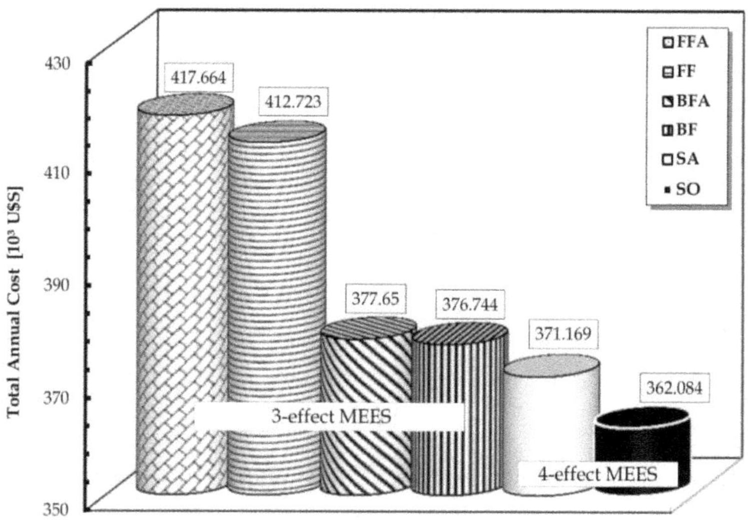

Fig. 30. Impact of the liquid and vapor flow pattern and area effects on the total annual cost of the MEES (Case III).

4.3.3 Comparison with conventional configurations

In example II, the best number of effects for the ideal MEES is four, which does not align with the traditionally suggested triple-effect countercurrent (BFA) from classical literature. The analysis of traditional structures with equal area effects (Table 2) reaffirms the superiority of the backward feed arrangement, indicating that three effects is the optimal quantity. FFA is 11.4% more costly than the structure BFA (Fig. 30). If both designs facilitate the optimization of transfer area distribution among various effects, the enhancement is negligible. The best choice (BF) is 4.1% costlier than SO. The optimal configuration for a MEES, constrained solely by the equality of areas of its effects (SA), is backward feed (BF).

5. Concise Summary

This chapter addresses the design of a MEES for caustic soda concentration by formulating a rigorous non-convex MINLP mathematical model and solving it with a mathematical optimizer. The new formulation proposal, unlike prior papers, includes as decision variables: (a) the trajectories of steam and liquid flows along the evaporator train, whose optimal selection, as evidenced by various authors and demonstrated in the example presented, influences the operational costs and investment of the MEES. The number of effects of the MEES, as demonstrated by the presented results, is a crucial design consideration, along with the heat transfer area in each effect, without relying on the assumption of equal areas across the effects, which often significantly escalates the total investment cost. We employed the total annual cost of the evaporation system, encompassing both fixed and operational expenses, as an evaluation criterion for alternative designs within the problem's solution space. Additional significant factors, often overlooked and rarely addressed concurrently with the pursuit of the optimal flow pattern, were examined, including: the elevation of the solution's boiling point and its dependence on temperature and concentration, the fluctuation of the overall heat transfer coefficient along the train in response to variations in concentration and temperature for each effect, and the functional relationship of the heat of a non-ideal solution with concentration and temperature. Furthermore, we examined alternative non-conventional design variants that emerge from permitting: the parallel feeding of the weak solution to two or more effects of the evaporator train, the introduction of two or more liquid streams (even of varying concentrations) to a specific effect, and the diversion of a liquid solution branch around an effect to circumvent its processing therein. A significant limitation identified in the mathematical model's resolution is the existence of numerous stationary points (local optima) and the considerable impact of the beginning point.It has been found that in order to obtain the optimal design of lower total cost annual: (i) not always steam flow pattern should be the traditional unifilar cascade, could be useful to feed steam live in more than one effect, (ii) the heat transfer areas do not necessarily have to be equal, (iii) the fresh feed stream should not always come in the last effect evaporation train. However, the synthesis of the optimal MEES only be achieved if you are optimized simultaneously structural, parametric, and operation variables.

6 . References

Foust, Alan S., Principles of Unit Operations, Second Ed. C. Wiley, New York, 1980.

Geankoplis, Ch. J., Transport Processes and Unit Operations, 2nd. Edition, 495-501, 1983.

Hillebrand Jr., J. B. & Westerberg, A.W., The Synthesis of Multiple-Effect Evaporator Systems Using Minimum Utility Insights - I. A Cascaded Heat Representation Computers & Chemical Engineering, Vol. 12, pp. 611-624, 1988.

Horvath, A. L., Handbook of Aqueous Electrolyte Solutions, Physical Properties, Estimation and Correlation Methods, Wiley, C. New York, 1985.

Irahola Ferreira, Jaime A., & Jaime Cerdá. Optimal Synthesis Of A Multiple-Effect Evaporator System. Fifth International Symposium On Process Systems Engineering (PSE), Korea, 1994.

Irahola Ferreira, Jaime A., & Jaime Cerdá. Síntesis Optima de un Sistema de Evaporación Multiefecto para la Concentración de Licor de Caña de Azúcar. IX Simposio Internacional en Aplicaciones de Informática. Infonor'96. Antofagasta. Chile. 1996.

Irahola Ferreira, Jaime A., Aplicación para Formular y Resolver Modelos Matemáticos de Sistemas de Evaporación Multiefecto. Información Tecnológica. Chile, 2008

Kern, D. Q. Process Heat Transfer. McGraw-Hill, New York .pp 375-452, 1999.

Maloney. James O. Perry's Chemical Engineers Handbook, 8th Edn. McGraw-Hill, New York 2008.

Nishitani, H. & Kunugita, E., The Optimal Flow-Pattern of Multiple Effect Evaporator Systems, Computers & Chemical Engineering, Vol. 3, pp. 261-268, 1979.

Peters, Max Stone Klaus D. Timmerhaus. Plantdesign and economics for chemical engineers. McGraw-Hill, Inc. Singapore, ISBN 0-07-100871-3, 1991.

Standiford, Ferris C. Jr., W. L. Badger, Evaporation. Chem. Engng, 70, 158-176, 1963.

Westerber, A.W. & Hillebrand Jr., J. B. The Synthesis of Multiple-Effect Evaporator Systems Using Minimum Utility Insights - II. Liquid Flow-Pattern Selection. Computers & Chemical Engineering, Vol. 12, pp. 625-636, 1988.

Index

www.ingramcontent.com/pod-product-compliance
Lightning Source LLC
LaVergne TN
LVHW081941200926
848484LV00009B/2486

9781787156357